Fortschritte *der Chemie organischer Naturstoffe*

Progress in the Chemistry of Organic Natural Products

66

Founded by L. Zechmeister
Edited by W. Herz, G. W. Kirby, R. E. Moore,
W. Steglich, and Ch. Tamm

Authors:
R. G. S. Berlinck, T. Hatano, T. Okuda, T. Yoshida

Springer-Verlag
Wien New York 1995

© 1995 by Springer-Verlag/Wien

Softcover reprint of the hardcover 1st edition 1995

Library of Congress Catalog Card Number AC 39-1015

Typesetting: Macmillan India Ltd., Bangalore-25

Printed on acid free and chlorine free bleached paper

With 6 Figures

ISSN 0071-7886
ISBN-13: 978-3-7091-9365-5   e-ISBN-13: 978-3-7091-9363-1
DOI: 10.1007/978-3-7091-9363-1

Contents

List of Contributors

BERLINCK, Dr. R. G. S., Departamento de Química e Física Molecular, Instituto de Química de São Carlos, Universidade de São Paulo, CP 780, 13560-970 São Carlos, SP, Brasil.

HATANO, Dr. T., Faculty of Pharmaceutical Sciences, Okayama University, Tsushima, Okayama 700, Japan.

OKUDA, Prof. Dr. T., Faculty of Pharmaceutical Sciences, Okayama University, Tsushima, Okayama 700, Japan.

YOSHIDA, Prof. Dr. T., Faculty of Pharmaceutical Sciences, Okayama University, Tsushima, Okayama 700, Japan.

List of Contributors

Hydrolyzable Tannins and Related Polyphenols

Takuo Okuda, Takashi Yoshida, and Tsutomu Hatano,
Faculty of Pharmaceutical Sciences, Okayama University, Tsushima,
Okayama 700, Japan

Contents

1. Introduction

Recent advances in isolation and structure elucidation of several hundred of new polyphenolic compounds of the tannin class using new techniques (*1*) have altered the definition of a tannin, at least that of a hydrolyzable tannin. Thus the term tannin now includes all polyphenolic compounds of defined chemical structure as well as the whole group of compounds whose chemical structures and properties are those of tannins. Tannins by the new definition are thus similar to other classes of natural products, such as alkaloids, terpenoids etc.

This modification in the definition of a tannin is due mainly to the isolation and structural characterization of a large number of hydrolyzable tannins in the decade which has elapsed since the publication of the last article dealing with hydrolyzable tannins in this series (*2*), although notable progress has been made in the chemistry of condensed tannins, too (*3*). The newly found hydrolyzable tannins comprise several new types of monomeric hydrolyzable tannins and their congeners and above all oligomeric hydrolyzable tannins which are the most remarkable new class of tannins (*4*).

Chemotaxonomic investigations have revealed that each hydrolyzable tannin is specifically associated with certain plant species in orders postulated to lie along certain lines of plant evolution (*5, 6*). Various biological and pharmacological activities of polyphenolic compounds, particularly those of hydrolyzable tannins, which often differ markedly from each other, have been discovered (*7–9*). These findings are also in favor of the above mentioned new definition of a tannin.

2. Monomeric Hydrolyzable Tannins

Among the monomeric hydrolyzable tannins newly found during the period covered in this review, the most notable ones are new types of ellagitannin analogs such as those having an oxidized hexahydroxydiphenoyl or HHDP group and ellagitannin-flavan condensates. However, some important new gallotannins have also been found.

2.1. Galloyl Esters

2.1.1. Galloylglucoses

The gallotannins in a plant species or a commercial product are generally a mixture of oligo- or polygalloyl esters of a sugar or a polyalcohol (*2, 3*). Although galloyl esters of glucose are widely distributed in dicotyledonous plants, various galloyl esters with hamamelose, quinic acid and some other sugar or polyalcohol residues have also been found in Dicotyledoneae. The mono-~pentagalloylglucoses (**1**)–(**6**), in which galloyl groups are directly attached to alcoholic oxygen of the glucose moiety are representatives of the most common type of gallotannin. These galloylglucoses, especially the tetra- and pentagalloylglucoses (**4** and **6**), are regarded as the precursors of polygalloylglucoses with depsidically linked galloyl groups, and also of various ellagitannins (*2, 3*). Some species of Euphorbiaceae, Fagaceae, Cornaceae and Hamamelidaceae contain dimeric hydrolyzable tannins (see Chapter 3) having some galloylglucose as a constituent monomer (*4*).

Galloylation in the biosynthetic pathway yielding pentagalloylglucose *via* intermediates with a smaller number of galloyl groups is effected by β-glucogallin (**1**) and 1,6-di-*O*-galloyl-β-D-glucose (**2**) as the galloyl donors. β-Glucogallin (**1**) is also considered to be a galloyl donor in the

(**1**) β-Glucogallin: $R^1 = R^2 = R^3 = R^4 = H$
(**2**) 1,6-Di-*O*-galloyl-β-D-glucose: $R^1 = R^2 = R^3 = H$, $R^4 = G$
(**3**) 1,2,6-Tri-*O*-galloyl-β-D-glucose: $R^1 = R^4 = G$, $R^2 = R^3 = H$
(**4**) 1,2,3,6-Tetra-*O*-galloyl-β-D-glucose: $R^1 = R^2 = R^4 = G$, $R^3 = H$
(**5**) 1,2,4,6-Tetra-*O*-galloyl-β-D-glucose: $R^1 = R^3 = R^4 = G$, $R^2 = H$
(**6**) 1,2,3,4,6-Penta-*O*-galloyl-β-D-glucose: $R^1 = R^2 = R^3 = R^4 = G$

formation of polygalloylglucoses with depsidically linked galloyl groups (*10*).

Several aspects of the chemical behavior of gallotannins including acyl migration were reported by 1970 (*2*). Synthetic studies permitted introduction of galloyl groups into specific hydroxyl groups of glucose, leading for example to 1,3,6-tri-*O*-galloyl-β-D-glucose (*11*).

2.1.2. Polygalloylglucoses with Depsidically Linked Galloyl Groups

Commercial products called "tannic acid" used as drugs or reagents are prepared mainly from galls of some *Rhus* or *Quercus* species and are mixtures of oligo-~polygalloylglucoses. Besides the galls of these plants, the leaves of various species of Anacardiaceae, Fagaceae, Hamamelidaceae, Paeoniaceae and Ericaceae also produce analogous galloylglucose mixtures (*12*).

The polygalloylglucoses from these plants contain a chain or a few chains of depsidically linked galloyl groups on a glucose core, in addition to several other galloyl groups directly bound to the sugar hydroxyl groups. The depsidic linkages in the polygalloylglucoses can be cleaved selectively at room temperature in weakly acidic solution, whereas the ester linkages on the alcoholic hydroxyl groups remain unaffected by this treatment (*13*).

NMR spectral analysis revealed that each depsidic linkage equilibrates by acyl migration between the *p*-depside form and the *m*-depside

(**7**) Trigallic acid

form. In the *m*-depside or *p*-depside form, the phenolic hydroxyl group *meta* or *para* to the carboxyl in the galloyl group is participating in the depsidic linkage. Thus, the ^{1}H-NMR spectrum of depsidic trigallic acid (**7**) shows the *m*-coupled doublets due to unequivalent protons characteristic of the *m*-depside form as well as two-proton singlets due to equivalent protons of the *p*-depside form, indicating the presence of four isomers arising from equilibration between two depside forms (the *m-m*, *m-p*, *p-m* and *p-p* forms) (*14*).

2.1.3. Galloyl Esters of Other Sugars or Polyalcohols

Hamamelitannin (*15*), first isolated from a *Hamamelis* species, is a digalloylhamamelose which is not responsible for the protein-binding activity of the extract from this plant. Several galloyl esters of hamamelose were later isolated from *Castanea* species of Fagaceae and *Sanguisorba* species of Rosaceae (*16, 17*). Galloyl esters of other sugars or polyalcohols, such as fructose, sucrose, sedoheptulose, *proto-* and *scyllo*quercitols, and glycerol, were also isolated from plants of various families including the Fagaceae (*18–20*).

Acertannin, a digalloyl ester of 1,5-anhydro-D-glucitol, was first isolated in 1912 from the leaf of *Acer ginnala* (*21*); its final structure was established later as 2,6-digalloyl-1,5-anhydro-D-glucitol (**8**) (*22*). Tri- and tetragalloyl derivatives (**9 and 10**) of anhydroglucitol were also isolated from the same plant species (*23, 24*). Although the binding activity of acertannin determined by the RMB (relative affinity to methylene blue) method was low (*25*), the tri- and tetragalloyl derivatives showed much stronger activity (*24*). Two isomeric trigalloyl anhydro-glucitols and three isomeric tetragalloylanhydroglucitols were isolated after methylation of each isomer mixture (*24*). In these tri- and tetragalloyl derivatives, the galloyl and/or the depsidically linked di- or trigalloyl groups are at O-2 and O-6 of the sugar residue while the hydroxyl groups at C-3 and C-4 are unacylated.

"Taratannin" from fruit pods of *Caesalpinia spinosa* is a mixture of polygalloylquinic acids, in which the hydroxyl groups at C-3, C-4 and C-5 are esterified with gallic acid or depsidically linked polygallic acids (*26*). Polygalloylquinic acids (**11**) were isolated from the leaf of *Koelreuteria paniculata* and the leaf of *Pistacia chinensis* (*27*). They have a polygalloyl chain attached exclusively at O-5 of quinic acid with the hydroxyl groups at C-1, C-3 and C-4 remaining unacylated. Analogous polygalloylshikimic acids were also obtained from a *Castanopsis* species (*28*).

(8) Acertannin

(9) Trigalloyl-1,5-anhydro-D-glucitols

(10) Tetragalloyl-1,5-anhydro-D-glucitols

(11) 5-O-Polygalloylquinic acids (n = 0 ~ 5)

2.1.4. Galloyl Esters of Phenolic Glycosides

Various gallates of O- and C-glucosides of flavonoids and several other types of phenolic compounds were isolated from many plant species, mostly from those belonging to those orders which contain species also producing hydrolyzable tannins. Some galloylated glycosides of these phenolics with a sugar residue other than glucose were also found. Examples are monogalloylated glycosides with disaccharides (29), and gallates of rhamnosides and galactosides of some flavonoids. There are also galloylated anthocyanins as exemplified by delphinidin 3-O-(2″-O-galloyl)-β-D-galactoside, by cyanidin 3-O-(2″-O-galloyl)-β-D-galactoside from *Victoria amazonica* (*30*), by cyanidin 3-*O*-(2″-*O*-galloyl)-β-D-glucoside and cyanidin 3-*O*-(2″-*O*-galloyl-6″-*O*-α-L-rhamnosyl)-β-D-glucoside from the leaf of *Acer* species (*31*).

It is noteworthy that some gallates of flavonoid glycosides, such as kaempferol 3-*O*-glucoside (astragalin) and quercetin 3-*O*-glucoside (isoquercitrin), exhibit appreciable binding activity to protein and basic substance (*24*) as well as inhibitory effects on the angiotensin converting enzyme (*32*) and xanthine oxidase (*33*), which are analogous to effects produced by several hydrolyzable tannins.

Gallates of *O*-glucosides of low molecular phenolics, such as arbutin, homoarbutin, stilbene, gallic acid, gentisic acid and trihydroxybenzyl alcohol, have been found in species of Ericaceae, Saxifragaceae, Pyrolaceae, Polygonaceae, Fagaceae, Betulaceae and Euphorbiaceae (*34–38*). Among these compounds, 2-*O*-galloyl-, 6-*O*-galloyl- and 4,6-di-*O*-galloyl-arbutin (**12–14**) (*34, 39*), and also an analog galloylated at the phenolic hydroxyl group of the aglycone (*p*-galloyloxyphenyl-β-D-glucoside of arbutin), were found in *Arctostaphyllos* (Ericaceae) and *Bergenia* (Saxifragaceae) species. Gallates of the *O*-glucosides possessing an aliphatic

(**12**) 2-*O*-Galloylarbutin: R^1 = G, R^2 = R^3 = H
(**13**) 6-*O*-Galloylarbutin: R^1 = R^2 = H, R^3 = G
(**14**) 4,6-Di-*O*-galloylarbutin: R^1 = H, R^2 = R^3 = G

(**15**) Salidroside-6″-gallate

(**16**) Cornuside

alcohol as the aglycone, such as lindleyin (gallate of hydroxy-phenylbutanone glucoside) and salidroside gallate (15) (40), were obtained from some species of Crassulaceae, Polygonaceae and Fagaceae. Gallates of oxypaeoniflorin and paeoniflorin (41), a monoterpene glucoside in which a glucose hydroxyl group is galloylated, were isolated from *Paeonia suffruticosa*, and cornuside (16), a gallate of secoiridoid glucoside (42), having a galloyl group on the aglycone, was obtained from *Cornus officinalis*.

Bergenin (17) and norbergenin (18), C-glucosides biogenetically producible from 2-O-galloylarbutin (39), are found in *Bergenia* species. Their gallates (19–25) were isolated from *Bergenia* (34) and *Mallotus* (Euphorbiaceae) species (43, 44).

Mono-~trigallates of benzophenone C-glucosides and xanthone C-glucosides were isolated from *Mangifera indica* (Anacardiaceae) (45).

(17) Bergenin: $R^1 = R^2 = R^3 = H, R^4 = Me$
(18) Norergenin: $R^1 = R^2 = R^3 = R^4 = H$
(19) 4-O-Galloylbergenin: $R^1 = G, R^2 = R^3 = H, R^4 = Me$
(20) 4-O-Galloylnorbergenin: $R^1 = G, R^2 = R^3 = R^4 = H$
(21) 11-O-Galloylbergenin: $R^1 = R^2 = H, R^3 = G, R^4 = Me$
(22) 11-O-Galloylnorbergenin: $R^1 = R^2 = R^4 = H, R^3 = G$
(23) 3,4-Di-O-galloylbergenin: $R^1 = R^2 = G, R^3 = H, R^4 = Me$
(24) 4,11-Di-O-galloylbergenin: $R^1 = R^3 = G, R^2 = H, R^4 = Me$
(25) 3,4,11-Tri-O-galloylbergenin: $R^1 = R^2 = R^3 = G, R^4 = Me$

2.1.5. Condensed Tannins with Galloyl Groups

Condensed tannins produced by several plants have partially galloylated structures. Kaki-tannin from immature fruits of *Diospyros kaki* is a mixture of partially galloylated proanthocyanidins. It is composed of prodelphinidin and procyanidin units as the monomers in a molar ratio of 1:2, and *ca.* 50% of the monomer units are galloylated at O-3. Its molecular weight is reported to be 5200 in the number average (M_n), and 11200 in the weight average (M_w) (46).

Condensed tannins obtained from rhubarb are also partially galloylated procyanidin oligomers and polymers; the extent of galloylation and oligomerization or polymerization of the proanthocyanidin units varies depending on the source or the method of preparation of the crude drugs from the plants (*47*). On the other hand, the condensed tannins from *Saxifraga stolonifera* leaves are highly galloylated (> 90%) procyanidins and are structurally more homogeneous (*47, 48*).

Higher oligomers of proanthocyanidins previously subjected to structure elucidation were often mixtures because of the difficulties associated with their purification. However, the structures of some lower oligomers of galloylated proanthocyanidins were determined after isolation of each component, and the biological activities of these compounds were screened. Among them, a tetramer of (−)-epicatechin 3-*O*-gallate (ECG) [ECG-(4β → 8)-ECG-(4β → 8)-ECG-(4β → 8)-ECG] (**26**), activated neutrophils and monocytes (*49*) and inhibited the proliferation of herpes simplex virus (HSV) (*50*).

2.1.6. Flavan-3-ol Gallates

The "tannin" of tea leaf is composed of polyphenols of low molecular weight represented by (−)-epigallocatechin gallate (EGCG) (**27**) and ECG (**28**). These galloylated flavan-3-ols exhibited binding activity to proteins and basic compounds comparable to that of average hydrolyzable tannins in spite of their low molecular weight (*25*). The main tea

(**27**) (-)-Epigallocatechin gallate

(**28**) (-)-Epicatechin gallate

(**29**) 3,3'-Di-*O*-galloyl-(-)-epicatechin

(**26**) 3-*O*-Galloylepicatechin-
(4β→8)-3-*O*-galloylepicatechin-
(4β→8)-3-*O*-galloylepicatechin-
(4β→8)-3-*O*-galloylepicatechin

(**30**) 3,4'-Di-*O*-galloyl-(-)-epicatechin

polyphenol, EGCG, also showed noticeable pharmacological activities such as antitumor and anticarcinogenic effects (51).

Flavan-3-ols with one or two galloyl groups have been obtained from species of Theaceae, Saxifragaceae and Leguminosae. Among them, a digalloyl derivative of (−)-epicatechin, isolated from the leaf of *Saxifraga stolonifera*, is equilibrated between two isomers, 3,3′-di-O-galloyl-(−)-epicatechin (29) and 3,4′-di-O-galloyl-(−)-epicatechin (30). This equilibration is due to migration of the galloyl group depsidically linked to the B-ring hydroxyl groups of flavan-3-ol (48).

2.2. Ellagitannins, Dehydroellagitannins and Naturally Occurring Oxidation Products

2.2.1. Ellagitannins

In a molecule of ellagitannin, hexahydroxydiphenoyl (HHDP) group(s) esterify the hydroxyl groups of a glucose (or polyalcohol) core, as exemplified by pedunculagin (31). Galloyl group(s) are often present in addition to the HHDP group (2, 3), as, for example, in casuarictin (32), the tellimagrandins I (34) and II (35), etc. (52, 53). They therefore liberate ellagic acid, and mostly gallic acid, too, upon hydrolysis with acid or alkali and also in the course of extraction and concentration of the extracts. Various ellagitannins possessing a trisgalloyl group, such as a valoneoyl, tergalloyl or macaranoyl group, etc., in which a galloyl group is bound to the HHDP group through an ether linkage, have been

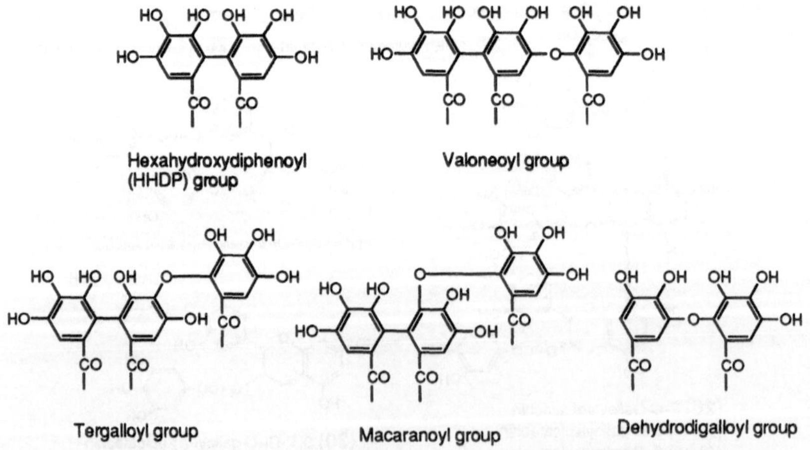

Hexahydroxydiphenoyl (HHDP) group

Valoneoyl group

Tergalloyl group

Macaranoyl group

Dehydrodigalloyl group

Chart 1

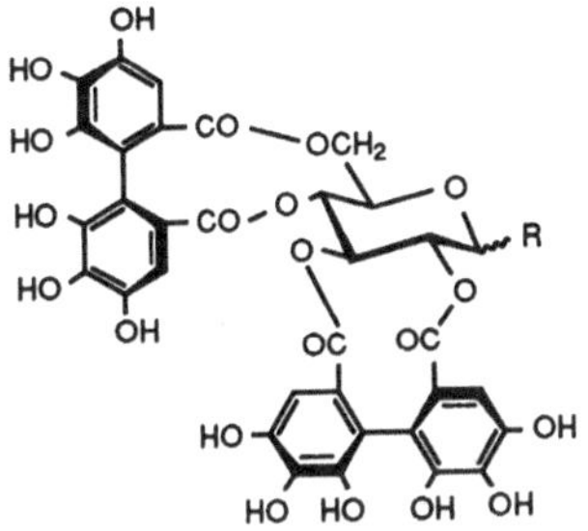

(31) Pedunculagin: R = OH
(32) Casuarictin: R = (β)-OG
(33) Potentillin: R = (α)-OG

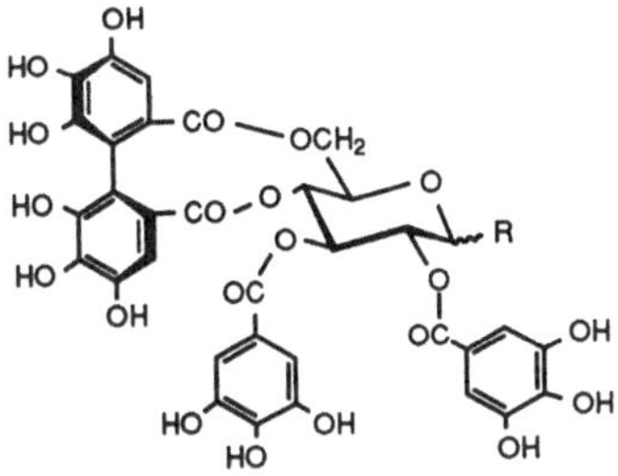

(34) Tellimagrandin I: R = OH
(35) Tellimagrandin II: R = (β)-OG

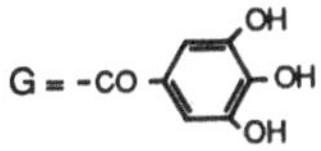

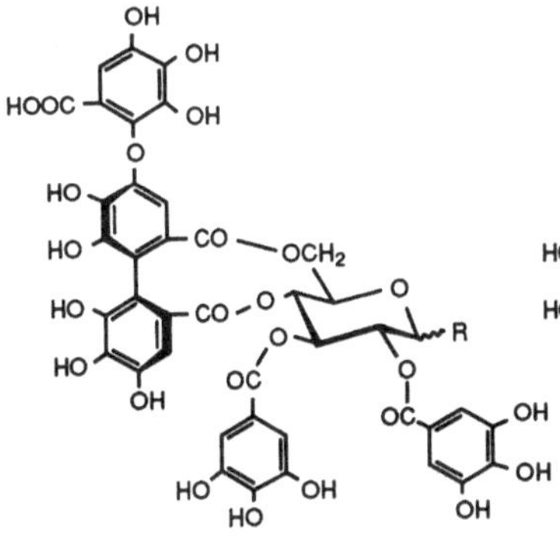

G = -CO

(36) Rugosin A: R = (β)-OG
(37) Rugosin B: R = OH

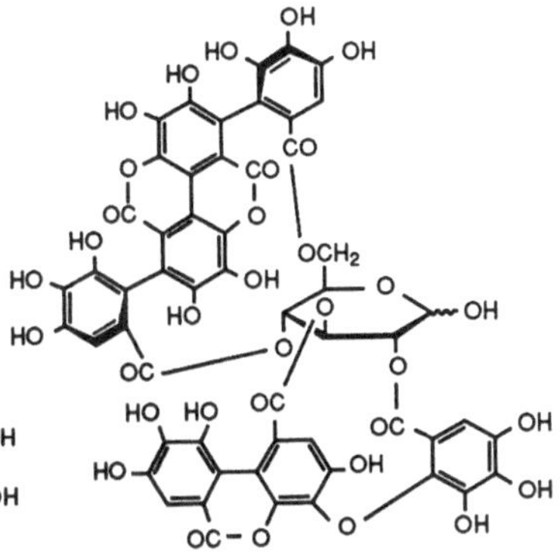

(38) Tergallagin

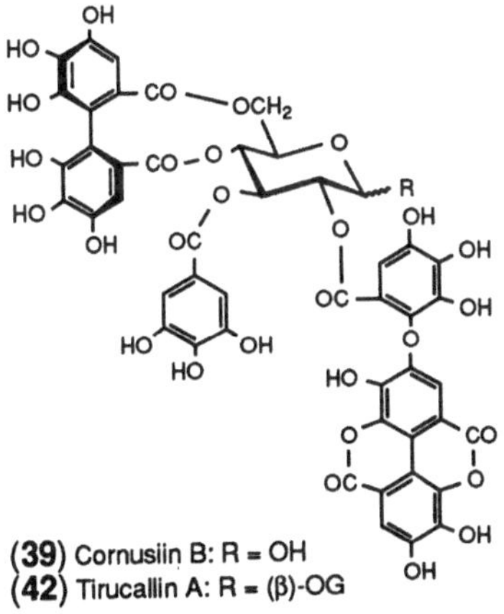

(39) Cornusiin B: R = OH
(42) Tirucallin A: R = (β)-OG

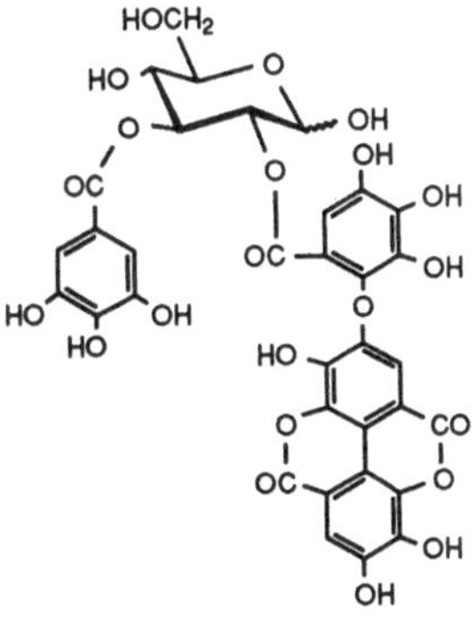

(40) Oenothein C

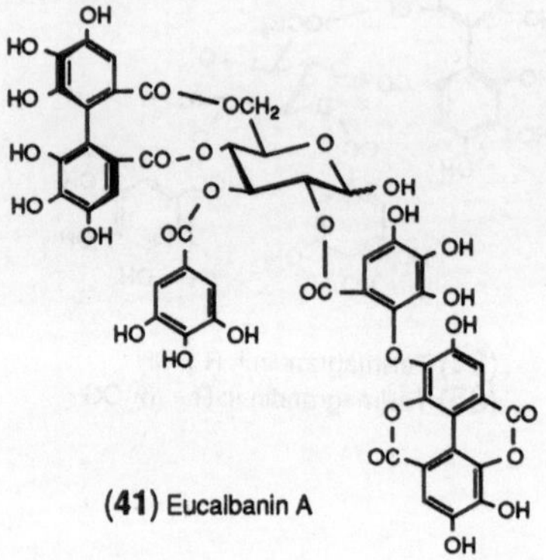

(41) Eucalbanin A

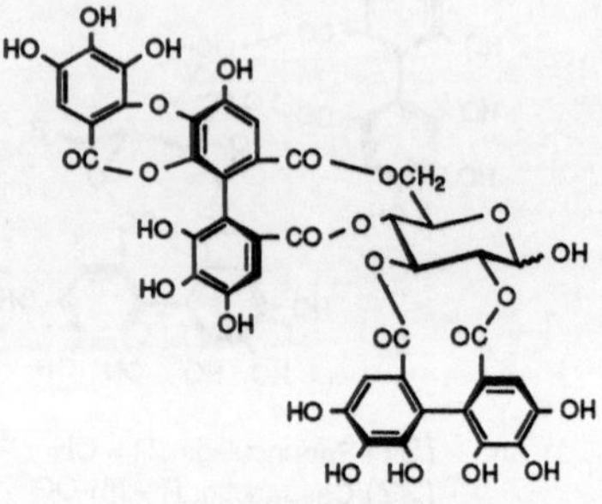

(43) Alnusiin

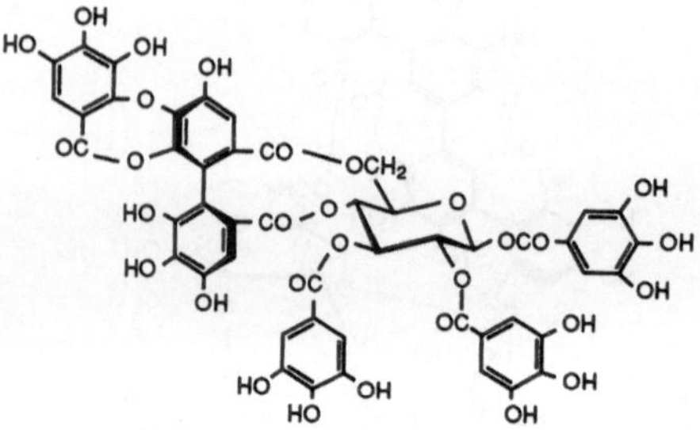

(44) Bicornin

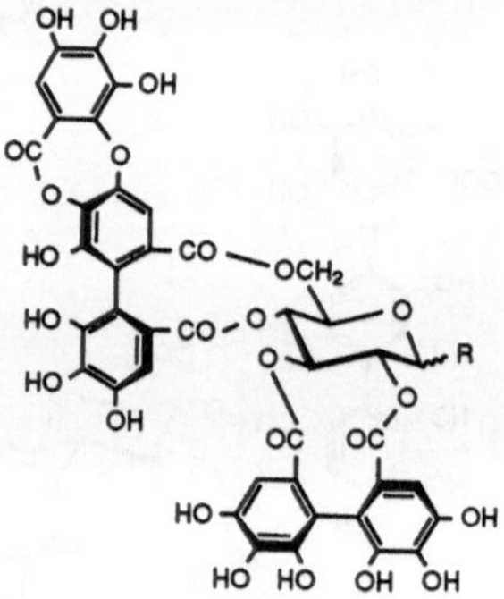

(45) Praecoxin C: R = (β)-OG
(46) Praecoxin D: R = OH

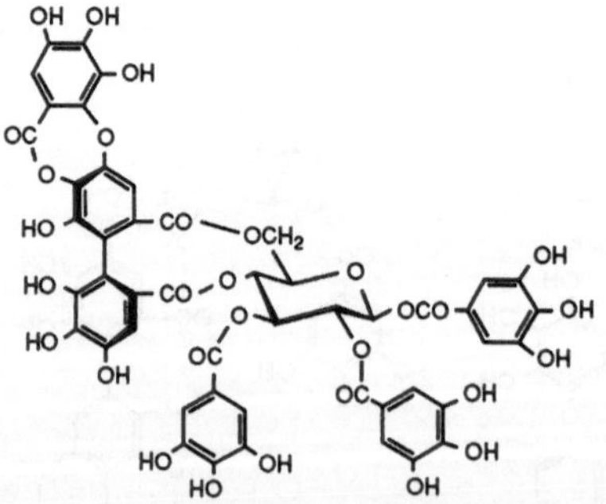

(47) Euprostin C

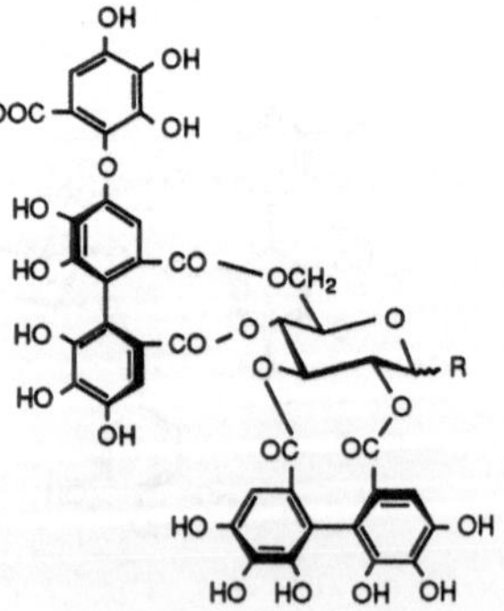

(48) Rugosin C: R = (β)-OG
(52) Praecoxin A: R = OH

isolated from various plant species. Examples are the rugosins A (**36**), B (**37**) (*54*) and tergallagin (**38**) (*55*), which did not give ellagic acid upon mild hydrolysis. These trisgalloyl groups present in some hydrolyzable tannins may form a dilactone or depsidone structure, *e.g.*, in cornusiin B (**39**) (*56*), oenothein C (**40**) (*56*), eucalbanin A (**41**) (*58*) and tirucallin A (**42**) (*57*) (dilactone form); alnusiin (**43**), bicornin (**44**) (*59*), praecoxins C (**45**) and D (**46**) (*60*) and euprostin C (**47**) (depsidone form) (*61*), *etc.* The labile depsidone linkage is easily hydrolyzed under mild conditions, as, for example, in the transformation of praecoxin C (**45**) to rugosin C (**48**) (*60*).

Ellagitannins containing a dehydrodigallic acid residue which is presumably formed in plants by oxidative coupling between two galloyl groups in a fashion similar to the formation of valoneic acid and its analogs (*4*) have been found in various plant species such as *Agrimonia pilosa* [agrimonic acids A (**55**) and B (**56**) (*62*)], *Coriaria japonica* [coriariins B (**57**) and J (depsidone) (**58**) (*63, 64*)] and *Reaumuria hirtella* [remurins A (**59**) and B (**60**) (*65*)], *etc.* The other ellagitannins (**49–54, 61** and **62**) found after 1982 are listed in Table 1.

2.2.2. Dehydroellagitannins: Equilibrated DHHDP Group in Geraniin and Analogs

Many hydrolyzable tannins containing a dehydrohexahydroxydiphenoyl (DHHDP) group which is an oxidatively modified congener of the HHDP group, have been isolated from various plants. Since the number of such tannins has increased considerably in recent years, it is convenient to classify them as dehydroellagitannins, although biogenetically they are metabolites of ellagitannins (*3*).

Geraniin (**63**) is the most notable member of this class because of its status as the first structurally defined dehydroellagitannin and because of its role in the biogenetic route to hydrolyzable tannins of various structures. This yellow crystalline tannin was originally isolated from *Geranium thunbergii*, an official anti-diarrheic in Japan (*70*). In aqueous solution the DHHDP group equilibrates between the six- (**63a**) and five-membered hemiacetal form (**63b**) as shown by NMR spectrometry (*70*). The ^{1}H-NMR spectrum measured immediately after solution of crystalline geraniin in acetone-d_6 exhibits signals due to the DHHDP group in the **63a**-form at $\delta 5.16$ (H-1′), 6.56 (H-3′) and 7.25 (H-3″), while upon equilibration in a D_2O solution containing acetone-d_6, these signals were accompanied by new signals at $\delta 4.72$ (d, $J = 2$ Hz, H-1′), 6.26 (d, $J = 2$ Hz, H-3′) and 7.28 (s, H-3″) which are ascribable to the **63b**-form. The ^{13}C-NMR spectrum showed, besides the signals of a methine group

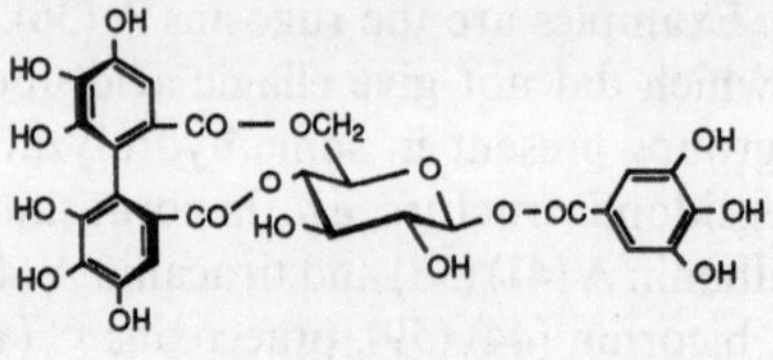

(49) Strictinin

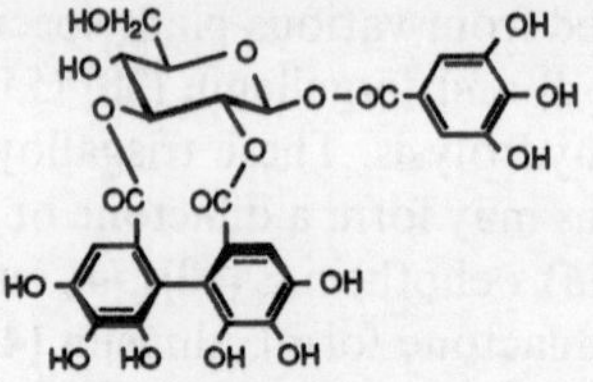

(50) Isostrictinin

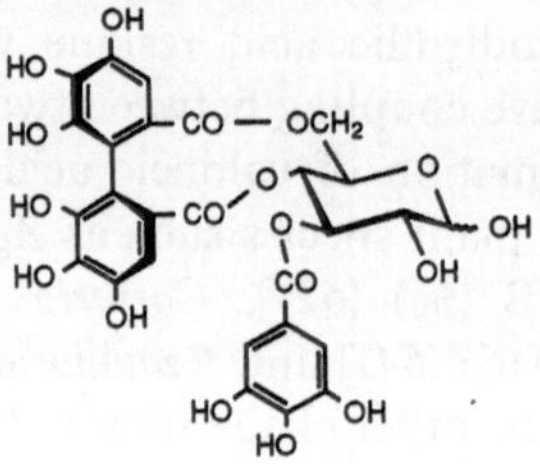

(51) Gemin D

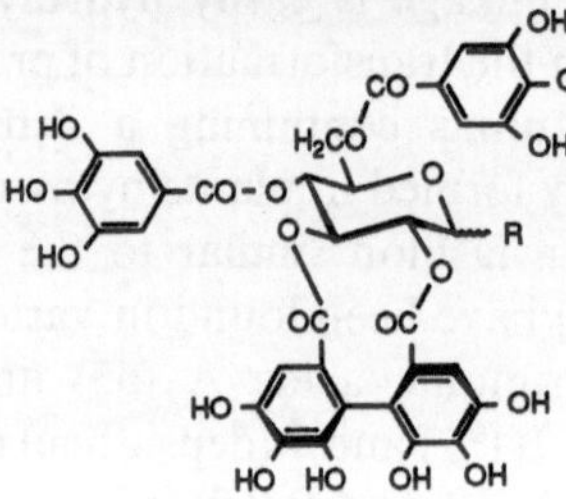

(53) Praecoxin B: R = OH
(54) Pterocaryanin C: R = (β)-OG

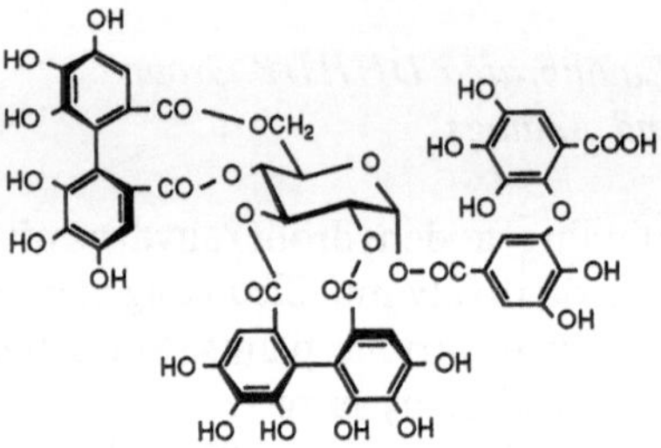

(55) Agrimonic acid A

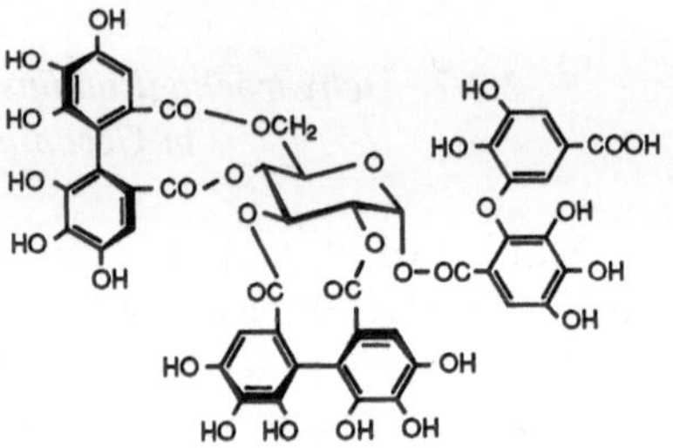

(56) Agrimonic acid B

[δ 46.2 and 51.9 (C-1′) of **63a** and **63b**)], the presence of an α,β-unsaturated ketonic function as revealed by signals of a ketonic carbon [δ 191.8, 194.8 (C-3′ of **63a** and **63b**)] and two olefinic carbons [δ 154.5, 149.2 (C-2′ of **63a** and **63b**); 128.6, 125.0 (C-3′ of **63a** and **63b**)]. The remaining two carbons in the A-ring were associated with four signals at δ 92.5–108.9, which are assignable to *gem*-diol carbon and/or acetal carbons in the **63a** and **63b** forms.

Condensation of geraniin with *o*-phenylenediamine yields a phenazine derivative (**64**) which provides additional evidence for the presence of a DHHDP group. The residual part of geraniin was proved to be corilagin (**65**) by partial hydrolysis of the phenazine derivative (**64**) in hot water (*70*).

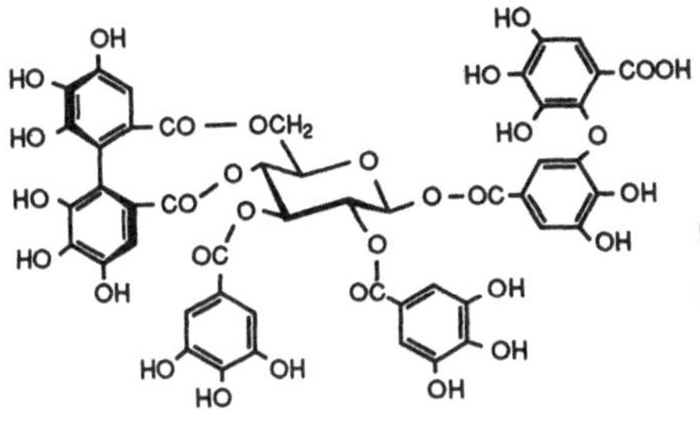

(57) Coriariin B

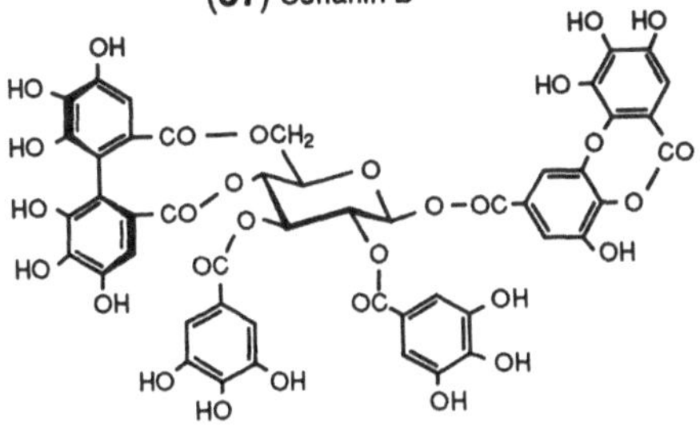

(58) Coriariin J

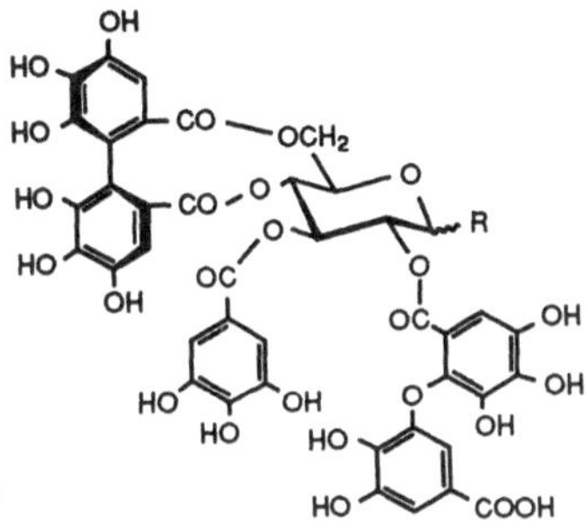

(59) Remurin A: R = (β)-OG
(60) Remurin B: R = OH

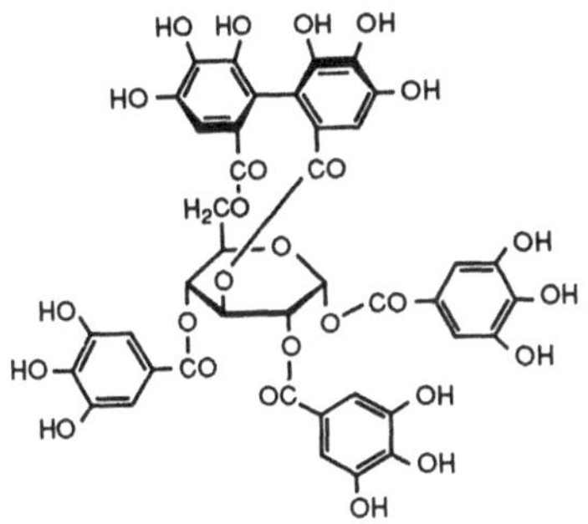

(61) Nupharin A

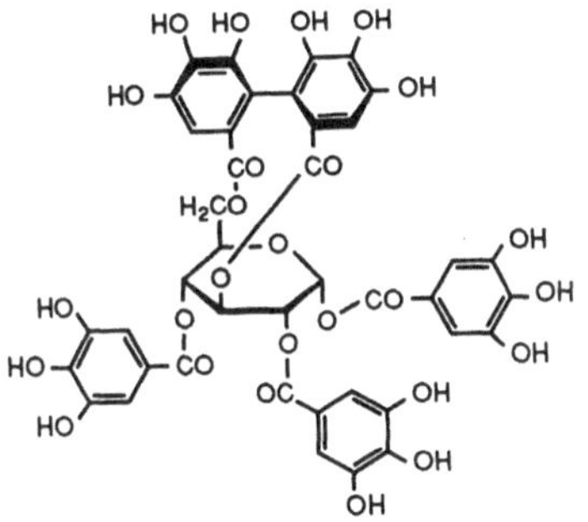

(62) Nupharin B

Alcoholic solutions of dehydroellagitannins show multiple and broad HPLC peaks due to formation of methanol adducts (63c–63f) at one of the carbonyl groups in the DHHDP groups as revealed by the NMR spectrum of geraniin in methanol (200).

Geraniin (63) was later found in species of *Geranium* (Geraniaceae), Euphorbiaceae, and some other families within Geraniales (71), and also in species of Cercidiphyllaceae, Aceraceae, Simaroubaceae, Rosaceae, Melastomataceae and Coriariaceae (3, 62, 72).

Table 1. *Monomeric Ellagitannins and Their Plant Sources*

Compound	Plant[a]	Family	Ref.
Tellimagrandin I (34)	*Tellima grandiflora*	Saxifragaceae	*(52, 53)*
Tellimagrandin II (35)	*T. grandiflora*		*(52, 53)*
Rugosin A (36)	*Rosa rugosa*	Rosaceae	*(54)*
Rugosin B (37)	*R. rugosa*		*(54)*
Rugosin C (48)	*R. rugosa*		*(54)*
Potentillin (33)	*Potentilla kleiniana*	Rosaceae	*(62, 107)*
Agrimonic acid A (55)	*Agrimonia pilosa*	Rosaceae	*(62, 107)*
Agrimonic acid B (56)	*A. pilosa*		*(62, 107)*
Casuarictin (32)	*Casuarina stricta*	Casuarinaceae	*(52)*
Strictinin (49)	*C. stricta*		*(52)*
Isostrictinin (50)	*Psidium guajava*	Myrtaceae	*(68)*
Gemin D (51)	*Geum japonicum*	Rosaceae	*(69)*
Pterocaryanin C (54)	*Pterocarya steroptera*	Fagaceae	*(66)*
Tergallagin (38)	*Terminalia catappa*	Combretaceae	*(55)*
Cornusiin B (39)	*Cornus officinalis*	Cornaceae	*(56)*
Oenothein C (40)	*Oenothera erythrosepala*	Onagraceae	*(56)*
Eucalbanin A (41)	*Eucalyptus alba*	Myrtaceae	*(58)*
Tirucallin A (42)	*Euphorbia tirucalli*	Euphorbiaceae	*(57)*
Euprostin C (47)	*Euphorbia prostrata*	Euphorbiaceae	*(61)*
Alnusiin (43)	*Alnus fordii*	Euphorbiaceae	*(59)*
Bicornin (44)	*Trapa bicornis*	Trapaceae	*(59)*
Praecoxin A (52)	*Stachyurus praecox*	Stachyuraceae	*(60)*
Praecoxin B (53)	*S. praecox*		*(60)*
Praecoxin C (45)	*S. praecox*		*(60)*
Praecoxin D (46)	*S. praecox*		*(60)*
Coriariin B (57)	*Coriaria japonica*	Coriariaceae	*(63)*
Coriariin J (58)	*C. japonica*		*(64)*
Remurin A (59)	*Reaumuria hirtella*	Tamaricaceae	*(65)*
Remurin B (60)	*R. hirtella*		*(65)*
Nupharin A (61)	*Nuphar japonicum*	Nymphaeaceae	*(67)*
Nupharin B (62)	*N. japonicum*		*(67)*

[a] Plant from which the tannin was first isolated

Furosinin (66) and didehydrogeraniin (= dehydrogeraniin) (67) are minor constituents in *G. thunbergii* and are structurally related to geraniin (73). The former exists as a mixture of eight isomers as a result of equilibration within two DHHDP groups and the anomeric center. The structure of terchebin (74), a component of myrobalans, was revised to 68 (75) which incorporates DHHDP group, based on spectral analogy to geraniin which also contains this group. Isoterchebin, initially isolated

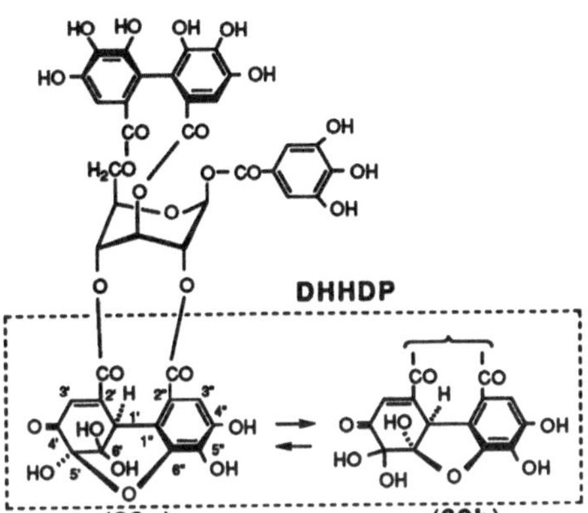

DHHDP

(63a)   (63b)
(63) Geraniin

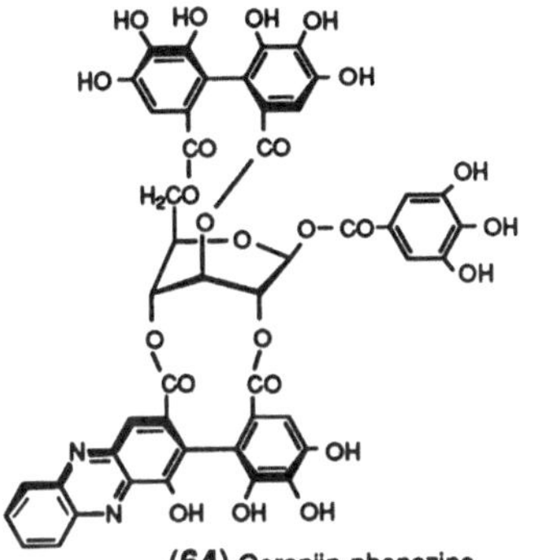

(64) Geraniin-phenazine

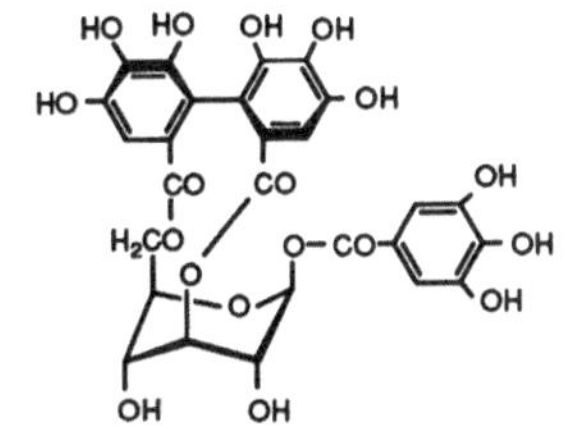

(65) Corilagin

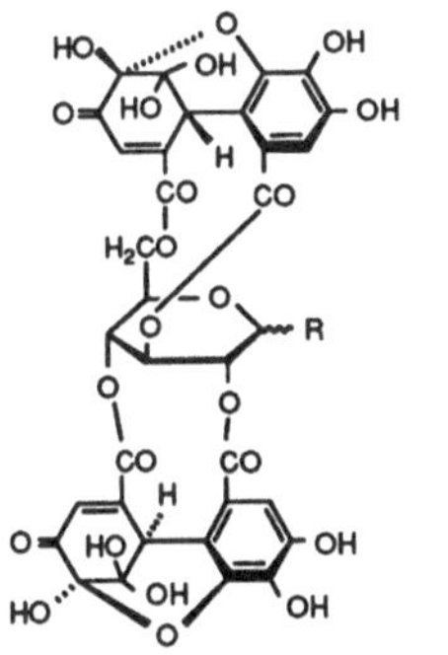

(66) Furosinin: R = OH
(67) Dehydrogeraniin: R = (β)-OG

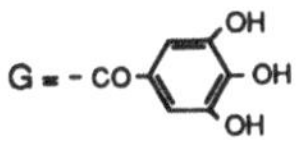

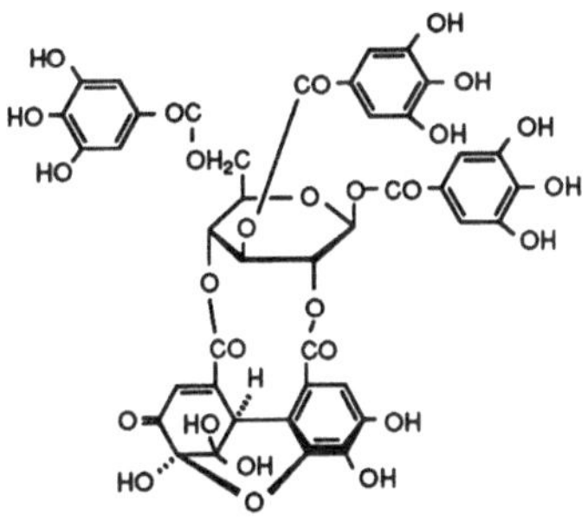

G = - CO

(68) Terchebin

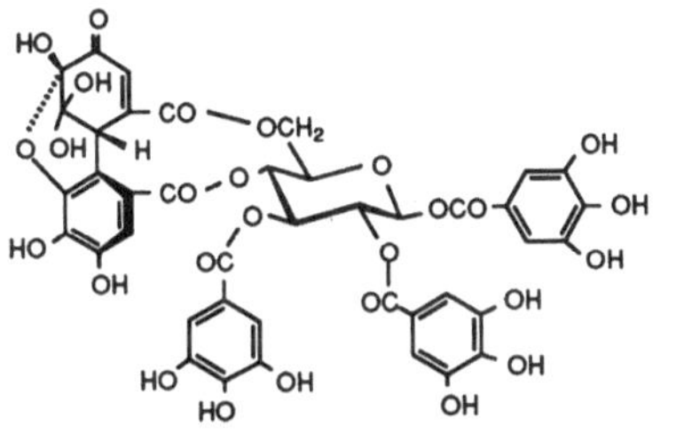

(69) Isoterchebin

from *Cytinus hypocistis* (*76*), and later from *Cornus officinalis*, was similarly formulated as **69** (*77*). The other examples (**70–79**) of monomeric dehydroellagitannins isolated to date are shown in Table 2.

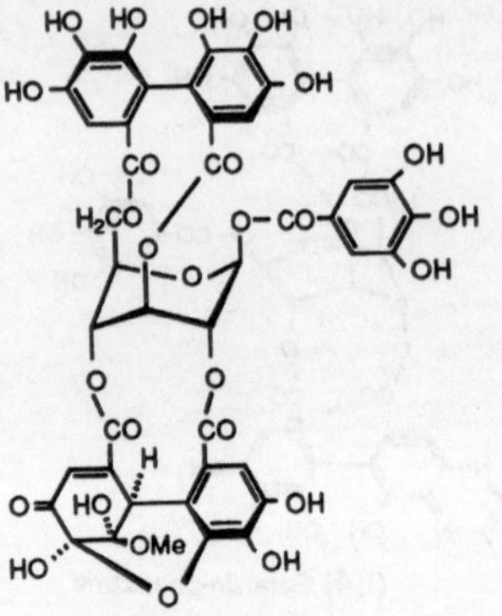

(63c)

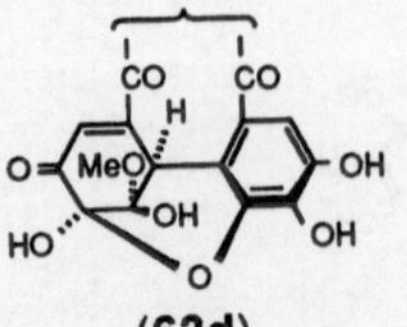

(63d)

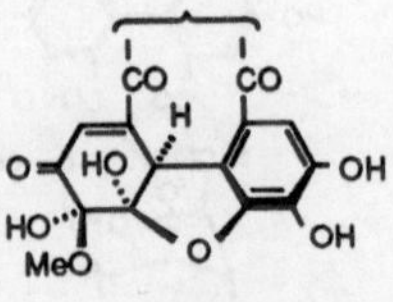

(63e)

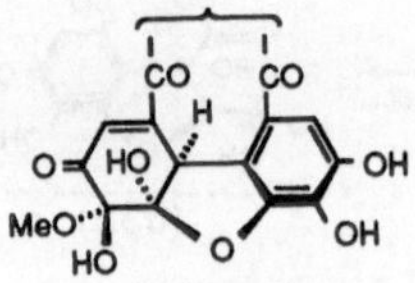

(63f)

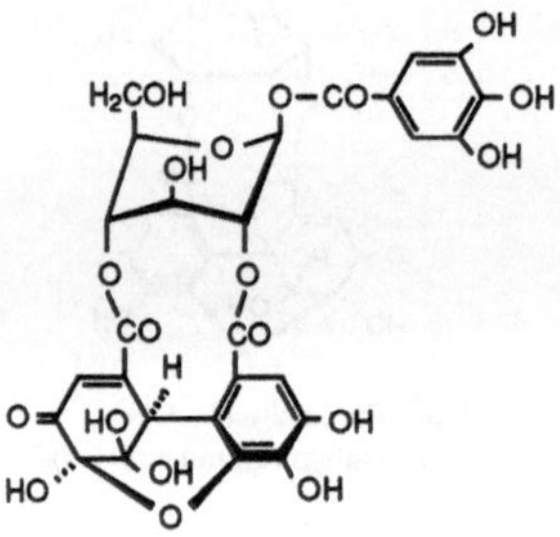

(70) Furosin

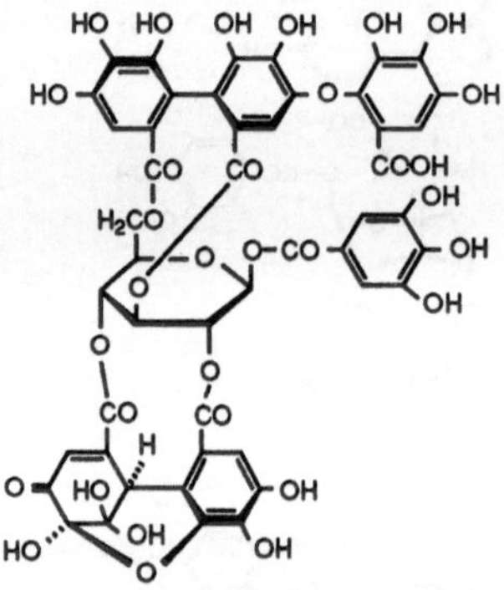

(71) Mallotusinic acid

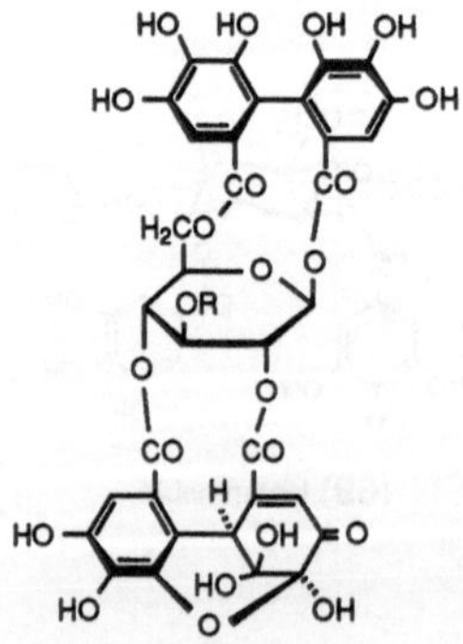

(72) Granatin A: R = H
(73) Helioscopinin A: R = G

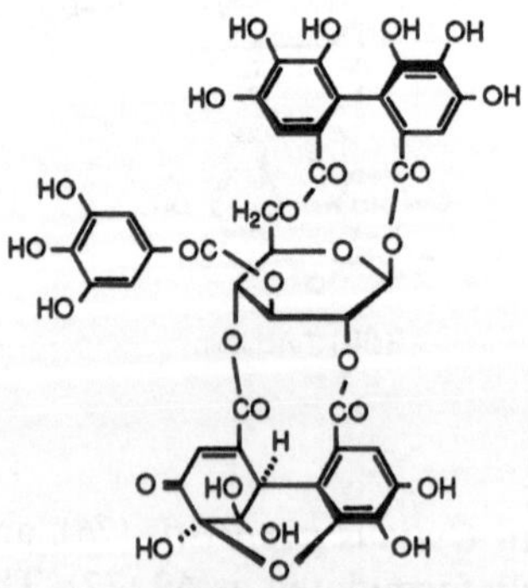

(74) Carpinusin

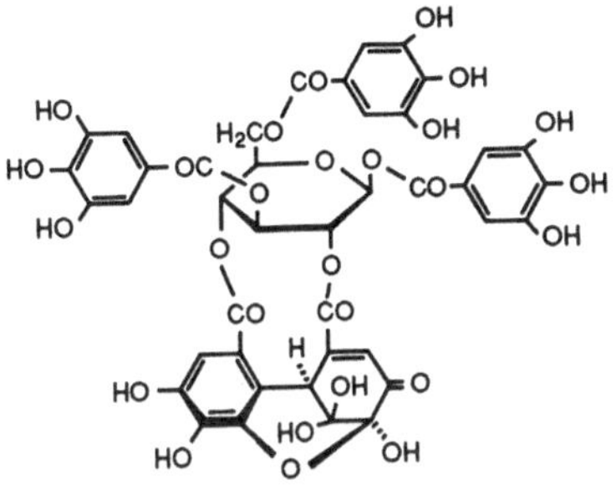

(**75**) Supinanin

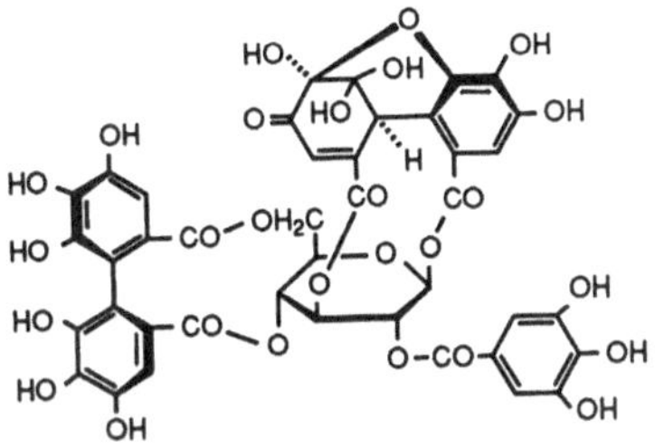

(**76**) Euphorscopin

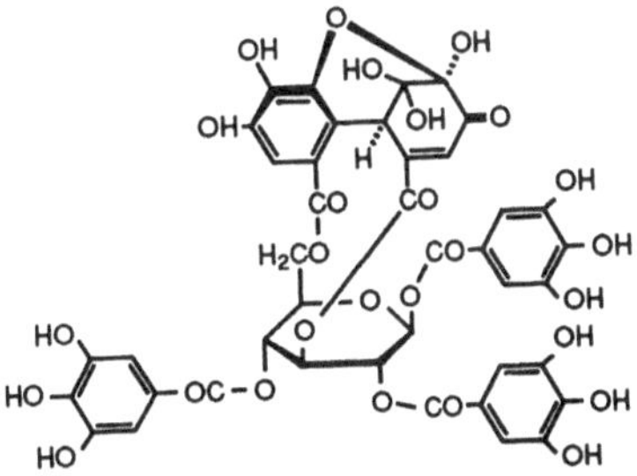

(**77**) Tanarinin

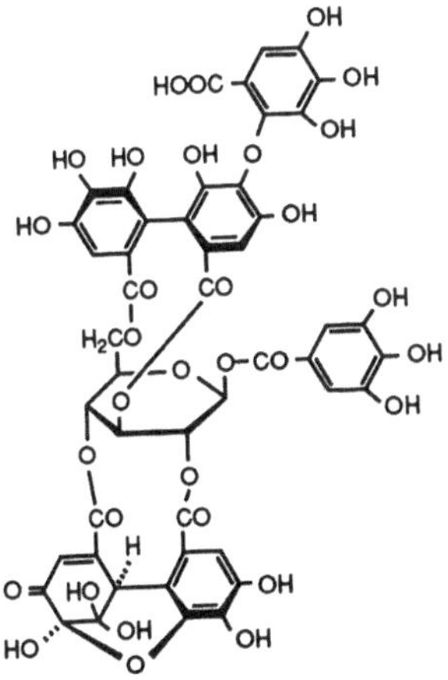

(**78**) Macarinin A

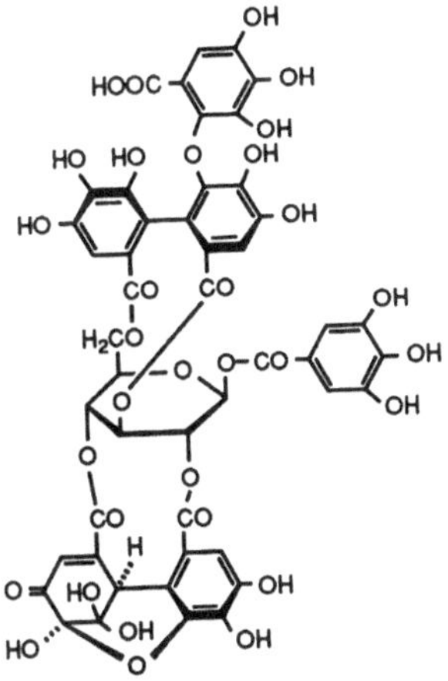

(**79**) Macaranin C

Table 2. *Monomeric Dehydroellagitannins*

Compound	Plant Sources[a]	Family	Ref.
Geraniin (63)	*Geranium thunbergii*	Geraniaceae	(70)
Furosinin (66)	*G. thunbergii*		(73)
Didehydrogeraniin (67)	*G. thunbergii*		(73)
Furosin (70)	*G. thunbergii*		(73)
Mallotusinic acid (71)	*Mallotus japonicus*	Euphorbiaceae	(79)
Granatin A (72)	*Punica granatum*	Punicaceae	(78)
Helioscopinin A (73)	*Euphorbia helioscopia*	Euphorbiaceae	(80)
Carpinusin (74)	*Carpinus laxiflora*	Betulaceae	(80, 81)
Supinanin (75)	*Euphorbia supina*	Euphorbiaceae	(82)
Euphorscopin (76)	*E. supina*		(82)
Tanarinin (77)	*Macaranga tanarius*		(83)
Macarinin A (78)	*Macaranga sinensis*		(84)
Macaranin C (79)	*M. sinensis*		(84)

[a] Plant from which the tannin was first isolated

2.2.3. *Optical Isomerism of the HHDP and DHHDP Groups and CD Spectra*

Atropisomerism of the HHDP Group. The atropisomerism of the HHDP group in the ellagitannins can be determined by measuring the specific optical rotation of chiral dimethyl hexamethoxydiphenate (80)

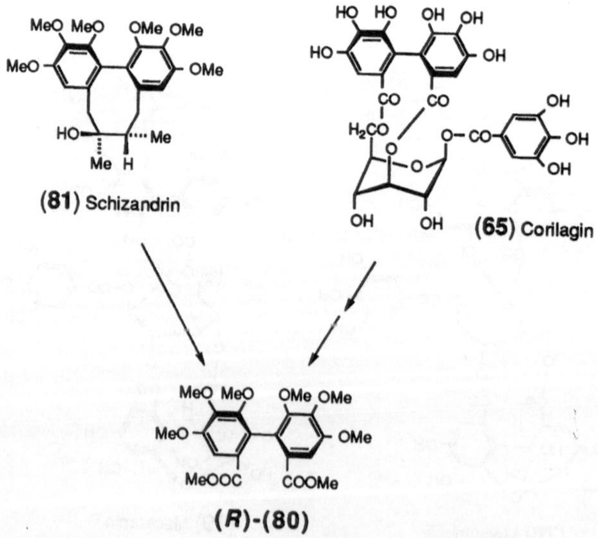

(81) Schizandrin

(65) Corilagin

(R)-(80)

Scheme 1

(Scheme 1) obtained by methanolysis of methylated ellagitannins (*70*). This determination is based on the optical identity of the dextrorotatory **80** obtained from corilagin (**65**) with that derived from schizandrin (**81**), a lignan from the fruits of *Schizandra chinensis*. The chirality of the biphenyl moiety in **81** was established as *R* based on an X-ray crystallographic analysis of the derivative of its antipodal congener, gomisin D (*85*). The *R*-configuration of (+)-**80** and the *S*-configuration of (−)-**80** are evidenced in the CD spectra in the form of split-type Cotton effects of opposite sign at 225 and 250 nm (*86*) (Fig. 1).

CD spectral data of a number of ellagitannins for which the absolute configuration of the HHDP group(s) was determined *via* methanolysis as

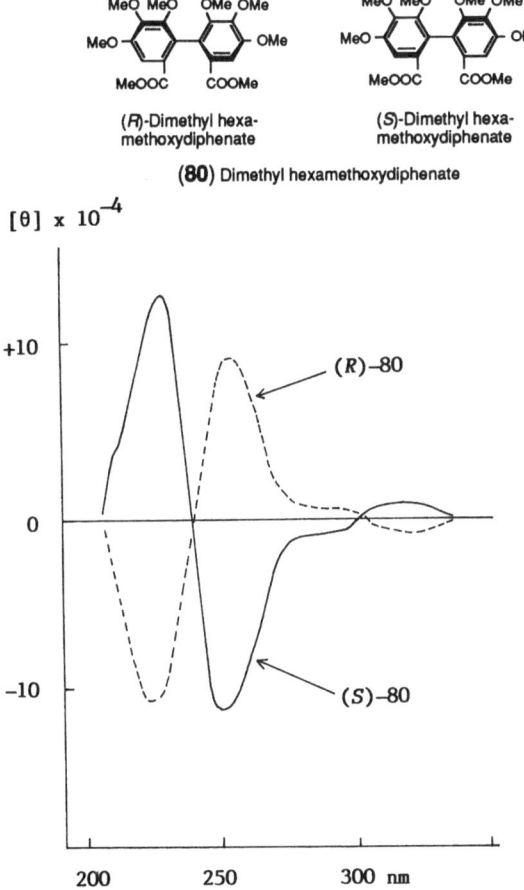

Fig. 1. Structures of chiroptical dimethyl hexamethoxydiphenates and their CD spectra

described above in the preceding paragraphs now allow one determine their absolute stereostructures by a convenient empirical method based on the CD spectra without the necessity of performing a chemical degradation (86).

HHDP-glucoses having an (R)- or (S)-HHDP group show Cotton effects near 235 and 265 nm, that near 265 nm when positive being due to (R)- and when negative being due to (S)-configuration of the HHDP group (see Fig. 1). On the other hand, galloylglucoses such as 1,2,6- and 1,2,3-tri-O-galloyl-β-D-glucoses exhibit a split Cotton effect centered at 270 nm (maxima at *ca.* 280 and 260 nm of opposite sign), which corresponds to an intramolecular charge-transfer transition (277 nm in the UV spectra) of the galloyl groups. The sign of the first Cotton effect near 280 nm is in agreement with the chirality between adjacent galloyl groups assigned by the exciton chirality rule (87). The sign and amplitude of the Cotton effects centered near 270 nm in the CD spectra of polygalloylglucoses are therefore the sum of the chiralities among the various galloyl groups (86).

The CD spectra of ellagitannins having both galloyl and HHDP groups show three Cotton effects at around 235, 265 and 285 nm. The Cotton effects near 265 nm result from overlapping of the longer-wavelength Cotton effect from the HHDP group with the shorter-wavelength Cotton effect from the galloyl groups (or the galloyl and HHDP groups). The Cotton effect at around 235 nm is therefore diagnostic for the absolute configuration of the HHDP group in an ellagitannin molecule [positive for the (S)-configuration and negative for the (R)-configuration] as shown in the CD spectra of 31, 34 and 65. The amplitude of this Cotton effect is also indicative of the number of HHDP groups in the molecule as shown by that of pedunculagin (31), the amplitude of whose CD curve near 235 nm is twice as large as that of tellimagrandin I (34) (Fig. 2). The empirical rule thus derived for the HHDP group is applicable to the absolute configuration of the valoneoyl group and analogs in both monomeric and oligomeric hydrolyzable tannins.

Absolute Configuration of the DHHDP Group. The absolute configuration of the asymmetric methine carbon (C-1′) within the DHHDP group was first established for geraniin (63) (70) whose CD spectrum is shown in Fig. 3. The R-configuration assigned to the phenylphenazine moiety in phenazine derivative (64) from geraniin, shown in Scheme 2, was based on the significant upfield shift of the glucose H-1 and the downfield shift of the H-5 signal upon transformation of 63 to 64. The configuration at C-1′ in geraniin was then assigned as R, since the R-phenylphenazine part could be formed only from a precursor having this

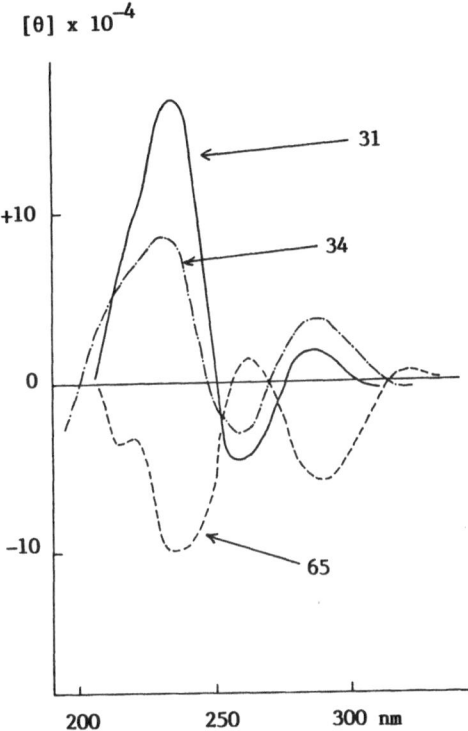

$[\theta] \times 10^{-4}$

+10

0

−10

31

34

65

200 250 300 nm

Fig. 2. CD spectra of pedunculagin (**31**), tellimagrandin I (**34**) and corilagin (**65**)

stereostructure. The sign of specific rotation of dimethyl ester (**82**), prepared by methylation of **64** followed by methanolysis, was positive (*70*).

The antipodal dimethyl ester, (−)-**82**, was obtained from isoterchebin [1,2,3-tri-*O*-galloyl-4,6-(1′*S*)-DHHDP-β-D-glucose] (**69**) (CD spectrum in Fig. 3) in a way similar to the preparation of (+)-**82** from **63**, which showed that **69** had the (1′*S*)-configuration. Additional evidence for this assignment was obtained by reduction of isoterchebin (**69**) with sodium dithionite or by catalytic hydrogenation, which afforded tellimagrandin II [1,2,3-tri-*O*-galloyl-4,6-*O*-(*S*)-HHDP-β-D-glucose] (**35**) (*70*). The *S*-configuration of the HHDP group in **35** was confirmed chemically and by the CD spectrum as described above.

The absolute stereochemistries of the DHHDP group in terchebin (**68**) (*75*), mallotusinic acid (**71**) (*79*) and the other dehydroellagitannins

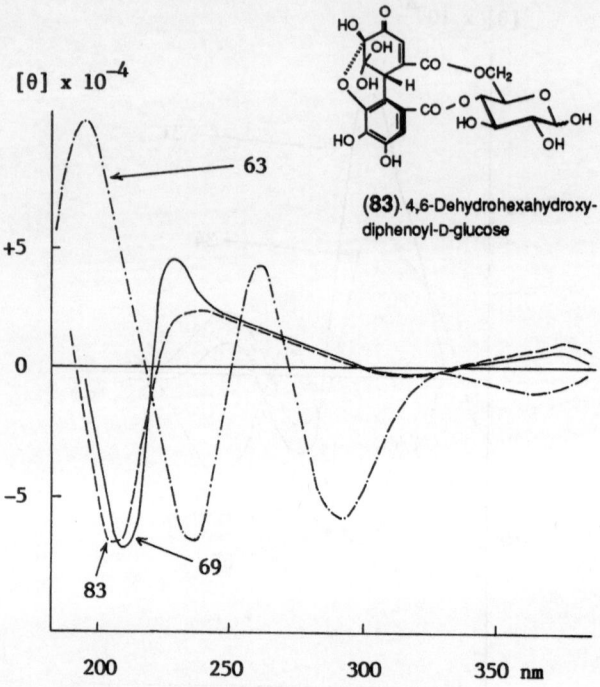

Fig. 3. CD spectra of geraniin (**63**), isoterchebin (**69**) and 4,6-DHHDP-glucose (**83**)

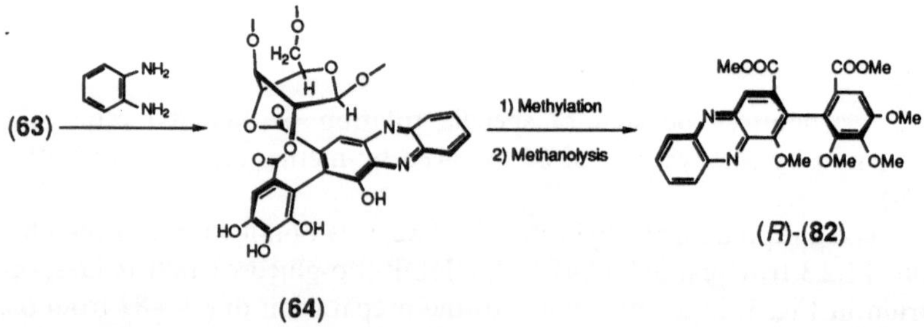

Scheme 2

were subsequently determined from the specific rotation of the phenyl-phenazine derivative (**82**). The CD spectrum of 4,6-DHHDP-glucose (**83**) (Fig. 3) showed Cotton effects characteristic of an α,β-unsaturated ketone at about 370, 235 and 210 nm due to the n-π* and the first and second π-π* transitions (*88*). However, in dehydroellagitannins with an HHDP group the Cotton effect at around 235 nm was overlapped by that of the HHDP group. The prominent peak at around 210 nm was thus correlated with the absolute configuration at C-1′ of the DHHDP group, being positive for the (1′R)- and negative for the (1′S)-configuration (see Fig. 3) (88). CD spectra of the phenazine derivatives of dehydroellagitannins show strong Cotton effects of opposite sign at around 280 nm and 250 nm which are useful for determining the absolute configuration at the phenylphenazine moiety (88).

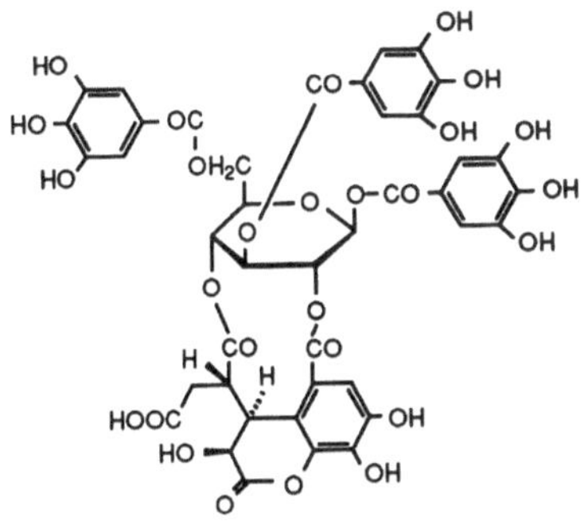

(84) Chebulinic acid

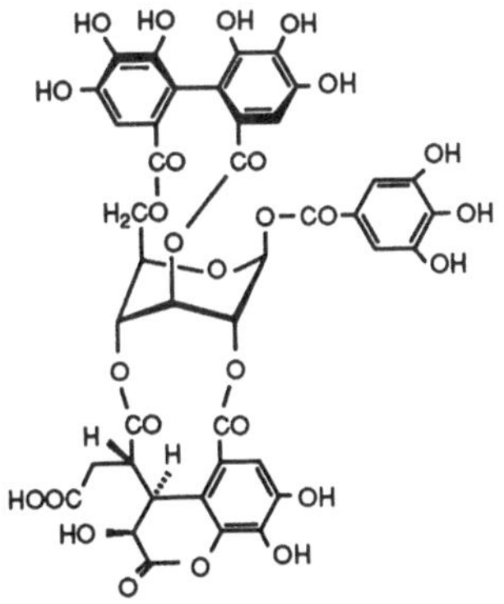

(85) Chebulagic acid

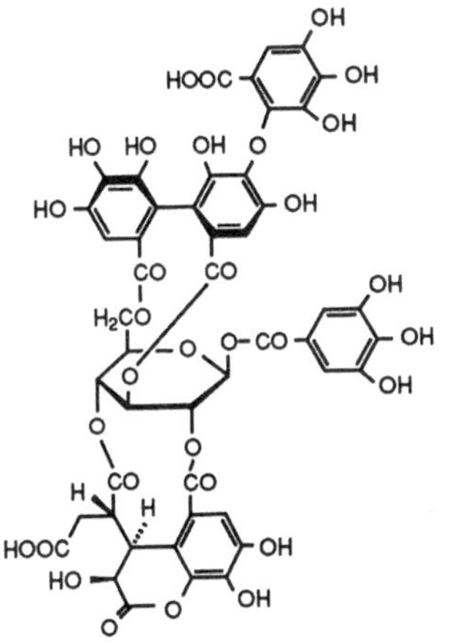

(86) Macarinin B

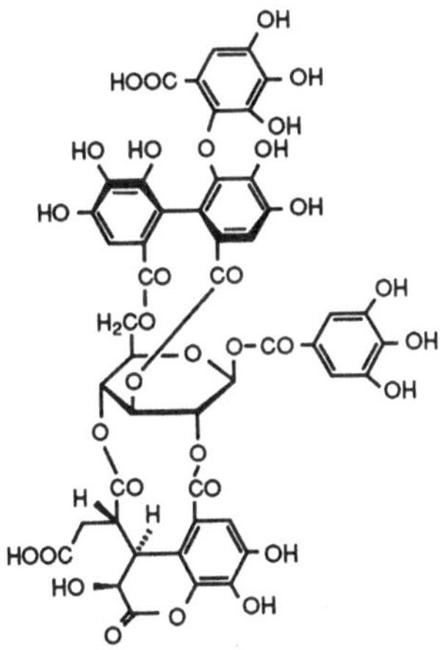

(87) Macarinin C

2.2.4. Substances Formed by Oxidation of the DHHDP Group

The structures of chebulinic acid and chebulagic acid from the myrobalans, an important traditional medicine in India, have been extensively studied since the middle of the 19th century (2, 3). The structures of these compounds, including the absolute configurations, were finally established as **84** and **85** (89) in 1982 after several revisions (90). Macarinins B (**86**) and C (**87**), analogs of **85** having a tergalloyl or a macaranoyl group in addition to the chebuloyl group, were isolated from

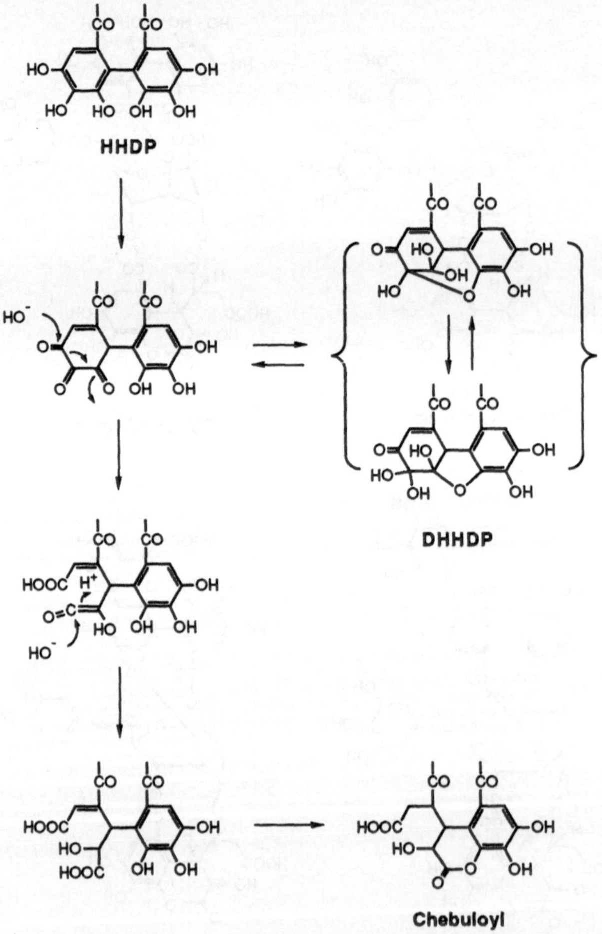

Scheme 3

Macaranga sinensis (*84*). The chebuloyl group in these compounds is considered to be the result of oxidation of the HHDP group (*2, 3*). Biogenetically the chebuloyl group can be formed from the HHDP group via the DHHDP group by a benzylic acid rearrangement-like C–C bond cleavage, as illustrated in Scheme 3.

Analogous metabolites, phyllanthusiins A (**88**), B (**89**) and C (**90**) and repandusinic acid (**91**), are produced by *Phyllanthus flexuosus* along with geraniin (**63**), corilagin (**65**) and putranjivain A (**97**) (see below) (*91*). Geraniinic acids B (**92**) and C (**93**) were isolated from *Geranium thunbergii* (*92*). These tannins are regarded as being formed by oxidative modifications of the DHHDP group of geraniin. Heterophylliin E (**94**) and repandusinin (**95**), isolated from *Corylus heterophylla* (*93*) and *Mallotus*

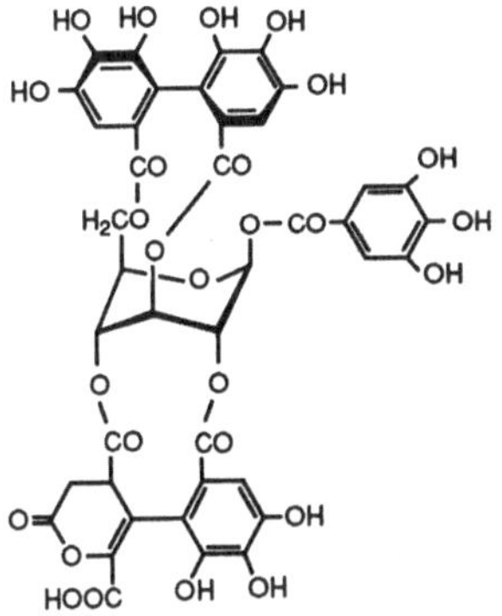

(**88**) Phyllanthusiin A

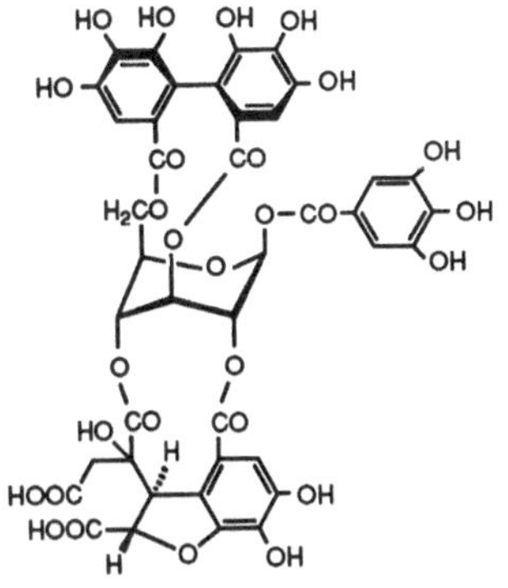

(**89**) Phyllanthusiin B

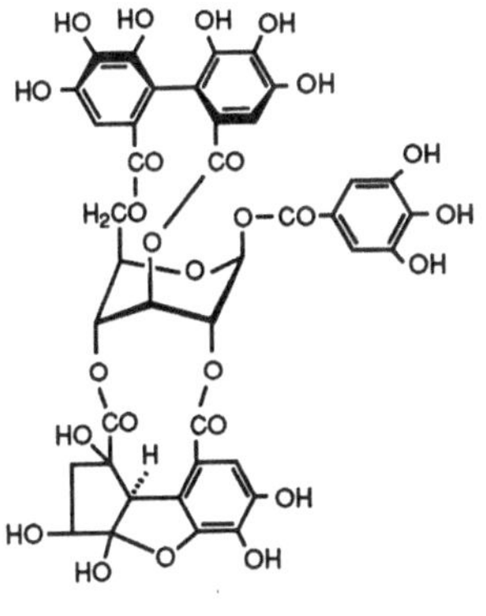

(**90**) Phyllanthusiin C

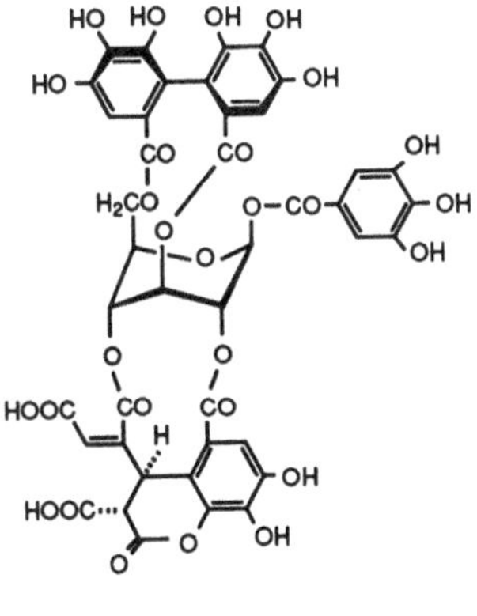

(**91**) Repandusinic acid

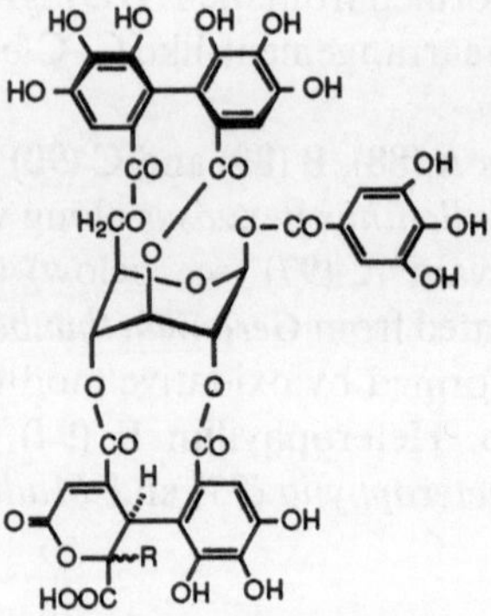

(92) Geraniinic acid B: R = (α)-H
(93) Geraniinic acid C: R = (β)-H

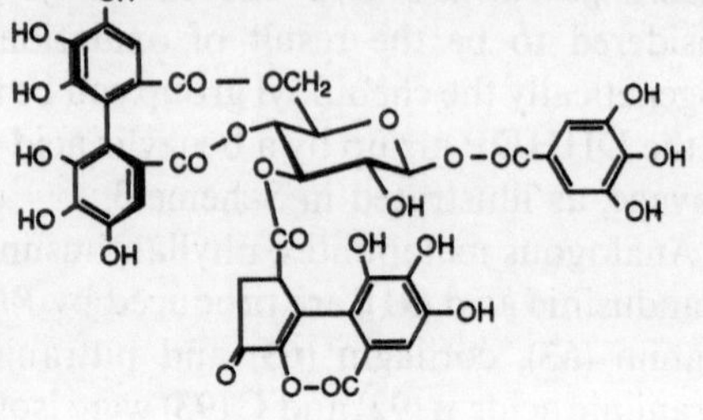

(94) Heterophylliin E

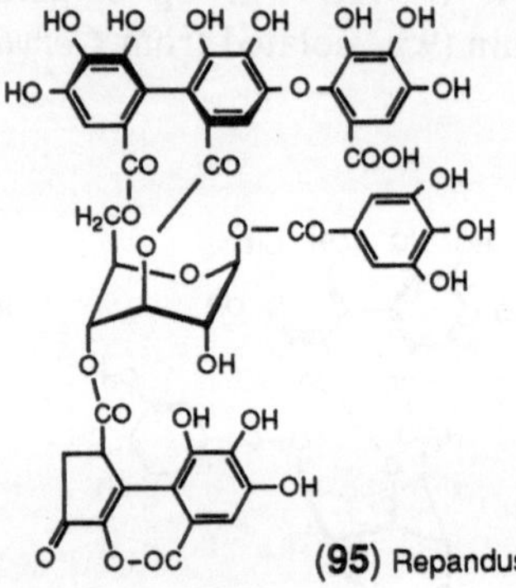

(95) Repandusinin

repandus (94), respectively, contain a brevifolincarboxyl group presumably formed from HHDP via a DHHDP group.

2.3. Condensates of Geraniin with Ascorbic Acid and Other Compounds

A condensate of geraniin with ascorbic acid named ascorgeraniin (elaeocarpusin) (96) was isolated from Geranium thunbergii (95–97), and also from Acer nikoense (96) and Elaeocarpus sylvestris (97). It co-occurs with geraniin (63) in Acer, Rhus, and Cercidiphyllum species (97) and was synthesized by condensation of 63 with ascorbic acid in moderately acidic aqueous solution or in aqueous methanol at room temperature (96, 98). An analog, putranjivain A (97), was isolated from several euphorbiaceous plants, Macaranga sinensis (84), Euphorbia tirucalli (57) and Phyllanthus flexuosus (91). Phyllanthusiin D (98), a condensate of geraniin with acetone, was isolated from acetone or aqueous acetone homogenates of P. flexuosus (91), P. amarus (= P. niruri) (99), and also from suspension

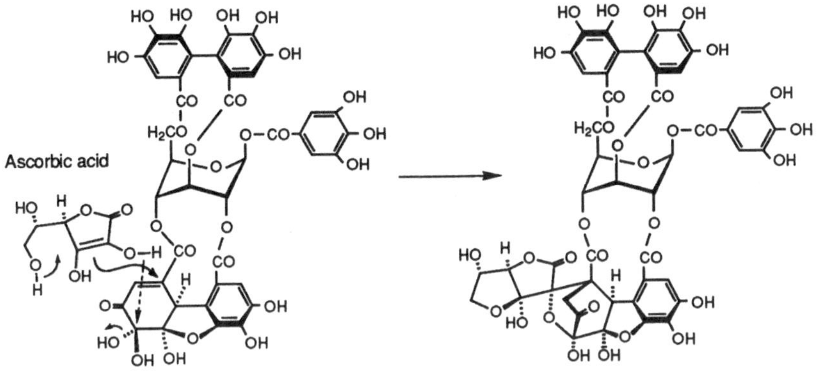

(63) Geraniin

(96) Ascorgeraniin
(= Elaeocarpusin)

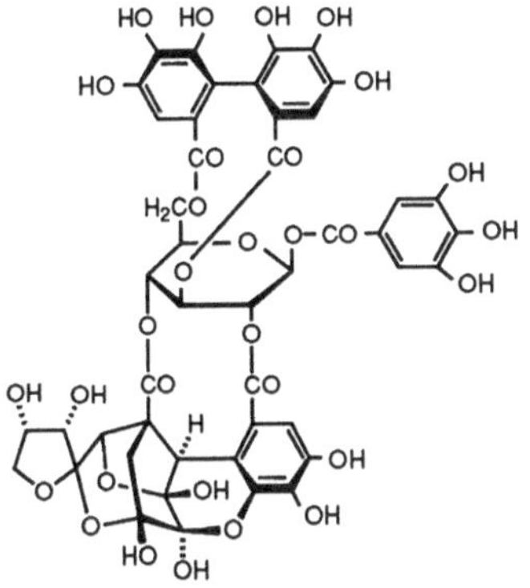

(97) Putranjivain A

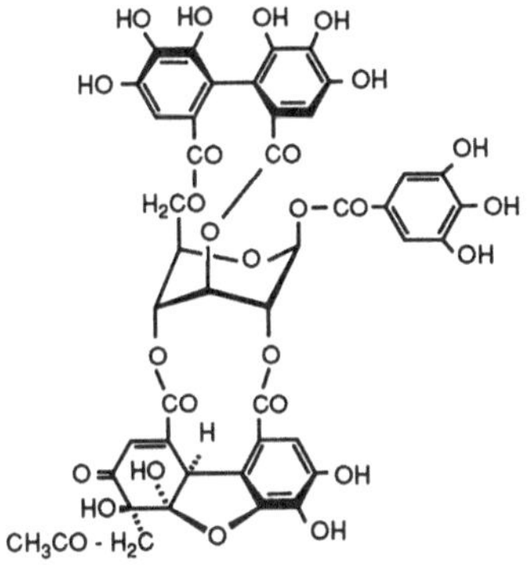

(98) Phyllanthusiin D

Chart 2

cultures of *Geranium thunbergii* (*100*). This compound was produced when a solution of geraniin in dry acetone containing a small amount of trifluoroacetic acid was refluxed. Therefore, it is likely that phyllanthusiin D (**98**) is an artefact formed during the extraction procedure.

2.4. *C*-Glucosidic Tannins and Complextannins

In addition at castalagin (**99**) and vescalagin (**100**), *C*-glucosidic monomers earlier found in the wood of *Castanea* and *Quercus* species (*101*), analogous *C*-glucosidic ellagitannins were isolated from *Casuarina* *stricta* (*52, 102*). These tannins, casuarinin (**101**), stachyurin (**102**) and casuariin (**103**), are widely distributed in various plant families such as the Casuarinaceae, Stachyuraceae, Myrtaceae (*68*), Betulaceae, Fagaceae, Hamamelidaceae, Lythraceae, Punicaceae, Melastomataceae, Rosaceae, Elaeagnaceae, Theaceae and Juglandaceae (*6*).

The complextannin monomers are usually composed of a *C*-glucosidic tannin monomer and flavan unit linked to each other through a C-C bond between C-1 of glucose and C-8 or C-6 of the flavan (*103*). They are widely distributed in the Fagaceae, Combretaceae and Myrtaceae and reflect broad distribution of the *C*-glucosidic tannin monomers and the flavans in these families (*6*). Some species of Theaceae and Melastomataceae also produce tannins of this type, such as the camelliatannins A (**104**), B (**105**) and F (**107**) in the leaf of *Camellia japonica* (*104, 105*), and the malabathrins A (**106**) and E (**108**) in *Melastoma malabathricum* (*106*). Some complextannins are made up of several oligomeric hydrolyzable tannin molecules which will be discussed in the sequel.

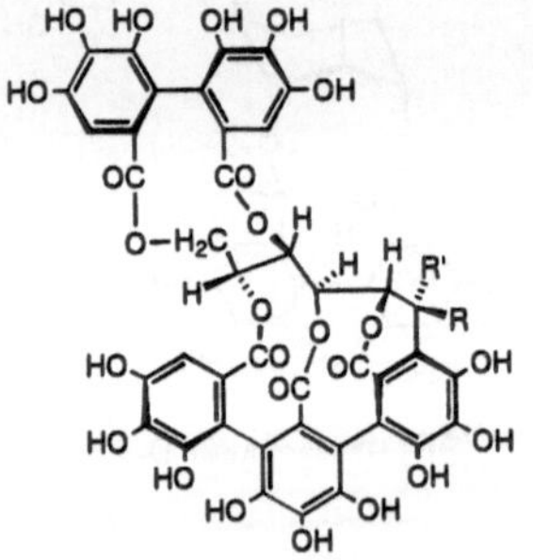

(**99**) Castalagin: R = H, R'= OH
(**100**) Vescalagin: R = OH, R'= H

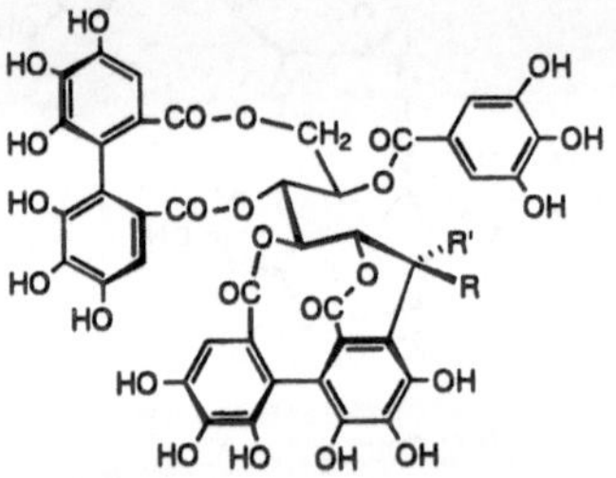

(**101**) Casuarinin: R = H, R'= OH
(**102**) Stachyurin: R = OH, R'= H

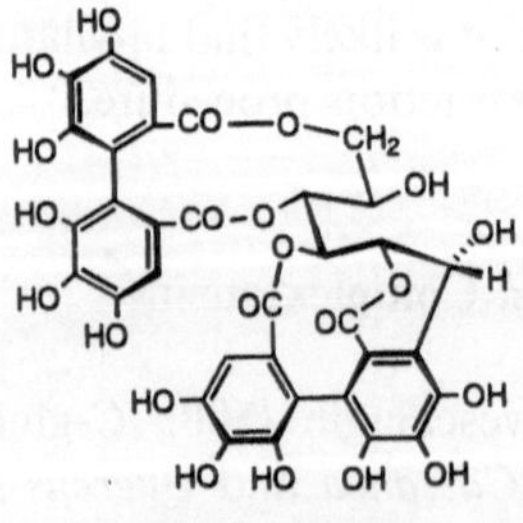

(**103**) Casuariin

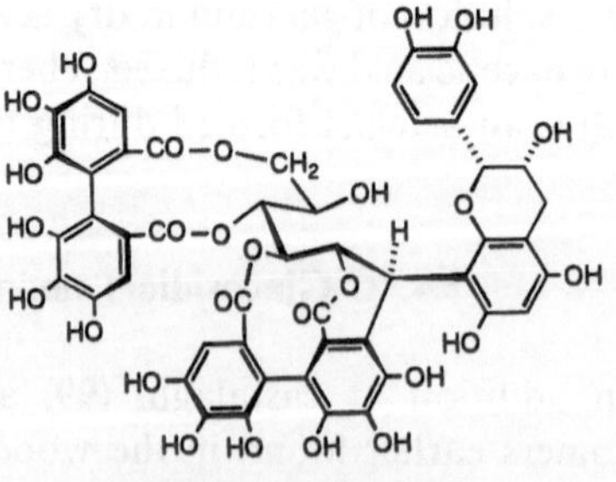

(**104**) Camelliatannin A

(105) Camelliatannin B: R = H
(106) Malabathrin A: R = G

G = - CO

(107) Camelliatannin F: R = H
(108) Malabathrin E: R = G

3. Oligomeric Hydrolyzable Tannins

The first dimeric ellagitannin to be reported was agrimoniin (109) (*107*) which is composed of two moles of potentillin (33), an α-glucosidic ellagitannin monomer (*62, 107*). Since the isolation of this dimer from *Agrimonia pilosa* in 1982, about 150 additional oligomers of various structures and molecular weights up to 3745 (nobotanin K, one of the tetramers from *Heterocentron roseum*) have been isolated by the end of 1992 and their structures elucidated (*6*). Their structural variety has been enhanced by multiplication of the monomer units which frequently differ from each other within a single oligomer molecule, and also by variations in the mode (linkage type) and extent of condensation among the mono-mers (*4, 6*). About 86% of these oligomers are dimers of diverse structure, about 11% are trimers, and five oligomers are tetramers.

3.1. Classification of Oligomeric Hydrolyzable Tannins According to the *O*-Donating Monomer

Although not many monomer structures are encountered often as constituents of oligomers, some monomers occurring in plants frequently and in large amount are also frequently found as constituents of various oligomer molecules. Examples are the tellimagrandins I (34) and II (35), pedunculagin (31) and casuarictin (32). Potentillin (33), strictinin (49), rugosin A (36), rugosin B (37), praecoxin B (53), pterocaryanin C (54), nupharin A (61), geraniin (63), corilagin (65), castalagin (99), vescalagin (100), casuarinin (101), stachyurin (102), sanguiin H-4 [1-O-galloyl-2,3-O-HHDP-α-D-glucose] and davidiin [1,6-O-(S)-HHDP-2,3,4-tri-O-galloyl-β-D-glucose] are also found comparatively often as constituent of oligomer molecules.

All oligomers found to date are listed in Table 3 and are arranged to the structural type of the monomer one of whose hydroxyl oxygens forms the ether linkage between the two monomer units participating in dimer formation. Some structural features of these O-donating monomers, within each structural type of oligomers in the classification of Table 3, are the following.

A) HHDP (or valoneoyl) group at O-2/O-3 of glucose
B) Galloyl groups at O-2 and O-3 of glucose.
C) HHDP group at O-3/O-6 of glucose.
D) C-Glucosidic monomer attached to a second monomer by a C–C bond.

Each structural type in Table 3 is further subdivided into several subtypes depending on the method of linkage between the monomers described in Subsect. 3.2. Clear correlations are generally observed between the distribution of oligomer types and plant taxonomy.

Several monomers of unusual structure are constituents of certain oligomers and will be discussed separately.

3.1.1. Geraniin in Oligomer Molecules

Geraniin (63), which has a DHHDP group and is a key compound in the metabolism of many hydrolyzable tannins having the $^1C_4$ glucose core (3, 6), has also been found to be the building block of several oligomeric hydrolyzable tannins isolated from some euphorbiaceous plants such as the euphorbins A–C (Euphorbia hirta) (57, 108, 109), euphorbin F (E. tirucalli) (57), antidesmin A (Antidesma pentandrum var. burbatum) (110), and bischofianin (Bischofia javanica) (111), etc. Another monomer [helioscopinin A (73) (80)] having the DHHDP group has been found in jolkianin, a dimer from E. jolkini (82). Such oligomers, having some dehydroellagitannin monomer as a monomeric structural unit, are specific for several species of Euphorbiaceae.

However, several monomers, which are presumed to be oxidative metabolites of geraniin (63), such as chebulagic acid (85) and the

Table 3. *Structural Units and Plant Sources of Oligomeric Hydrolyzable Tannins*

Compound	Composition	Linking unit	Extent of oligomerization	Main plant source	Plant family	Ref.
A) Oligomers having HHDP group at O-2/O-3						
A1) Oligomers having *m*-GOG (DHDG) unit						
Agrimoniin (**109**)	(**33**)-[*m*-GOG(1-1')]-(**33**)	*m*-GOG	Dimer	*Agrimonia pilosa* Koidz.	Rosaceae	*107*
Laevigatin B (**110**)	(U1)-[*m*-GOG(1-1')]-(**33**)	*m*-GOG	Dimer	*Rosa laevigata* Michx.	Rosaceae	*114*
Laevigatin C (**111**)	(**33**)-[*m*-GOG(1-1')]-(U1)	*m*-GOG	Dimer	*R. laevigata*	Rosaceae	*114*
Laevigatin D (**112**)	(U2)-[*m*-GOG(2-1')]-(**33**)	*m*-GOG	Dimer	*R. laevigata*	Rosaceae	*114*
Laevigatin E (**113**)	(U1)-[*m*-GOG(1-1')]-(U1)	*m*-GOG	Dimer	*R. laevigata*	Rosaceae	*115*
Laevigatin F	(**33**)-[*m*-GOG(1-1')]-(U3)	*m*-GOG	Dimer	*R. laevigata*	Rosaceae	*115*
Laevigatin G	(U1)-[*m*-GOG(1-1')]-(U3)	*m*-GOG	Dimer	*R. laevigata*	Rosaceae	*115*
Davuriciin D2 (**114**)	(U4)-[*m*-GOG(1-1')]-(**33**)	*m*-GOG	Dimer	*Rosa davurica* Pall.	Rosaceae	*116*
Davuriciin T1	{(**33**)-[*m*-GOG(1-1')]-(**33**)}-[*m*-DOG (4',6'-1'')]-(**49**)	*m*-GOG + *m*-DOG	Trimer	*R. davurica*	Rosaceae	*116*
A2) Oligomers having *m*-GOD (sanguisorboyl) unit						
Sanguiin H-3	(**50**)-[*m*-GOD(1-4',6')]-(**33**)	*m*-GOD	Dimer	*Sanguisorba officinalis* L.	Rosaceae	*112*
Sanguiin H-6 (**153**)	(**32**)-[*m*-GOD(1-4',6')]-(**33**)	*m*-GOD	Dimer	*S. officinalis*	Rosaceae	*112*
Sanguiin H-8	(**50**)-[*m*-GOD(1-4',6')]-(U3)	*m*-GOD	Dimer	*S. officinalis*	Rosaceae	*112*
Sanguiin H-9	(**49**)-[*m*-GOD(1-4',6')]-(U3)	*m*-GOD	Dimer	*S. Officinalis*	Rosaceae	*112*
Sanguiin H-10	(**49**)-[*m*-GOD(1-4',6')]-(**33**)	*m*-GOD	Dimer	*S. officinalis*	Rosaceae	*112*
Sanguiin H-11 (**158**)	(**32**)-[*m*-GOD(1-4',6')]-(**33**)-[*m*-GOD(1-4',6')]-(**32**)-[*m*-GOD(1-4',6')]-(**33**)	*m*-GOD × 3	Tetramer	*S. officinalis*	Rosaceae	*112*
Roshenin A (**154**)	(U5)-[*m*-GOD(1-4',6')]-(**33**)	*m*-GOD	Dimer	*Rosa henryi* Boul	Rosaceae	*140*
Roshenin B (**155**)	(**32**)-[*m*-GOD(1-4',6')]-(**31**)	*m*-GOD	Dimer	*R. henryi*	Rosaceae	*140*
Roshenin C	(**32**)-[*m*-GOD(1-4',6')]-(U3)	*m*-GOD	Dimer	*R. henryi*	Rosaceae	*140*

Table 3 *(continued)*

Compound	Composition	Linking unit	Extent of oligomerization	Main plant source	Plant family	Ref.
Roshenin D	(U6)-[*m*-GOD(1-4',6')]-(33)	*m*-GOD	Dimer	*R. henryi*	Rosaceae	*140*
Roshenin E	(U2)-[*m*-GOD(1-4',6')]-(32)	*m*-GOD	Dimer	*R. henryi*	Rosaceae	*140*
Lambertiannin C (156)	(32)-[*m*-GOD(1-4',6')]-(32)-[*m*-GOD(1-4',6')]-(33)	*m*-GOD × 2	Trimer	*Rubus lambertianus* Seringe	Rosaceae	*141*
Lambertiannin D (157)	(32)-[*m*-GOD(1-4',6')]-(32)-[*m*-GOD(1-4',6')]-(32)-[*m*-GOD(1-4',6')]-(32)	*m*-GOD × 3	Tetramer	*R. lambertianus*	Rosaceae	*141*
A3) Oligomers having *m*-DOG (valoneoyl) unit at O-2/O-3						
Camelliatannin D (141)	(U7)-[*m*-DOG(3,2-2')]-(34)	*m*-DOG	Dimer	*Camellia japonica* L.	Theaceae	*105*
Camelliatannin H	(34)-[*m*-DOG(3,2-2')]-(34)	*m*-DOG	Dimer	*C. japonica*	Theaceae	*105*
Nobotanin B (136)	(32)-[*m*-DOG(3,2-4')]-(54)	*m*-DOG	Dimer	*Tibouchina semidecandra* Cogn.	Melastomataceae	*128*
Nobotanin G	(32)-[*m*-DOG(3,2-4')]-(U8)	*m*-DOG	Dimer	*Heterocentron roseum* A. Br. et Bouch.	Melastomataceae	*72*
Nobotanin H	(32)-[*m*-DOG(3,2-4')]-(U9)	*m*-DOG	Dimer	*H. roseum*	Melastomataceae	*72*
Nobotanin I	(32)-⟨*m*-DOG(3,2-4')]-(U10)	*m*-DOG	Dimer	*H. roseum*	Melastomataceae	*72*
Nobotanin J	(32)-[*m*-DOG(3,2-4')]-(54)-[*m*-DOG(2',3'-1'')]-(32)	*m*-DOG × 2	Trimer	*H. roseum*	Melastomataceae	*148*
Malabathrin B	(32)-[*m*-DOG(3,2-4')]-(53)	*m*-DOG	Dimer	*Melastoma malabathricum* L.	Melastomataceae	*148*
Malabathrin C	(31)-[*m*-DOG(3,2-4')]-(54)	*m*-DOG	Dimer	*M. malabathricum*	Melastomataceae	*148*
Reginin D (145)	(31)-[*m*-DOG(3,2-5' or 2,3-5')]-(101)	*m*-DOG	Dimer	*Lagerstroemia flos-regina* Retz.	Lythraceae	*137*

A4) Oligomers having *m*-DOG (valoneoyl) unit at O-4/O-6

Compound	Structure	Unit	Type	Species	Family	Ref.
Rugosin F (128)	(32)-[*m*-DOG(4,6-1')]-(35)	*m*-DOG	Dimer	*Rosa rugosa* Thunb.	Rosaceae	*121*
Degalloylrugosin F	(31)-[*m*-DOG(4,6-1')]-(35)	*m*-DOG	Dimer	*Corylus heterophylla* Fisch.	Betulaceae	*138*
Heterophylliin D	(32)-[*m*-DOG(4,6-1')]-(32)	*m*-DOG	Dimer	*C. heterophylla*	Betulaceae	*138*
Davuricin D1	(32)-[*m*-DOG(4,6-1')]-(49)	*m*-DOG	Dimer	*Rosa davurica* Pall.	Rosaceae	*116*
Roxbin A	(31)-[*m*-DOG(4,6-1')]-(32)	*m*-DOG	Dimer	*Rosa roxburghii* Tratt.	Rosaceae	*149*
Calamanin B	(33)-[*m*-DOG(4,6-1')]-(35)	*m*-DOG	Dimer	*Terminalia calamansanai* Rolfe	Combretaceae	*150*
Calamanin C	(33)-[*m*-DOG(4,6-1')]-(35) – [*m*-DOG(4',6'-1'')]-(35)	*m*-DOG × 2	Trimer	*T. calamansanai*	Combretaceae	*150*
Camelliin A (172)	(31)-[*m*-DOG(4,6-2')]-(34)	*m*-DOG	Dimer	*Camellia japonica* L.	Theaceae	*134*
Nobotanin A (134)	(32)-[*m*-DOG(4,6-4')]-(53)	*m*-DOG	Dimer	*Tibouchina semidecandra* Cogn.	Melastomataceae	*127*
Nobotanin F (135)	(32)-[*m*-DOG(4,6-4')]-(54)	*m*-DOG	Dimer	*T. semidecandra*	Melastomataceae	*127*
Medinillin B	(31)-[*m*-DOG(4,6-4')]-(53)	*m*-DOG	Dimer	*Medinilla magnifica* Lindle.	Melastomataceae	*166*
Nobotanin C	{(32)-[*m*-DOG(4',6'-4)]-(53)]-[*m*-DOG(3',2'-4'')]-(54)	*m*-DOG × 2	Trimer	*Tibouchina semidecandra* Cogn.	Melastomataceae	*128*
Nobotanin E	{(32)-[*m*-DOG(4',6'-4)]-(54)-[*m*-DOG(3',2'-4'')]-(54)	*m*-DOG × 2	Trimer	*T. semidecandra*	Melastomataceae	*128*
Nobotanin K (148)	{(32)-[*m*-DOG(4',6'-4)]-(54)]-[*m*-DOG(3',2'-4'')]-(54)-[*m*-DOG(2'',3''-1''')]-(32)	*m*-DOG × 3	Tetramer	*Heterocentron roseum* A. Br. et Bouch.	Melastomataceae	*139*
Nobotanin L	{(32)-[*m*-DOG(4',6'-4)]-(53)]-[*m*-DOG(3',2'-4'')]-(53)	*m*-DOG × 2	Trimer	*Tibouchina semidecandra* Cogn.	Melastomataceae	*151*
Nobotanin M	{(32)-[*m*-DOG(3',2'-4'')]-(53)}-[*m*-DOG(4',6'-4)]-(U8)	*m*-DOG × 2	Trimer	*T. semidecandra*	Melastomataceae	*151*
Nobotanin N	{(32)-[*m*-DOG(4',6'-4)]-(53)}-[*m*-DOG(3',2'-4'')]-(U11)	*m*-DOG × 2	Trimer	*T. semidecandra*	Melastomataceae	*151*
Barringtin A	(U12)-[*m*-DOG(6,4-2')]-(U13)	*m*-GOD	Dimer	*Barringtonia asiatica* Kurz.	Lecythidaceae	*152*
Reginin A (142)	(31)-[*m*-DOG(6,4-5')]-(101)	*m*-DOG	Dimer	*Lagerstroemia flos-regina* Retz.	Lythraceae	*136*

Table 3 (continued)

Compound	Composition	Linking unit	Extent of oligomerization	Main plant source	Plant family	Ref.
Reginin B (143)	(31)-[m-DOG(6,4-5')]-(102)	m-DOG	Dimer	L. flos-regina	Lythraceae	136
Reginin C (144)	(31)-[m-DOG(6,4-5')]-(U14)	m-DOG	Dimer	L. flos-regina	Lythraceae	137
B) Oligomers having galloyl groups at O-2 and O-3						
B1) Oligomers having GOG unit						
Gemin A (115)	(35)-[m-GOG(1-1')]-(33)	m-GOG	Dimer	Geum japonicum Thunb.	Rosaceae	113
Gemin B	(35)-[m-GOG(1-1')]-(U1)	m-GOG	Dimer	G. japonicum	Rosaceae	113
Gemin C	(U15)-[m-GOG(1-1')]-(33)	m-GOG	Dimer	G. japonicum	Rosaceae	113
Coriariin A (116)	(35)-[m-GOG(1-1')]-(35)	m-GOG	Dimer	Coriaria japonica A. Gray	Coriariaceae	63
Coriariin C	(35)-[m-GOG(1-1')]-(36)	m-GOG	Dimer	C. japonica	Coriariaceae	117
Hirtellin A (121)	(35)-[m-GOG(1-2')]-(35)	m-GOG	Dimer	Reaumuria hirtella Jaub. et Sp.	Tamaricaceae	65
Hirtellin B (124)	(35)-[m-GO-m-GOG(1,2-2')]-(35)	m-GO-m-GOG	Dimer	R. hirtella	Tamaricaceae	120
Hirtellin E	(35)-[m-GO-m-GOG(1,2-2')]-(U15)	m-GO-m-GOG	Dimer	R. hirtella	Tamaricaceae	119
Hirtellin T1	m-GOG(1-2')]-(35)-[m-GO-m-GOG(1'-2'')]-(35)	m-GOG × 2	Trimer	R. hirtella	Tamaricaceae	153
Tamarixinin A (125)	(35)-[m-GO-m-GOG(1,2-2')]-(35)	m-GO-m-GOG	Dimer	Tamarix pakistanica Qaiser.	Tamaricaceae	65
Tamarixinin C (175)	(35)-[p-GOG(1-2')]-(35)	p-GOG	Dimer	T. pakistanica	Tamaricaceae	154
Hirtellin C (122)	cyclic{(35)-[p-GOG(1-2') + m-GOG(1'-2)]-(35)}	p-GOG + m-GOG	Dimer	Reaumuria hirtella Jub. et Sp.	Tamaricaceae	119
Hirtellin F (123)	cyclic{(35)-[p-GOG(1-2') + m-GOG(1'-2)]-(U15)}	p-GOG + m-GOG	Dimer	R. hirtella	Tamaricaceae	119
Tamarixinin B (174)	cyclic{(35)-[p-GOG(1-2') + m-GOG(1'-2)]-(35)}	p-GOG × 2	Dimer	Tamarix pakistanica Qaiser.	Tamaricaceae	154

B2) Oligomers having DOG unit

Rugosin D (126)	(35)-[*m*-DOG(4,6-1')]-(35)	*m*-DOG	Dimer	*Rosa rugosa* Thunb.	Rosaceae	*121*
Rugosin E (127)	(34)-[*m*-DOG(4,6-1')]-(35)	*m*-DOG	Dimer	*R. rugosa*	Rosaceae	*121*
Rugosin G (129)	(35)-[*m*-DOG(4,6-1')]-(35)-[*m*-DOG(4,6-1')]-(35)	*m*-DOG × 2	Trimer	*R. rugosa*	Rosaceae	*121*
Eusupinin A	(35)-[*m*-DOG(4,6-1')]-(36)	*m*-DOG	Dimer	*Euphorbia supina* Rafin.	Euphorbiaceae	*155*
Coriariin D	(34)-[*m*-DOG(4,6-1')]-(36)	*m*-DOG	Dimer	*Coriaria japonica* A. Gray	Coriariaceae	*117*
Coriariin E	(51)-[*m*-DOG(4,6-1')]-(35)	*m*-DOG	Dimer	*C. japonica*	Coriariaceae	*117*
Euprostin B	(34)-[*m*-DOG(4,6-1')]-(35)-[*m*-DOG(4',6'-1'')]-(35)	*m*-DOG × 2	Trimer	*Euphorbia prostrata* Ait.	Euphorbiaceae	*61*
Heterophylliin C (147)	(101)-[*m*-DOG(4,6-1')]-(32)	*m*-DOG	Dimer	*Corylus heterophylla* Fisch.	Betulaceae	*138*
Heterophylliin B (146)	(101)-[*m*-DOG(4,6-1')]-(35)	*m*-DOG	Dimer	*C. heterophylla*	Betulaceae	*138*
Woodfordin A	(35)-[*m*-DOG(4,6-2')]-(4)	*m*-DOG	Dimer	*Woodfordia fruticosa* Kurz.	Lythraceae	*132*
Woodfordin B	(U16)-[*m*-DOG(4,6-2')]-(34)	*m*-DOG	Dimer	*W. fruticosa*	Lythraceae	*132*
Woodfordin G	(39)-[*m*-DOG(4,6-2')]-(U13)	*m*-DOG	Dimer	*W. fruticosa*	Lythraceae	*143*
Woodfordin H	(39)-[*m*-DOG(4,6-2')]-(34)	*m*-DOG	Dimer	*W. fruticosa*	Lythraceae	*143*
Loropetalin A	(35)-[*m*-DOG(4,6-2')]-(3)	*m*-DOG	Dimer	*Loropetalum chinense* Oliv.	Hamamelidaceae	*156*
Eucalbanin B (176)	(34)-[*m*-DOG(4,6-2')]-(34)	*m*-DOG	Dimer	*Eucalyptus alba* Reinw.	Myrtaceae	*58*
Eucalbanin C (137)	(34)-[*p*-DOG(4,6-2')]-(34)	*p*-DOG	Dimer	*E. alba*	Myrtaceae	*58*
Isorugosin D (169)	(35)-[*m*-DOG(6,4-1')]-(35)	*m*-DOG	Dimer	*Liquidambar formosana* Hance	Hamamelidaceae	*157*
Isorugosin E (170)	(34)-[*m*-DOG(6,4-1')]-(35)	*m*-DOG	Dimer	*L. formosana*	Hamamelidaceae	*120*
Isorugosin G (171)	(35)-[*m*-DOG(6,4-1')]-(35)-[*m*-DOG(6',4'-1'')]-(35)	*m*-DOG × 2	Trimer	*L. formosana*	Hamamelidaceae	*158*
Cornusiin A (130)	(34)-[*m*-DOG(6,4-2')]-(34)	*m*-DOG	Dimer	*Cornus officinalis* Sieb. et Zucc.	Cornaceae	*123*
Cornusiin C (131)	(34)-[*m*-DOG(6,4-2')]-(34)-[*m*-DOG(6',4'-2'')]-(34)	*m*-DOG × 2	Trimer	*C. officinalis*	Cornaceae	*123*
Cornusiin D	(34)-[*m*-DOG(6,4-2')]-(35)	*m*-DOG	Dimer	*C. officinalis*	Cornaceae	*124*
Cornusiin E	(35)-[*m*-DOG(6,4-2')]-(35)	*m*-DOG	Dimer	*C. officinalis*	Cornaceae	*124*
Cornusiin F	(51)-[*m*-DOG(6,4-2')]-(34)-[*m*-DOG(6',4'-2'')]-(34)	*m*-DOG ×'2	Trimer	*C. officinalis*	Cornaceae	*124*

Table 3 *(continued)*

Compound	Composition	Linking unit	Extent of oligomerization	Main plant source	Plant family	Ref.
Camptothin A	(51)-[*m*-DOG(6,4-2')]-(34)	*m*-DOG	Dimer	*Camptotheca acuminata* Decne.	Nyssaceae	*125*
Camptothin B	(35)-[*m*-DOG(6,4-2')]-(34)	*m*-DOG	Dimer	*C. acuminata*	Nyssaceae	*125*
Trapanin A (132)	(35)-[*m*-DOG(6,4-2')]-(34)-[*m*-DOG(6',4'-2'')]-(34)	*m*-DOG × 2	Trimer	*Trapa japonica* Flerov.	Trapaceae	*126*
TrapaninB (133)	(34)-[*m*-DOG(6,4-2')]-(34)-[*m*-DOG(6',4'-2'')]-(34)-[*m*-DOG(6'',4''-2'')]-(34)	*m*-DOG × 3	Tetramer	*T. japonica*	Trapaceae	*126*
Schimawalin B	(U17)-[*m*-DOG(6,4-2')]-(U13)	*m*-DOG	Dimer	*Schima wallichii* Korth.	Theaceae	*135*
Phillyraeoidin A	(35)-[*m*-DOG(6,4-4' or 4,6-4')]-(6)	*m*-DOG	Dimer	*Quercus phillyraeoides* A. Gray	Fagaceae	*159*
Phillyraeoidin B	(34)-[*m*-DOG(6,4-4' or 4,6-4')]-(6)	*m*-DOG	Dimer	*Q. phillyraeoides*	Fagaceae	*159*
Phillyraeoidin C	(49)-[*m*-DOG(6,4-4' or 4,6-4')]-(6)	*m*-DOG	Dimer	*Q. phillyraeoides*	Fagaceae	*159*
Phillyraeoidin D	(49)-[*m*-DOG(6,4-4' or 4,6-4')]-(U18)	*m*-DOG	Dimer	*Q. phillyraeoides*	Fagaceae	*159*
Lagerstronin	(100)-[*m*-DOG(6,4-5' or 4,6-5')]-(101)	*m*-DOG	Dimer	*Lagerstroemia indica* L.	Lythraceae	*160*
Cornusiin G	(35)-[*m*-DOG(6,4-6')]-(4)	*m*-DOG	Dimer	*Cornus officinalis* Sieb. et Zucc.	Cornaceae	*42*
Camellin B (140)	cyclic-{(34)-[*m*-DOG(4,6-1') + *m*-DOG(6',4'-2)]-(35)}	*m*-DOG × 2	Dimer	*Camellia japonica* L.	Theaceae	*134*

Compound	Structure		Type	Species	Family	Ref.
Loropetalin C	cyclic{{(34)-[m-DOG(4,6-1')] + m-DOG(6',4'-2")]-(35)}-[m-DOG(4',6'-2")]-(51)	m-DOG + m,m"-D(OG)$_2$	Trimer	Loropetalum chinense Oliv.	Hamamelidaceae	*156*
Oenothein A (159)	cyclic{{(34)-[m-DOG(4,6-2')] + m-DOG(6',4'-2)]-(34)}-[m-DOG(4',6'-2")]-(34)	m-DOG + m,m"-D(OG)$_2$	Trimer	Oenothera biennis L.	Onagraceae	*142*
Oenothein B (138)	cyclic{(34)-[m-DOG(4,6-2')] + m-DOG(6',4'-2)]-(34)}	m-DOG × 2	Dimer	Oenothera erythrosepala Borbas	Onagraceae	*129*
Woodfordin C (139)	cyclic{(U16)-[m-DOG(4,6-2')] + m-DOG(6',4'-2)]-(34)}	m-DOG × 2	Dimer	Woodfordia fruticosa Kurz.	Lythraceae	*132, 133*
Woodfordin D (160)	cyclic{(U16)-[m-DOG(4,6-2')] + m-DOG(6',4'-2)]-(34)}-[m-DOG(4',6'-2")]-(34)	m-DOG + m,m"-D(OG)$_2$	Trimer	W. fruticosa	Lythraceae	*142*
Woodfordin E	cyclic{(34)-[m-DOG(4,6-2')] + m-DOG(6',4'-2)]-(34)}-[m-DOG(4',6'-2")]-(U13)	m-DOG + m,m"-D(OG)$_2$	Trimer	W. fruticosa	Lythraceae	*143*
Woodfordin F	cyclic{(U16)-[m-DOG(4,6-2')] + m-DOG(6',4'-2)]-(34)}-[m-DOG(4',6'-2")]-(34)-[m-DOG(4",6"-2"')]-(34)	m-DOG × 2 + m,m"-D(OG)$_2$	Tetramer	W. fruticosa	Lythraceae	*143*
Woodfordin I	cyclic{(U16)-[m-DOG(4,6-2')] + m-DOG(6',4'-2)]-(37)}	m-DOG × 2	Dimer	W. fruticosa	Lythraceae	*143*

C) Oligomers having glucose $^1C_4$ or skew-boat conformation

C1) Oligomers having m-GOG unit

Compound	Structure		Type	Species	Family	Ref.
Nupharin D (118)	(62)-[m-GOG(1-2')]-(U19)	m-GOG	Dimer	Nuphar japonicum DC.	Nymphaeaceae	*118*
Nupharin C (117)	(62)-[m-GOG(1-2')]-(61)	m-GOG	Dimer	N. japonicum	Nymphaeaceae	*118*
Nupharin E (119)	(62)-[m-GOG(1-2')]-(62)	m-GOG	Dimer	N. japonicum	Nymphaeaceae	*118*
Nupharin F (120)	(62)-[m-GOG(1-2')]-(62)-[m-GOG(1-2')]-(61)	m-GOG × 2	Trimer	N. japonicum	Nymphaeaceae	*118*
Jolkianin	(U20)-[m-GOG(2-3')]-(73)	m-GOG	Dimer	Euphorbia jolkini Boiss.	Euphorbiaceae	*82*

Table 3 (continued)

Compound	Composition	Linking unit	Extent of oligomerization	Main plant source	Plant family	Ref.
C2) Oligomers having *m*-DOG unit						
Euphorbin A (**149**)	(**63**)-[*m*-DOG(3,6-2')]-(**6**)	*m*-DOG	Dimer	*Euphorbia hirta* L.	Euphorbiaceae	*57, 108*
Euphorbin B (**151**)	(**53**)-[*m*-DOG(6,3-2')]-(**6**)	*m*-DOG	Dimer	*E. hirta*	Euphorbiaceae	*57, 108*
Euphorbin C (**161**)	(**63**)-[*m,p*-D(OG)$_2$(3,6-2',1')]-(**35**)	*m,p*-D(OG)$_2$	Dimer	*E. hirta*	Euphorbiaceae	*109*
Euphorbin D (**162**)	(**63**)-[*m,p*-D(OG)$_2$-(3,6-2',1')]-(**6**)	*m,p*-D(OG)$_2$	Dimer	*E. hirta*	Euphorbiaceae	*161*
Euphorbin E (**163**)	(**63**)-[(dehydro)*m,p*-D(OG)$_2$(3,6-1',2')]-(**U21**)	dehydro-*m,p*-D(OG)$_2$	Dimer	*E. hirta*	Euphorbiaceae	*144*
Euphorbin F (**150**)	(**63**)-[*m*-DOG(3,6-2')]-(**35**)	*m*-DOG	Dimer	*Euphorbia tirucalli* L.	Euphorbiaceae	*57*
Antidesmin A (**152**)	(**63**)-[*m*-DOG(3,6-2')]-(**U20**)	*m*-DOG	Dimer	*Antidesma pentandrum* Merr. var *barbatum* Merr.	Euphorbiaceae	*110*
Excoecarianin	(**63**)-[*m*-DOG(3,6-2')]-(**U22**)	*m*-DOG	Dimer	*Excoecaria kawakamii* Hayata	Euphorbiaceae	*162*
Euphorhelin	(**63**)-[*m*-DOG(3,6-2')]-(**3**)	*m*-DOG	Dimer	*Euphorbia helioscopia* L.	Euphorbiaceae	*82*
Eumaculin A	(**65**)-[*m*-DOG(3,6-2')]-(**6**)	*m*-DOG	Dimer	*Euphorbia maculata* L.	Euphorbiaceae	*155*
Tirucallin B	(**65**)-[*m*-DOG(3,6-2')]-(**U22**)	*m*-DOG	Dimer	*Euphorbia tirucalli* L.	Euphorbiaceae	*57*
Excoecarinin A	(**U23**)-[*m,p*-D(OG)$_2$(3,6-2')]-(**34**)	*m,p*-D(OG)$_2$	Dimer	*Excoecaria kawakamii* Hayata	Euphorbiaceae	*162*
Excoecarinin B	(**U24**)-[*m,p*-D(OG)$_2$-(3,6-2')]-(**34**)	*m,p*-D(OG)$_2$	Dimer	*E. kawakamii*	Euphorbiaceae	*162*

Name	Structure	DOG	Type	Species	Family	Ref.
Bischofianin	(63)-[m-DOG(3,6-4' or 6,3-4')]-(6)	m-DOG	Dimer	Bischofia javanica Blume	Euphorbiaceae	111
Eumaculin B	(65)-[m-DOG(6,3-2')]-(6)	m-DOG	Dimer	Euphorbia maculata L.	Euphorbiaceae	163
Mallotannin A	(65)-[m-DOG(6,3-6')]-(21)	m-DOG	Dimer	Mallotus japonicus Muell.-Arg.	Euphorbiaceae	164
Mallotannin B	(65)-[m-DOG(6,3-6')]-(U25)	m-DOG	Dimer	M. japonicus	Euphorbiaceae	164

D) CD, CG and CFC oligomers having C–C bond at open-chain glucose C-1

Name	Structure	DOG	Type	Species	Family	Ref.
Alienanin A (164)	(102)-[C-D(6')]-(31)	CD	Dimer	Quercus aliena Blume	Fagaceae	145
Alienanin B (165)	(102)-[C-D(6')]-(101)	CD	Dimer	Q. aliena	Fagaceae	145
Casuglaunin A	(102)-[C-G(5')]-(101)	CG	Dimer	Casuarina glauca Sieber	Casuarinaceae	165
Casuglaunin B	(102)-[C-F(6)]-Catechin-[F(8)-C]-(102)	CFC	Dimer	C. glauca	Casuarinaceae	165
Roburin A	(100)-[C-D(6')]-(100)	CD	Dimer	Quercus robur L.	Fagaceae	146
Roburin B	(100)-[C-D(6')]-(U26)	CD	Dimer	Q. robur	Fagaceae	146
Roburin C	(100)-[C-D(6')]-(U27)	CD	Dimer	Q. robur	Fagaceae	146
Roburin D	(100)-[C-D(6')]-(99)	CD	Dimer	Q. robur	Fagaceae	146
Anogeissusin A (167) (= castamolinin)	(100)-[C-D(6')]-(U28)	CD	Dimer	Anogeissus acuminata Guill. et Perr. var. lanceolata Wall. ex Clarke	Combretaceae	147

Table 3 (continued)

Compound	Composition	Linking unit	Extent of oligomerization	Main plant source	Plant family	Ref.
Anogeissusin B (168)	(100)-[C-D(6')]-(U29)	CD	Dimer	A. acuminata var. lanceolata	Combretaceae	147
Anogeissinin (166)	(100)-[C-F(6')]-Catechin-[F(8)-C]-(100)	CFC	Dimer	A. acuminata var. lanceolata	Combretaceae	147

U1: Sanguiin H-4
U2: 2-O-galloyl-4,6-O-HHDP-D-glucose
U3: 1-O-galloyl-4,6-O-HHDP-α-D-glucose
U4: 1α-O-galloyl-praecoxin A
U5: Sanguiin H2
U6: Nobotanin D
U7: Camelliatannin E
U8: 1,4,6-Tri-O-galloyl-β-L-glucose
U9: 1,4,6-Tri-O-galloyl-2,3-O-valoneoyl-β-D-glucose
U10: 1,4,6-Tri-O-galloyl-2,3-O-(monolactonized-valoneoyl)-β-D-glucose
U11: 4,6-Di-O-galloyl-D-glucose
U12: 1β-O-(Dilactonized-valoneoyl))-pedunculagin
U13: 2,3-Di-O-galloyl-D-glucose
U14: Pterocarinin A
U15: 1,2,3-Tri-O-galloyl-β-D-glucose
U16: 1α-O-galloyl-tellimagrandin I
U17: Euprostin A
U18: 2,3,4,6-Tetra-O-galloyl-D-glucose
U19: 1,2,4-Tri-O-galloyl-α-D-glucose
U20: Davidiin
U21: Dehydroisoterchebin
U22: Punicafolin
U23: 1-O-Galloyl-2,4-O-DHHDP-3,6-O-tergalloyl-β-D-glucose
U24: 1-O-Galloyl-2,4-O-elaeocarpunsinoyl-3,6-O-tergalloyl-β-D-glucose
U25: 6-O-Galloyl-D-glucose
U26: Roburin E
U27: Grandinin
U28: Eugenigrandin A
U29: Actissimin A
HHDP: hexahydroxydiphenoyl
DHHDP: dehydrohexahydroxydiphenoyl

phyllanthusiins A (**88**)–C (**90**), from *Terminalia chebula* and *Phyllanthus flexuosus*, have so far not been found as constituents of any oligomer.

3.1.2. *C-Glucosidic Ellagitannins and Complextannin Monomers in Oligomer Molecules*

Several *C*-glucosidic ellagitannins make up the molecules of more than 10 oligomeric ellagitannins, and these oligomers are more specifically correlated with plant systematics than the monomers. Castalagin (**99**) and vescalagin (**100**), casuarinin (**101**) and stachyurin (**102**), exemplify these *C*-glucosidic monomers. Some complextannin monomers are also found as the monomeric units of oligomers.

3.1.3. *α-Glucosidic Ellagitannins in Oligomer Molecules*

The α-glucosidic ellagitannin monomers potentillin (**33**) and sanguiin H-4 (*112*) make up some oligomer molecules isolated from several species of Rosaceous plants (*Agrimonia*, *Geum* and *Rosa* species *etc.*). It is noteworthy that both β-glucosidic monomers such as casuarictin (**32**) and α-glucosidic monomers exemplified by potentillin (**33**) make up several dimers, *e.g.*, gemins A–C (*113*), while only α-glucosidic monomers [potentillin (**33**) and sanguiin H-4] occur in the dimers agrimoniin and laevigatins C and E–G (*114, 115*) as well as in davuriciin D$_2$ (*116*).

3.2. Subsidiary Classification of Oligomers Based on the Type of Linkage

The classification of oligomeric hydrolyzable tannins in Table 3 is accompanied by a subsidiary system which is based on the type of linking unit between monomers. As shown in the third column of the table, the type of linking unit generally correlates well with the classification based on the type of monomer and also with plant taxonomy. In trimers and tetramers, the unit linking the first and second monomer is generally the same as that linking the second and third and the third and fourth.

3.2.1. **GOG**- *and* **GOGOG**-*Type Oligomers*

Oligomers of the **GOG** type (Scheme 4) contain a linking unit composed of two galloyl (**G**) groups mutually bound through an ether

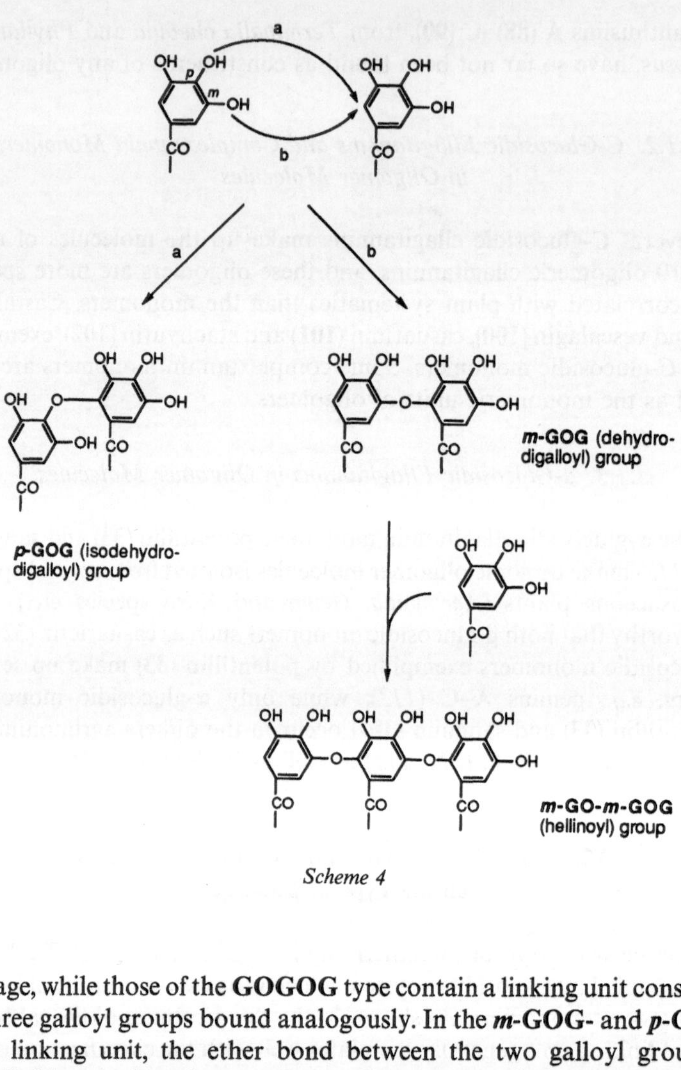

p-GOG (isodehydro-
digalloyl) group

m-GOG (dehydro-
digalloyl) group

m-GO-m-GOG
(hellinoyl) group

Scheme 4

linkage, while those of the **GOGOG** type contain a linking unit consisting of three galloyl groups bound analogously. In the *m*-**GOG**- and *p*-**GOG**-type linking unit, the ether bond between the two galloyl groups is supplied by a hydroxyl group *meta* and *para*, respectively, to the carboxyl group-bearing carbon in a galloyl group.

Frequently encountered monomers such as casuarictin (**32**), potentillin (**33**) and the tellimagrandins I (**34**) and II (**35**), often make up oligomers having a **GOG**- or **GOGOG**-type linkage. In most of the dimers having the **GOG**-type linkage, two carboxyl groups of the **GOG**-unit esterify the anomeric hydroxyl group of each glucose core, except for laevigatin D (*114*) in which one of the two ester linkages from the **GOG** group is at O-2 of a glucose core.

Oligomers of these two types are found in the plants of Rosaceae, Coriariaceae, Nymphaeaceae, Tamaricaceae and Fagaceae (Table 3). Between the two types of oligomers, *m*- and *p*-GOG types, the *m*-GOG (dehydrodigalloyl; DHDG) type oligomers are more frequently encountered in various plant families. In the *m*-GOG dimers found in rosaceous and coriariaceous plants the linking unit is attached to C-1 of both glucose cores [agrimoniin (**109**) (*62, 107*) and its analogs; laevigatins (**110–113**) (*114, 115*), davuriciin D$_2$ (**114**) (*116*), gemin A (**115**) (*113*) and

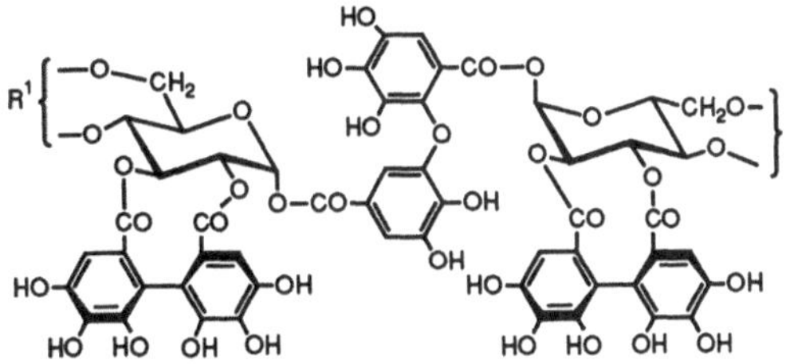

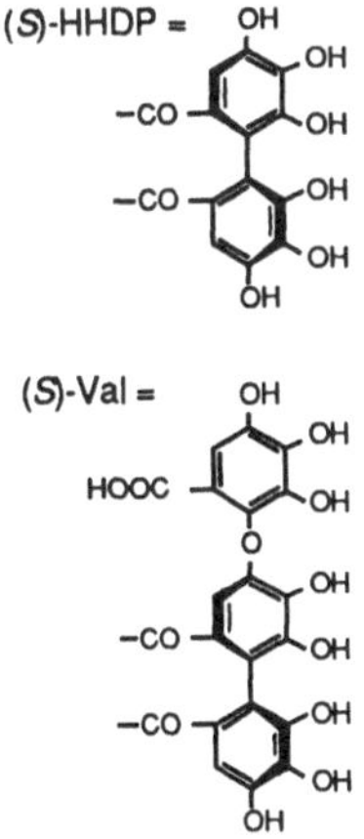

(**109**) Agrimoniin: R^1= R^2= (*S*)-HHDP
(**110**) Laevigatin B: R^1= H,H; R^2= (*S*)-HHDP
(**111**) Laevigatin C: R^1= (*S*)-HHDP; R^2= H,H
(**113**) Laevigatin E: R^1= R^2= H,H
(**114**) Davuriciin D2: R^1= (*S*)-Val; R^2= (*S*)-HHDP

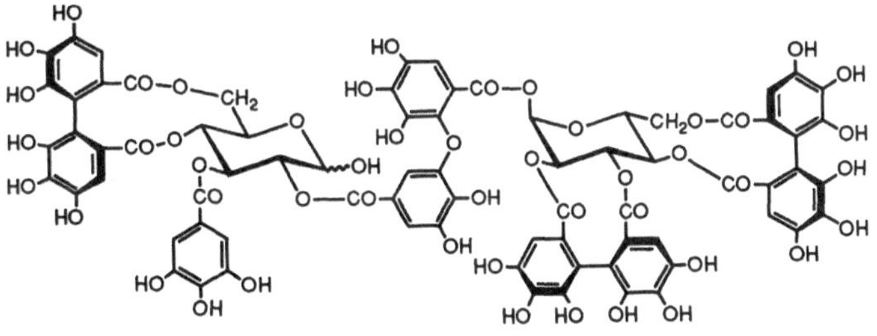

(**112**) Laevigatin D

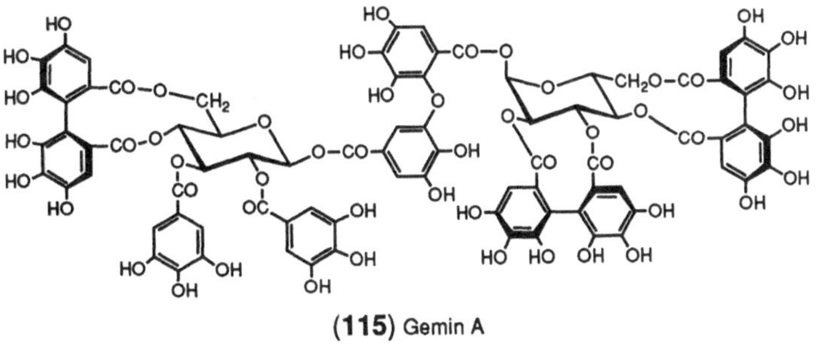

(**115**) Gemin A

coriariins A (116) (63) and C (117), etc.]. Among them, agrimoniin occurs in *Agrimonia, Rosa, Potentilla* and some other genera of Rosaceae (5) while gemin A (115) has been found only in *Geum* species (5). These oligomers may be useful as chemotaxonomic markers because of the specificity of their occurrence (5).

Nupharins C (117), D (118), E (119) and F (120) which have the *m*-GOG unit bound to glucose C-1 of a monomer and to C-2 of another monomer were isolated from *Nuphar japonicum* (Nymphaeaceae) (118). They are composed of nupharin A (1,2,4-tri-*O*-galloyl-3,6-*O*-(*S*)-hexahydroxydiphenoyl-α-D-glucose) (61), nupharin B (62) and/or their analogs. Tamaricaceous plants also produce oligomers with a linking mode similar to that of the nupharins [*e.g.* hirtellin A (121), a dimer of tellimagrandin II (35), and its analogs isolated from *Reaumuria hirtella* and *Tamarix pakistanica* (65)]. These two series of oligomers seem to be characteristic constituents of Tamaricaceae and Nymphaeaceae, respectively.

Some dimers which contain both a *p*-GOG (isodehydrodigalloyl) unit and a *m*-GOG unit [hirtellins C (112) and F (123), etc.] were isolated from

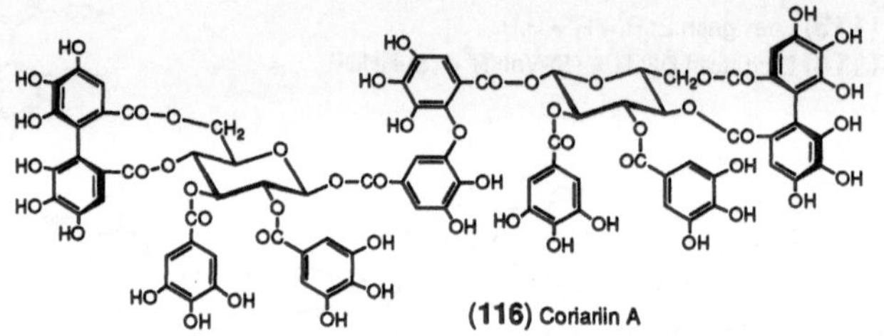

(116) Coriariin A

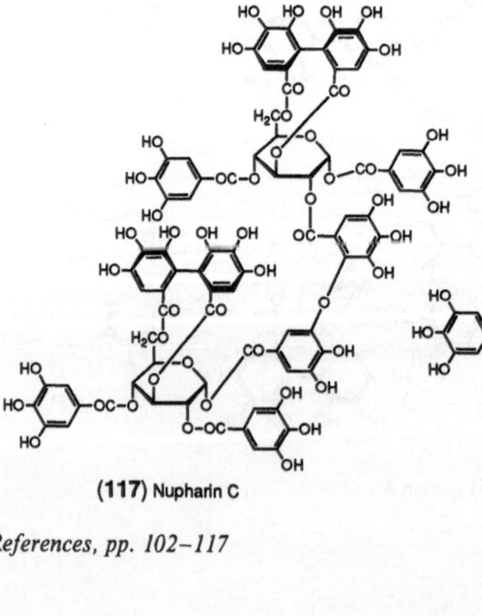

(117) Nupharin C

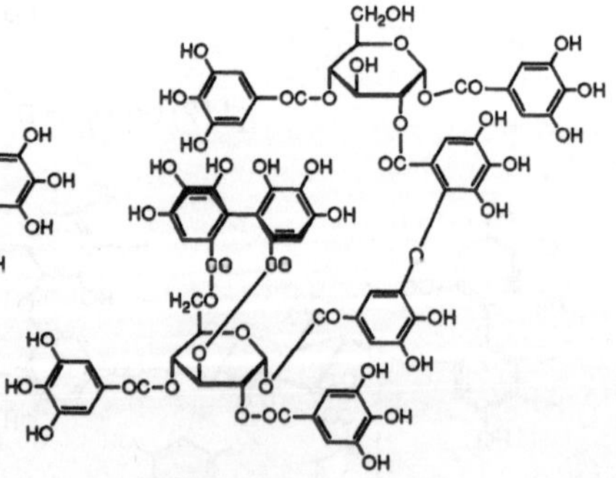

(118) Nupharin D

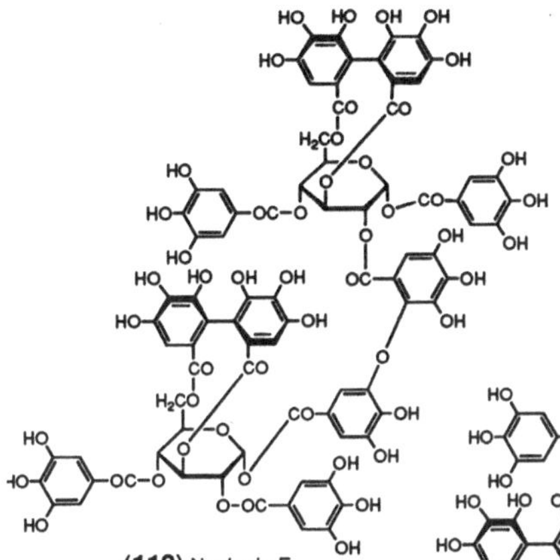

(119) Nupharin E

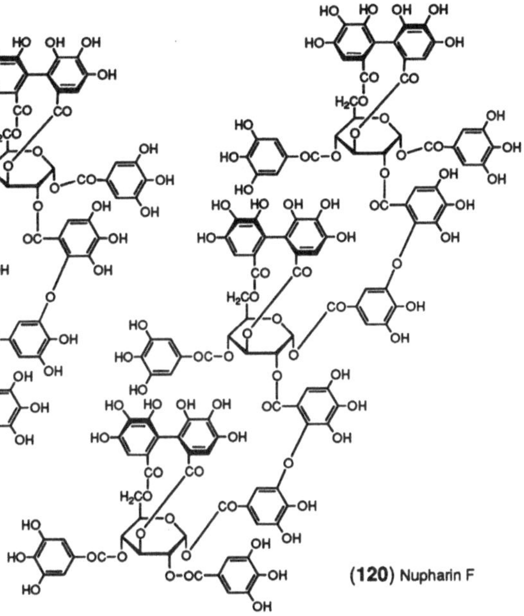

(120) Nupharin F

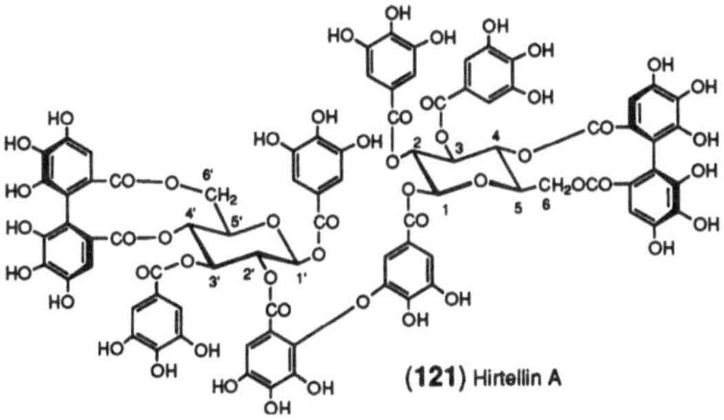

(121) Hirtellin A

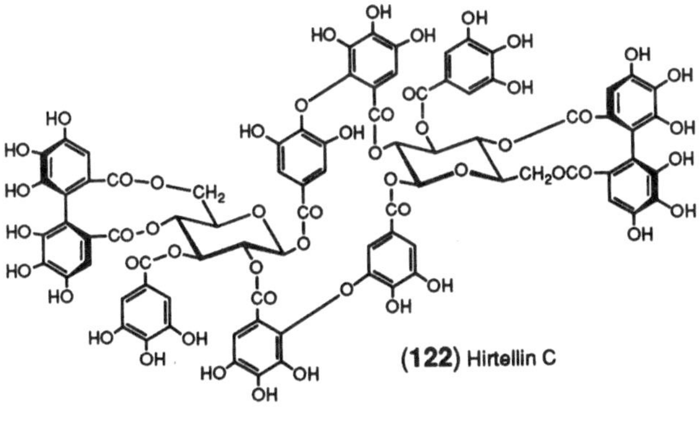

(122) Hirtellin C

Reaumuria hirtella (Tamaricaceae) (*119*). *R. hirtella* and *Tamarix paki-stanica* of Tamaricaceae produce also **GOGOG**-type dimers [hirtellin B (**124**) and tamarixinin A (**125**)], having a trisgalloyl group as the linkage unit in which an additional galloyl (**G**) group binds with its *m*-oxygen to a *m*-**GOG** unit (*119, 120*). These two **GOGOG**-type dimers can therefore be classified as *m*-**GO**-*m*-**GOG**-type dimers.

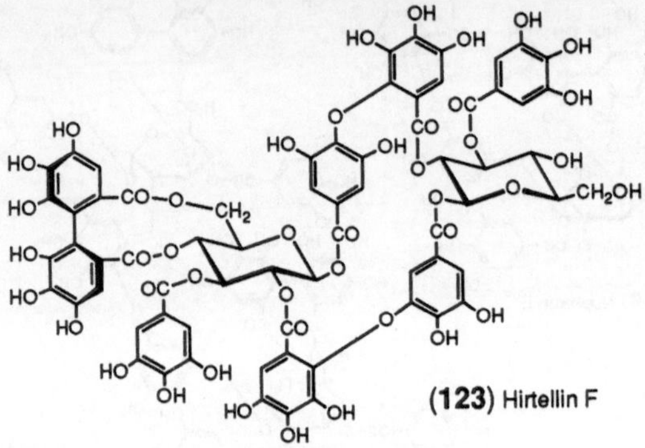

(**123**) Hirtellin F

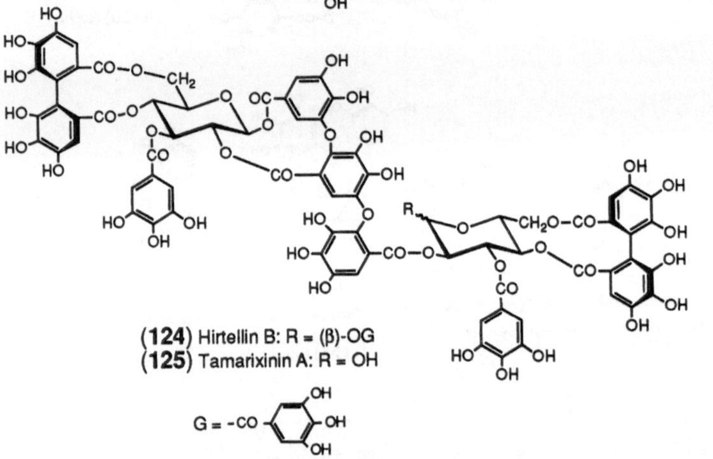

(**124**) Hirtellin B: R = (β)-OG
(**125**) Tamarixinin A: R = OH

G = -CO

3.2.2. **DOG**-*Type Oligomers*

In **DOG**-type oligomers the linking unit is the valoneoyl or tergalloyl group as exemplified by structures **126–136** and **137**, respectively. These linking units can be regarded as products resulting from oxidative coupling between the HHDP (abbreviated to **GG** or **D**) oxygen of a monomer and the galloyl (**G**) carbon of another monomer and are hence abbreviated to **DOG**. The *m*- and *p*-**DOG** units have been called the valoneoyl and tergalloyl groups. Table 3 shows that *m*-**DOG**-type

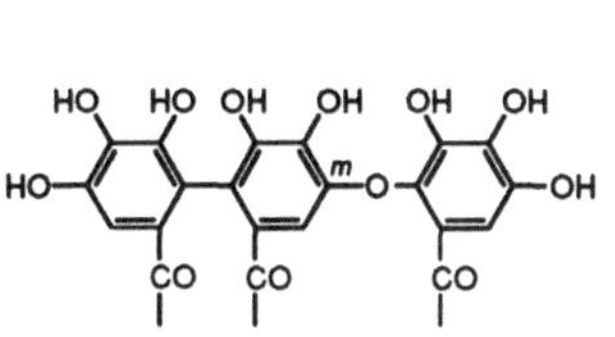

m-DOG (valoneoyl)
group

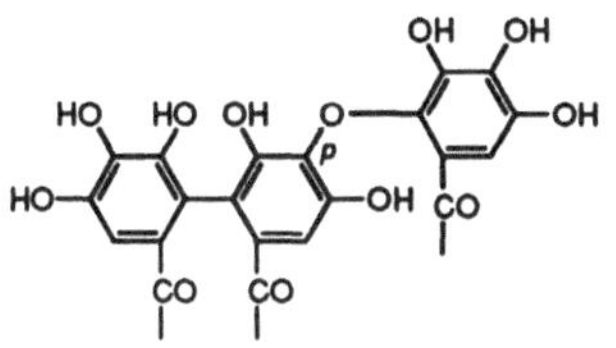

p-DOG (tergalloyl)
group

Chart 3

oligomers, among all types of oligomers, are the most frequently encountered type in various plant families.

The monomers tellimagrandins I (**34**) and II (**35**), pedunculagin (**31**),
casuarictin (**32**), praecoxin B (**53**), pterocaryanin C (**54**), and galloylglucoses and several *C*-glucosidic and complex tannins, have been found to
make up oligomers having the **DOG**-type linkage(s).

The rugosins D (**126**), E (**127**), F (**128**) and G (**129**), of the *m*-**DOG** type
were initially isolated from *Rosa rugosa* (*121*), but occur widely in other
genera of Rosaceae [*Filipendula* (*12*, *122*)] and also in Coriariaceae,
Trapaceae, Stachyuraceae, Hamamelidaceae, Fagaceae and Euphorbi-
aceae (*4*) (Table 3). The cornusiins A (**130**) and C (**131**)–F (*123*, *124*), and

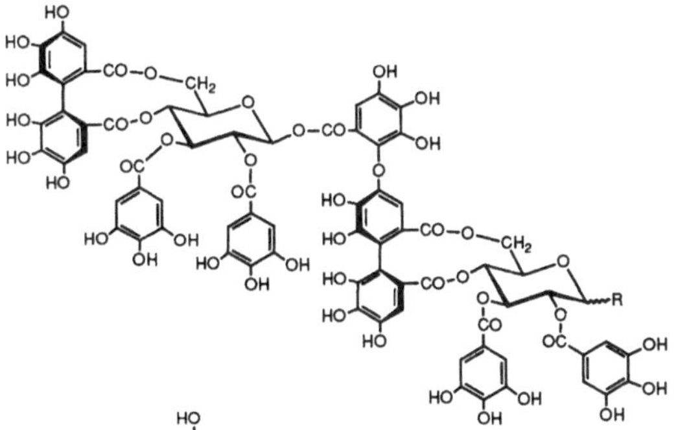

(**126**) Rugosin D: R = (β)-OG
(**127**) Rugosin E: R = OH

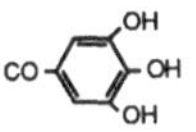

G = - CO

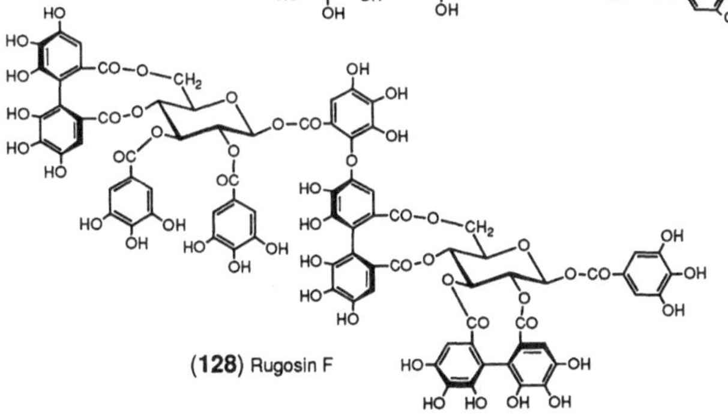

(**128**) Rugosin F

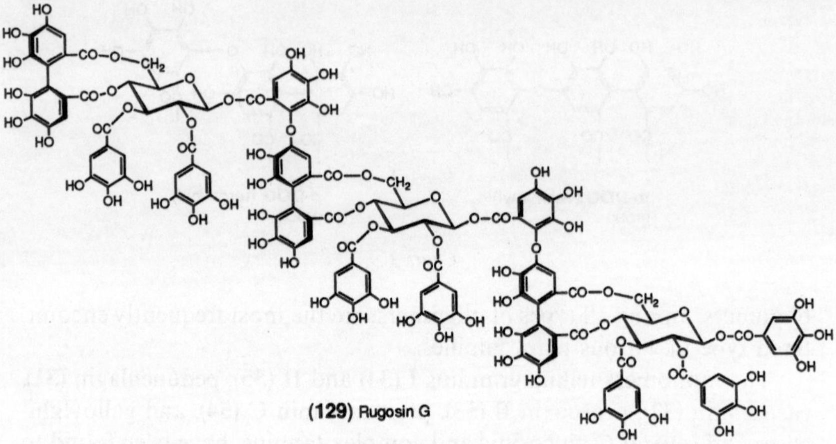

(129) Rugosin G

the camptothins A and B (*125*), **DOG** type oligomers isolated from Cornaceae and Nyssaceae within Cornales, are made up of tellimagrandin I (**34**) and its analogs, having the *m*-**DOG** unit at O-4/O-6 of one monomer and at O-2′ of a second. Each of these compounds exists as an equilibrium mixture of several anomers as a result of anomerization of the glucose residues. The cornusiins and their analogs, trapanins A (**132**) and B (**133**) (*126*), occur in *Trapa japonica* (Trapaceae, Myrtales) which is taxonomically close to Cornales.

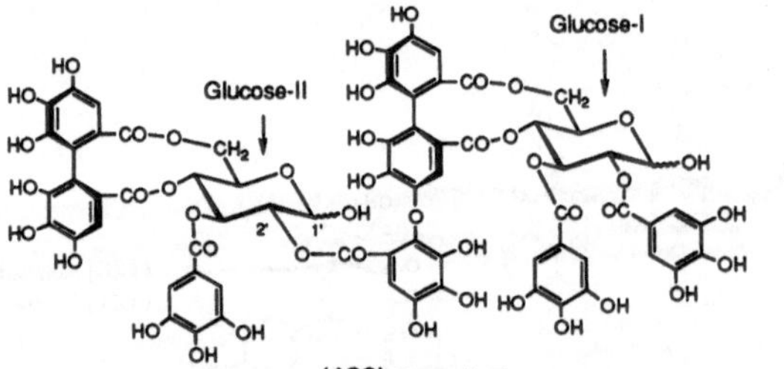

(130) Cornusiin A

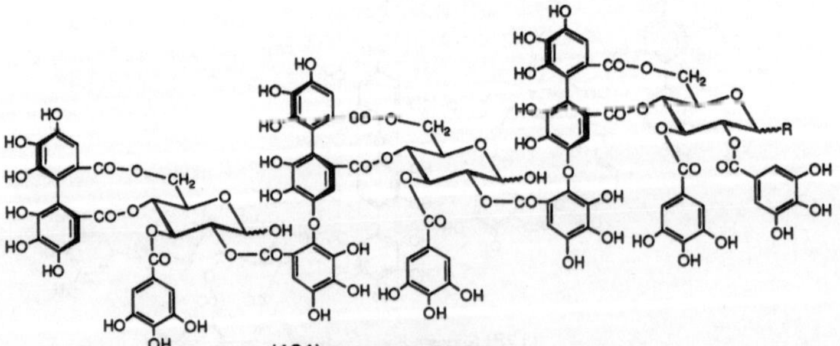

(131) Cornusiin C: R = OH
(132) Trapanin A: R = (β)-OG

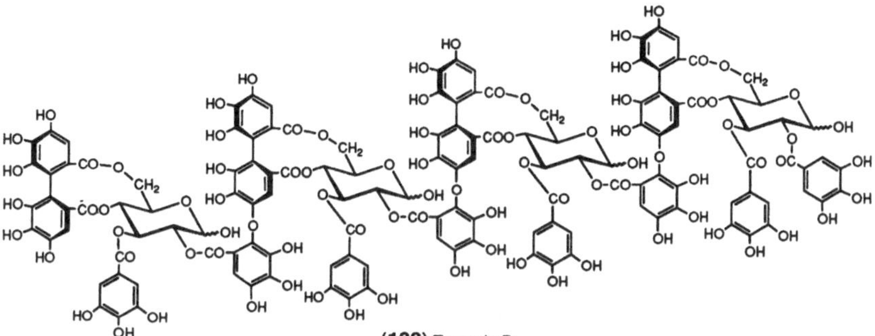

(133) Trapanin B

In the nobotanins A (134), F (135) (127) and B (136) (128), the galloyl part (G) in the *m*-DOG unit is at glucose O-4 of one of the monomer units; this location is characteristic of the oligomer structures in Melastomataceae. Many plants of Myrtales which comprise the Melastomataceae and

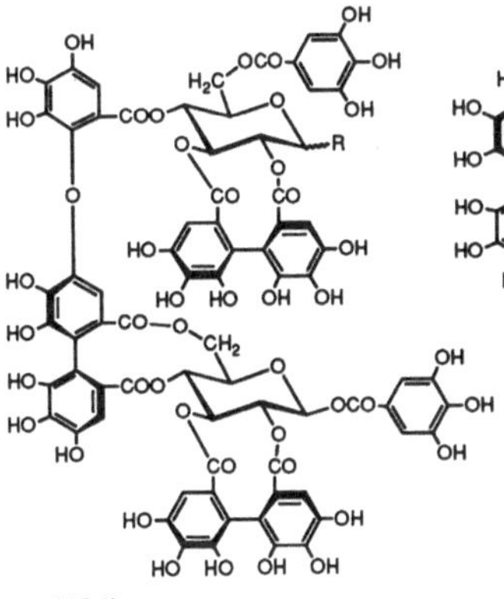

(134) Nobotanin A: R = OH
(135) Nobotanin F: R = (β)-OG

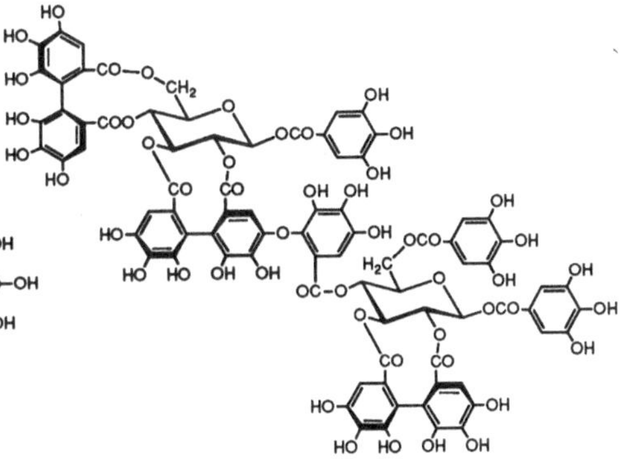

(136) Nobotanin B

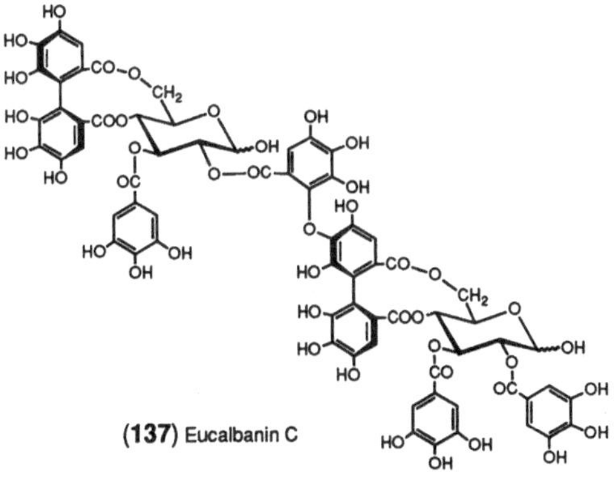

(137) Eucalbanin C

also the Trapaceae, Onagraceae, Myrtaceae and Lythraceae, according to Cronquist's system of classification, are generally rich in hydrolyzable tannins, particularly in the oligomers. A dimer having a *p*-DOG (tergalloyl) group as the linking unit, eucalbanin C (137), was found in *Eucalyptus alba* (Myrtaceae) (58).

Oenothein B (138), a dimer of unique macrocyclic structure, has been isolated from *Oenothera* (129) and *Epilobium* species of Onagraceae (130) and from *Lythrum anceps* (131). It is also found in *Woodfordia fruticosa* of Lythraceae along with woodfordin C (139) (132) which is an analog of oenothein B. These dimers contain two *m*-DOG units of mutually reverse orientation on the two glucose cores. A macrocyclic dimer of analogous structure, camelliin B (140), is found in the flowers of several species of Theaceae, *Camellia japonica, C. sasanqua* (134) and *Schima wallichii* (135)

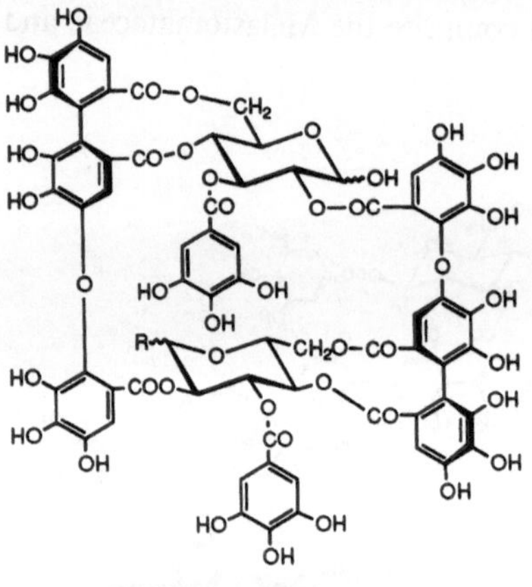

(138) Oenothein B: R = OH
(139) Woodfordin C: R = (α)-OG

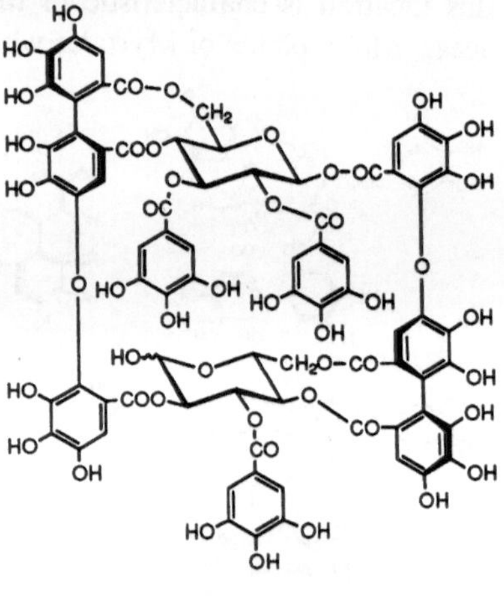

(140) Camelliin B

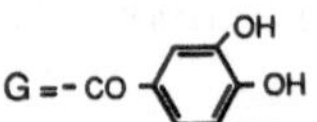

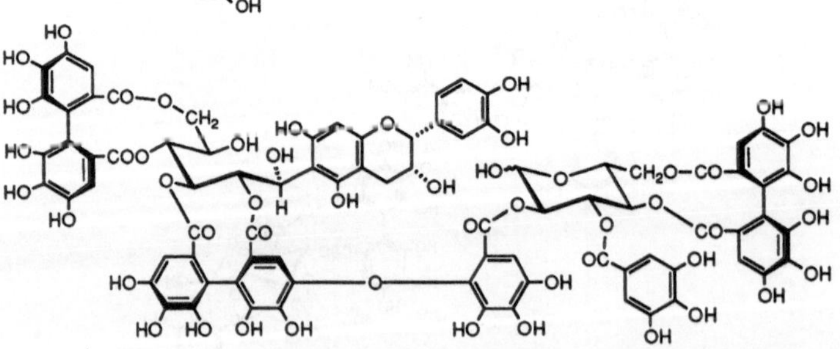

G = - CO [structure with OH, OH, OH]

(141) Camelliatannin D

(Table 3). However, the leaves of *C. japonica* lack camelliin B (**140**) and produce camelliatannin D (**141**), a dimer of a somewhat different type. One of its monomers is a complex tannin with an open-chain glucose core (*105*).

There exists also another group of **DOG**-type dimers which may be regarded as products resulting from C-O oxidative coupling between *C*-glucosidic tannin monomer and an ellagitannin monomer with a $^4C_1$ glucose core are found in species of Lythraceae and Betulaceae. Examples are the reginins A (**142**), B (**143**), C (**144**) and D (**145**) from *Lagerstroemia flos-regina* (Lythraceae) (*136, 137*). Analogous dimers, the heterophylliins B (**146**) and C (**147**), are produced in *Corylus heterophylla* (Betulaceae) (*138*).

(**142**) Reginin A: R = OH, R' = H
(**143**) Reginin B: R = H, R' = OH
(**144**) Reginin C: R = H, R' = A

(**145**) Reginin D

(**146**) Heterophylliin B: R = R' = G
(**147**) Heterophylliin C: R ~ R' = (*S*)-HHDP

Tetramers of the **DOG** type, trapanin B (**133**) (*126*) and nobotanin K (**148**) (*139*), are found in trapaceous and melastomataceous plants.

Two sub-types of **DOG**-type oligomers have been found in species of Euphorbiaceae. Oligomers with the **DOG** unit at O-4/O-6 on the $^4C_1$ glucopyranose core, such as rugosin D (**126**), and those with the **DOG** unit at O-3/O-6 on the glucose core in the $^1C_4$ or boat conformation, exemplified by the euphorbins A (**149**), B (**151**) and F (**150**) and antidesmin A (**152**) (*57, 108, 110*). The latter sub-type has a dehydroellagitannin as one of the components and has been found only in Euphorbiaceae by the end of 1993.

(**148**) Nobotanin K

3.2.3. GOD-*Type Oligomers*

The **GOD**-type linking unit results from oxidative coupling between one of the galloyl (**G**) hydroxyl groups of a monomer and an HHDP (**D**) carbon of a second monomer, and the *m*-**GOD** unit corresponding to the so-called sanguisorboyl group (*4*). Several monomers, tellimagrandin II (**35**), potentillin (**33**) and sanguiin H-4, are constituents of oligomers having a **GOD**-type linkage.

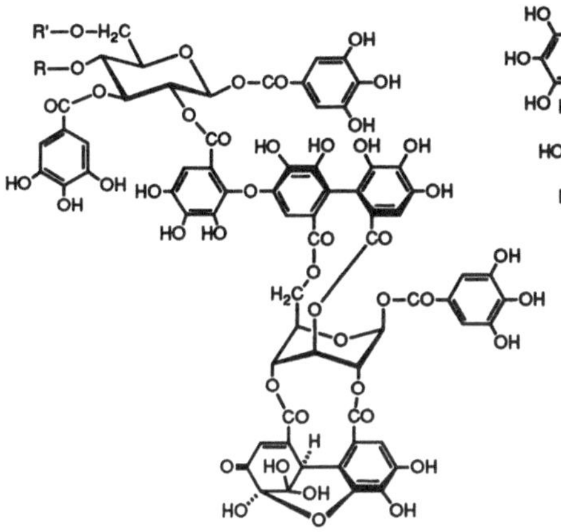

(149) Euphorbin A: R = R'= G
(150) Euphorbin F: R ~ R'= (S)-HHDP

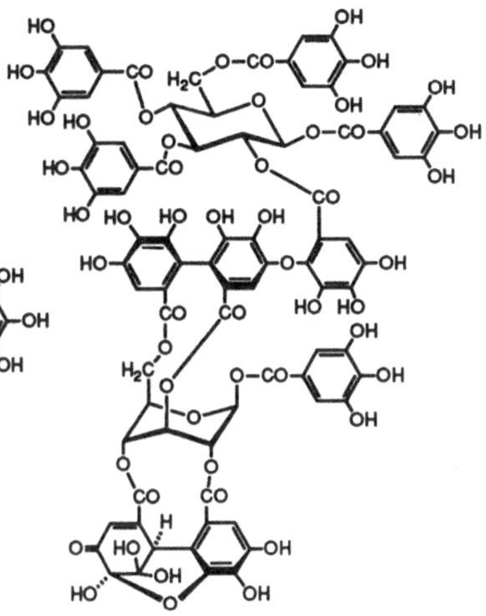

(151) Euphorbin B

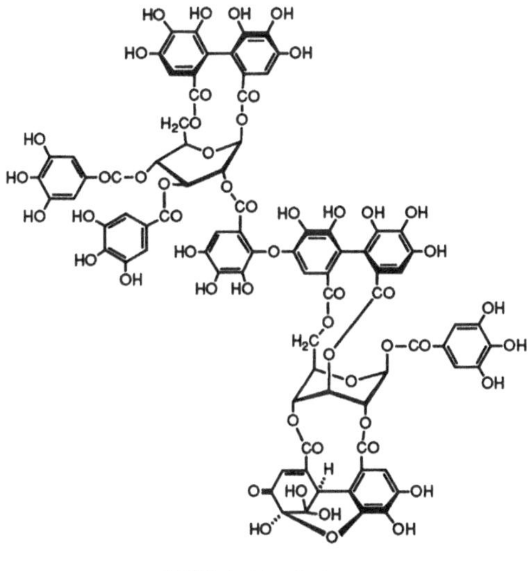

(152) Antidesmin A

Only a small number of **GOD** type oligomers has so far been found and all of these belong to the ***m*-GOD**-type (Table 3). These ***m*-GOD**-type oligomers, sanguiin H-6 (**153**) (*82*) and roshenin B (**155**) (*140*) (dimers), lambertianins C (**156**) (trimer) and D (**157**) (tetramer) (*141*), and sanguiin H-11 (**158**) (tetramer) (*112*), are mostly found in the genera *Sanguisorba* and *Rubus* (*5*), although some ***m*-GOD** dimers, roshenins A (**154**) and B (**155**), were isolated from *Rosa henryi* (*140*).

(153) Sanguiin H-6: R^1= (α)-OG, R^2~ R^3= (S)-HHDP
(154) Roshenin A: R^1= (α)-OG, R^2~ R^3= (S)-Sang
(155) Roshenin B: R^1= OH, R^2~ R^3= (S)-HHDP

(156) Lambertianin C

3.2.4. $D(OG)_2$-Type Oligomers

In a $D(OG)_2$-type linking unit, two hydroxyl oxygens of an HHDP (D) group participate in forming ether linkages with two galloyl (G) groups. In these galloyl groups the carbon *ortho* to the carboxyl groups takes part in the ether linkage. Two isomeric $D(OG)_2$ units, the m, m''-$D(OG)_2$ type and m, p-$D(OG)_2$ type (Scheme 5) which differ in the location of one of these ether linkages on the D group are found in oligomers isolated from members of the Onagraceae, Euphorbiaceae and Lythraceae. The monomers tellimagrandins I (34) and II (35) and geraniin (63), have so far been found as constituents of $D(OG)_2$ type oligomers.

m, m''-$D(OG)_2$-*Type Oligomers.* The linking unit [the woodfordinoyl group (4, 132)] in this type of oligomer has two ether linkages *meta* to the

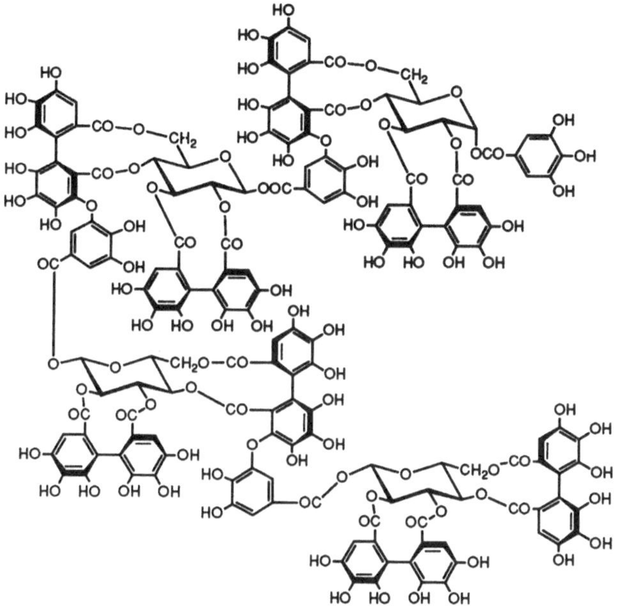

(157) Lambertianin D

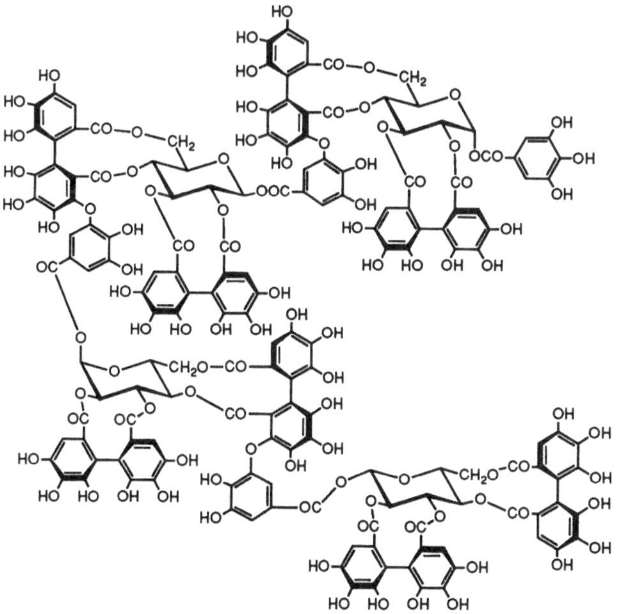

(158) Sanguiin H-11

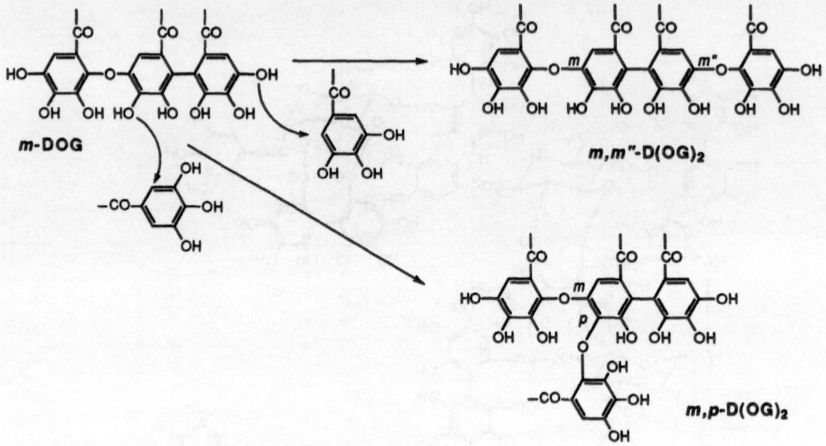

Scheme 5

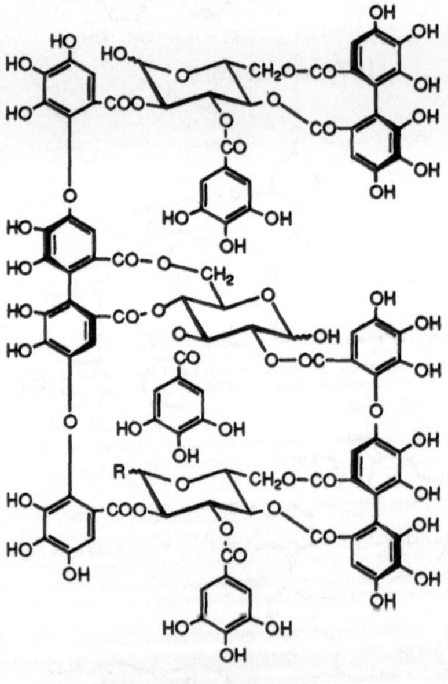

(**159**) Oenothein A: R = OH
(**160**) Woodfordin D: R = (α)-OG

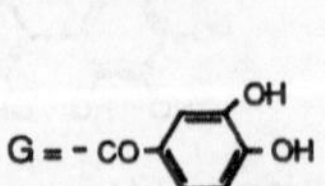

G = – CO ⟨OH, OH, OH⟩

carboxyls of an HHDP group (see Addendum). Oenothein A **(159)** *(142)* and the woodfordins D **(160)** *(142)* and F *(143)* have macrocyclic structures with this linkage characteristic of the oligomers in plants of the Onagraceae and Lythraceae.

(161) Euphorbin C: R ~ R'= (*S*)-HHDP
(162) Euphorbin D: R = R'= G

(163) Euphorbin E

m, p-D(OG)$_2$-*Type Oligomers.* The linking unit [the euphorbinoyl group (*4*)] in this type of oligomer has two ether linkages *meta* and *para* to the carboxyl of a phenyl ring in an HHDP (**D**) group (Scheme 5). Dimers of this type, structurally correlated with euphorbin A (**149**), are specific to certain Euphorbiaceae (*111, 57, 108*).

Euphorbin E (**163**) (*144*), produced together with euphorbins A (**149**), B (**151**), C (**161**) and D (**162**) in *Euphorbia hirta*, is regarded as a highly oxygenated metabolite of euphorbin C with a dehydroeuphorbinoyl group as the linking unit.

3.2.5. Oligomers Composed of C-Glucosidic Ellagitannin Monomers Mutually Linked through C–C Bond

In oligomers of this type, the *C*-glucosidic monomers are attached to each other through a C–C bond. Monomers most often found in oligomers of this type are stachyurin (**102**), casuarinin (**101**), castalagin (**99**) and vescalagin (**100**). The oligomers of this type can be further classified into a **CD**, a **CG** and a **CFC** type.

CD-*and* **CG**-*Type Oligomers.* In the dimer molecules of this type, the *C*-glucosidic monomers, vescalagin (**100**) and castalagin (**99**), are directly bound with each other *via* a C–C bond between glucose C-1 of a monomer and the HHDP or galloyl group of the other monomer. Examples are the alienanins A (**164**, a dimer composed of **31** and **100**) and B (**165**, a dimer composed of **101** and **102**), from *Quercus aliena* (*145*), and the roburins A–D from *Q. robur* (*146*).

CFC-*Type Oligomers.* In **CFC** dimers, the *C*-glucosidic monomers are bound with each other through a flavan 3-ol residue (**F**). An example is anogeissinin (**166**) isolated from a combretaceous plant, in which two vescalagin (**100**) units are connected to the C-6 and C-8 positions of the (+)-catechin unit (*147*). Some **CD**-type oligomers such as the anogeissusins A (**167**) and B (**168**) (*147*) have a flavan unit (**F**) bound to C-1 of one of the *C*-glucosidic monomers which is not participating in the *C*-glucosidic dimer formation.

3.3. Detection and Isolation of Oligomers from Plants

The oligomeric hydrolyzable tannins in plant extracts can be characterized by high-performance GPC (*1*), the color reactions of ellagitannins (*167, 168*) and quantitation of tannins based on the binding potency of

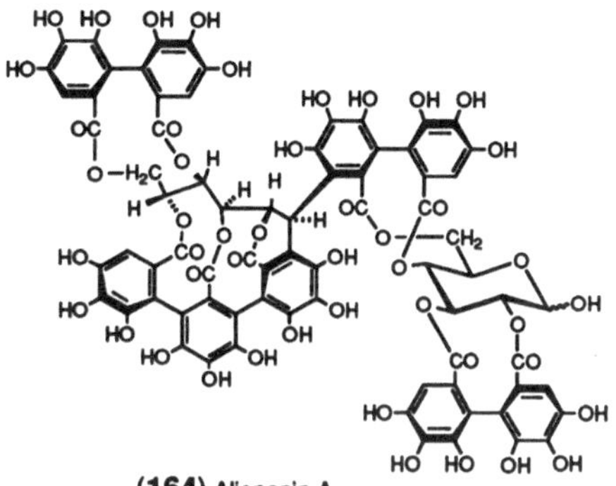

(164) Alienanin A

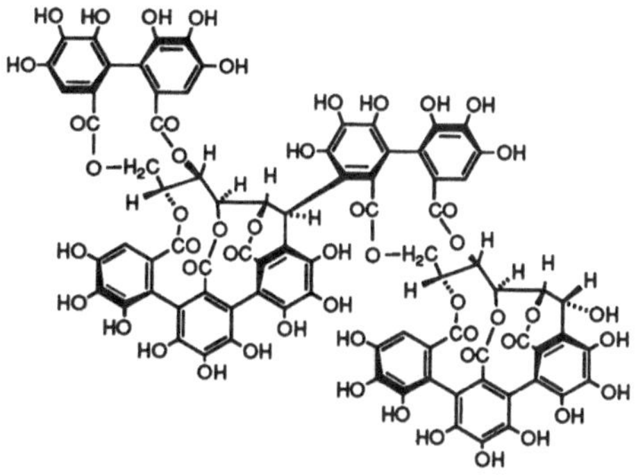

(166) Anogeissinin

(165) Alienanin B

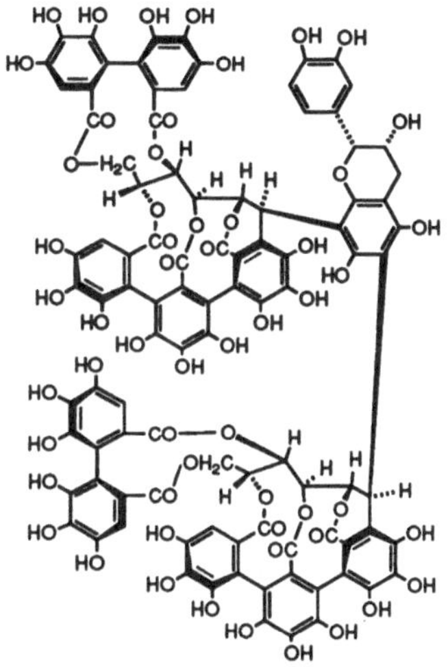

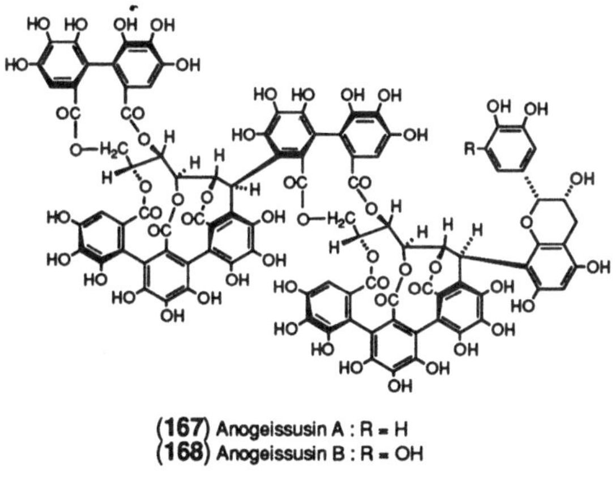

(167) Anogeissusin A : R = H
(168) Anogeissusin B : R = OH

tannins with proteins or basic compounds such as methylene blue (*168*). A more convenient method for preliminary analysis of oligomeric hydrolyzable tannins is use of normal-phase HPLC, which permits an approximate estimate of the degree of oligomerization, based on the retention time which increases with extent of oligomerization (*1*).

Some labile oligomers such as nobotanin K (**148**) from *Heterocentron roseum* (*4*) are apt to be hydrolyzed on gel columns and their separation and purification by column chromatography on solid supports is impracticable. Centrifugal partition chromatography (CPC), which does not require any solid support, permits isolation and purification of such labile oligomers (*131, 139, 169, 170*).

3.4. Spectroscopic Analysis of Oligomeric Hydrolyzable Tannins

Structure elucidation of hydrolyzable tannins, especially of oligomeric hydrolyzable tannins, has been greatly facilitated by remarkable advances in NMR and mass spectrometry during the last decade.

3.4.1. Determination of Molecular Weight

Fast-atom bombardment mass spectroscopy (FABMS) has often been applied successfully to determination of the molecular weight of hydrolyzable tannin monomers and oligomers which are highly polar, poorly volatile and thermally labile. The molecular weights of various oligomers up to woodfordin F (*143*) which is a tetramer (m.w. *ca.* 3200) have been determined by FABMS measurement in the presence of NaCl and KCl, which is usually effective in discriminating and confirming the molecular ion. In positive FABMS the species, $[M + Na]^+$ and/or $[M + K]^+$ in addition to $[M + H]^+$, are generally observed if a small amount of NaCl or KCl is added to a matrix agent such as *m*-nitrobenzyl alcohol. The negative FABMS of hydrolyzable tannins gives a prominent peak due to the $[M-H]^-$ ion.

3.4.2. Characterization of Monomer Structures in Oligomers by NMR

NMR analysis employing a variety of two dimensional measurements of high magnetic field is the most reliable and convenient tool for structure elucidation of tannins of this class (*1, 171*). The nature and number of polyphenolic acyl groups (galloyl, HHDP and valoneoyl, *etc.*) in hydrolyzable tannins are usually determined by characteristic signals in the ^{1}H- and ^{13}C-NMR spectra and by acid hydrolysis which yields free polyphenolcarboxylic acids or their dilactones. The ^{13}C-NMR signals of the sugar residues frequently found in the oligomers have also been fully assigned (Table 4). Therefore, the monomeric constituents of the oligomers can be easily determined by comparing the glucose carbon resonances of the oligomers with those of the presumed monomers (*172*). For example, the conclusion that rugosin D (**126**) is composed of two moles of tellimagrandin II (**35**) is easily arrived at when the glucose carbon resonances of **126** and **35** are compared. Table 4 lists the glucose carbon resonances in commonly encountered monomers.

This convenient way of structure eludication is also applicable to oligomers which exist as an equilibrium mixture of anomers (*173*). Thus while the ^{1}H NMR spectrum of trapanin A (**132**), a trimer forming an equilibrium mixture of four anomers, is exceedingly complicated due to

Table 4. ^{13}C NMR Data^a for the Glucose Moieties in the Main Monomers Having $^4C_1$ Glucose Core

Compound	C-1	C-2	C-3	C-4	C-5	C-6
Pentagalloylglucose (**6**)	93.4	71.9	73.5	69.5	74.1	62.9
1,2,4,6-Tetragalloylglucose (**5**)	93.4	73.8	73.3	71.7	74.0	63.1
Tellimagrandin II (**35**)	93.8	71.8	73.3	70.8	73.1	63.1
Tellimagrandin I (**34**)						
α-anomer	91.2	72.9	71.1	71.1	67.2	63.5
β-anomer	96.7	74.1	73.5	71.1	72.0	63.5
Rugosin A (**36**)	94.2	72.2	73.6	71.2	73.6	63.6
Casuarictin (**32**)	92.4	76.0	77.3	69.3	73.5	63.1
Potentillin (**33**)	90.7	74.1	76.0	69.1	71.0	63.2
Pedunculagin (**31**)						
α-anomer	91.8	75.6	75.8	69.9	67.4	63.6
β-anomer	95.4	78.3	77.6	69.6	72.5	63.6
Pterocaryanin C (**54**)	91.9	75.3	77.4	69.2	73.9	62.7
Praecoxin B (**53**)						
α-anomer	91.3	75.1	75.4	68.7	68.4	63.1
β-anomer	94.9	77.7	77.6	68.5	73.1	63.2
Gemin D (**51**)						
α-anomer	94.0	72.1	74.0	71.0	67.5	63.8
β-anomer	98.4	74.8	75.9	71.3	72.0	63.8
Strictinin (**49**)	95.9	74.7	75.6	72.8	73.2	63.7
Isostrictinin (**50**)	92.9	71.1	72.9	73.4	69.2	65.9
Nupharin A (**61**)	89.5	70.0	73.5	74.9	77.9	65.1

^{a} Measured in acetone-d$_6$ + D$_2$O

T. OKUDA, T. YOSHIDA, and T. HATANO

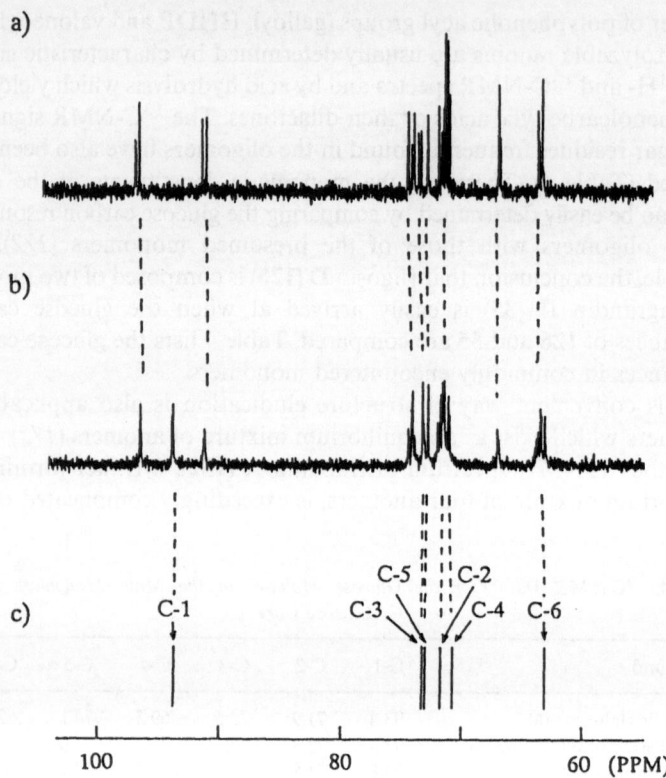

Fig. 4. ^{13}C-NMR spectral comparison of the glucose moieties of trapanin A (**132**) (b) with cornusiin A (**130**) (a) and tellimagrandin II (**35**) (c)

Table 5. ^{13}C NMR Data^a for the Glucose Moieties in the Most Common Monomers Having $^1C_4$ Glucose and Open-Chain Glucose Core

Compound	C-1	C-2	C-3	C-4	C-5	C-6
Geraniin (**63**)						
63a-form	90.8	69.9	63.3	65.9	72.6	63.6
63b-form	91.8	70.4	62.3	66.8	73.1	63.8
Corilagin (**65**)	94.2	68.8	70.4	62.2	75.5	64.3
Chebulagic acid (**85**)	91.5	70.5	61.7	66.1	73.5	63.9
Phyllanthusiin A (**88**)	91.7	69.7	63.2	66.5	73.3	64.0
Phyllanthusiin B (**89**)	91.8	70.6	61.5	67.8	73.6	64.0
Casuarinin (**101**)	67.6	76.7	69.8	74.2	71.2	64.6
Casuariin (**103**)	68.5	77.1	70.8	77.1	68.5	67.2
Stachyurin (**102**)	65.5	81.0	70.9	73.3	72.0	64.5

^{a} Measured in acetone-d_6 + D_2O

the presence of four lines for each aromatic proton singlet and their overlap, it was easily shown to be made up of tellimagrandins I (**34**) and II (**35**), since the glucose carbon signals of the trimer corresponded well with those of tellimagrandin II (**35**) and cornusiin A (**130**), the latter being a dimer of tellimagrandin I (**34**). The correspondence is shown in Fig. 4. Similar information about the monomeric constituents in oligomers possessing $^1C_4$ glucopyranose and/or open-chain glucose cores has been obtained from the data accumulated for monomers of various types (Table 5).

3.4.3. Orientation of the Valoneoyl Group in *m*-**DOG** Type Oligomers

The orientation of the linking unit in an oligomer can be determined by partial hydrolysis (see Sect. 4) and by ^{1}H-^{13}C long-range COSY NMR spectrometry. The latter technique has been applied to several *m*-**DOG** type tannins such as rugosin D (**126**) and isorugosin D (**169**) which differ only in the relative orientation of the valoneoyl groups (*157,158*). A comparison of the valoneoyl proton signals (H_A-H_C) of rugosin D (**126**) and isorugosin D (**169**) showed significant differences in the chemical shifts of the H_A signal (δ 6.46 in **126** and δ 6.66 in **169**). This difference in chemical shifts of the H_A signal was also observed in several analogs of **126** [rugosin E (**127**)–G (**129**) (δ 6.42–6.53)] and **169** [isorugosin E (**170**) and G (**171**) (δ 6.58–6.66)] (*121,157,158*). The H_A signals in other rugosin- and isorugosin-type oligomers experience analogous shifts except in macrocyclic oligomers. Thus, the chemical shift of the valoneoyl H_A signal permits a straightforward assignment for the orientation of the valoneoyl group at O-4/O-6 of the glucose core of such oligomers. This convenient method was also applied to other classes of *m*-**DOG**-type oligomers which contain the galloyl part of the valoneoyl group on O-2, O-4 or O-6 of a glucose core.

The orientation of the valoneoyl group in oligomers of this class can also be assigned from the chemical shift of the H-3 signal of glucose-I, which is *ca.* 0.3 ppm upfield (δ *ca.* 5.5) in isorugosin-type dimers (or larger oligomers) compared with rugosin-type oligomers.

A similar difference allowing assignment of the orientation of a valoneoyl group has also been found among those oligomers which contain the HHDP part of a valoneoyl group on O-3/O-6 of a $^1C_4$ glucopyranose core as exemplified by the euphorbins A (**149**) and B (**151**) (*57,108*). The valoneoyl H_B signal, which resonates at a higher field in

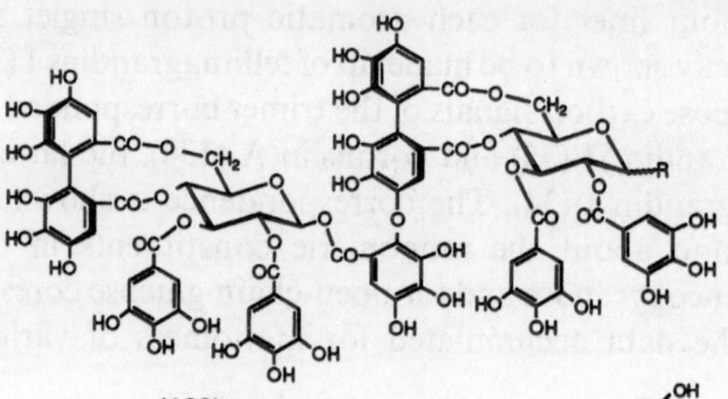

(169) Isorugosin D: R = (β)-OG
(170) Isorugosin E: R = OH

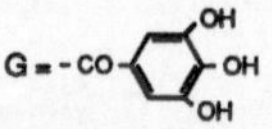

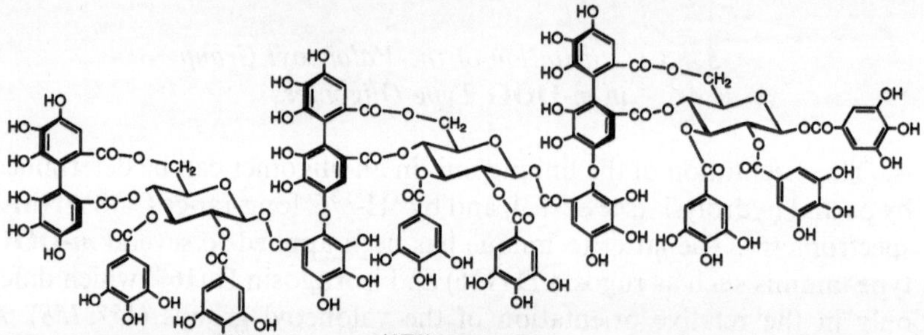

G = -CO

(171) Isorugosin G

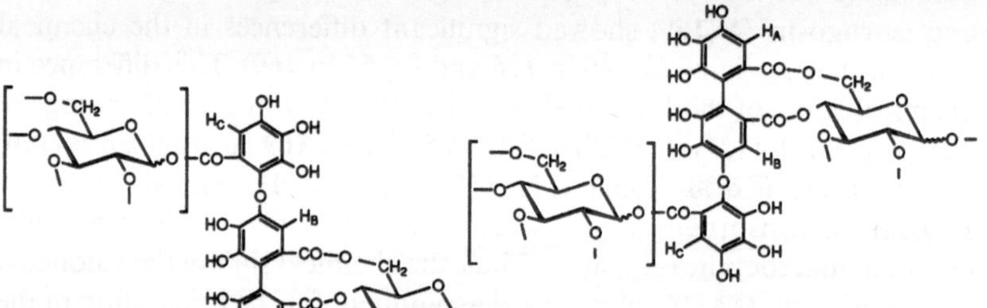

Rugosin-type
H_A δ 6.42 - 6.53

Isorugosin-type
H_A δ 6.58 - 6.66

Chart 4

euphorbin A-type dimers (δ 6.15–6.26) than in euphorbin B-type dimers (δ 6.40–6.42), can be used to differentiate these isomers.

3.4.4. Other ¹H-NMR Spectral Features of Oligomers

In the ¹H-NMR spectra of *m*-DOG [4, 6(or 6, 4) − 2′] type oligomers having the galloyl part of the valoneoyl unit at O-2′, such as cornusiin A (130) and camelliin A (172), the α-oriented H-1′ (β-anomer) on glucose-II

(172) Camelliin A

resonates at remarkably higher field ($\Delta\delta\,0.4$–0.8 ppm) than H-1' of analogs having a galloyl group on O-2' (δ ca. 5.2). This upfield shift is interpreted as being due to the anisotropic effect of the HHDP part of the valoneoyl group (129, 174). A similar upfield shift of the α-oriented anomeric proton signal is observed in m-GOG (1-2') type oligomers [hirtellin A (121) etc.] in which the H-1 signal resonates at higher field ($\delta\,5.35$) than that ($\delta\,6.20$) of tellimagrandin II (35). The chemical shift of the anomeric proton in the spectra of these types of oligomers is therefore

Table 6. Chemical Shifts of the Anomeric Proton in Tannins Having a Galloyl Part of Valoneoyl (m-DOG), Dilactonized Valoneoyl Group or Dehydrodigalloyl (m-GOG) Group on O-2 of the Glucose Core

Compound	H-1 of β-anomer	
Tannins having a free hydroxyl group on the anomeric center		Shift difference from 34
Tellimagrandin I (34)	5.15	
Oenothein C (40)	4.68	− 0.47
Cornusiin A (130)	4.50, 4.53	− 0.65, − 0.62
Oenothein B (138)	4.42	− 0.73
Camelliin A (167)	4.37, 4.36	− 0.78, − 0.79
Eucalbanin B (171)	4.37, 4.36	− 0.78, − 0.79
Tannins having a (β)-acyl group on the anomeric center		Shift difference from 35
Tellimagrandin II (35)	6.20	
Tirucallin A (42)	5.69	− 0.51
Hirtellin A (121)	5.35	− 0.85
Hirtellin B (124)	5.60	− 0.6

useful for assigning the location of the galloyl ester linkage within the valoneoyl (*m*-**DOG**) group or dehydrodigalloyl (*m*-**GOG**) group attached to glucose (Table 6).

3.5. Significance of Oligomers as Taxonomic Markers

An HPLC survey of rosaceous plants, using agrimoniin (**109**), gemin A (**115**), rugosin D (**126**), sanguiins H-6 (**153**) and H-11 (**158**) and several monomers as markers, demonstrated that several oligomers occurred

Table 7. *Distribution of Oligomeric Hydrolyzable Tannins in Dicotyledonae*

Order	Family	Type	Number of oligomers found
Nymphaeales	Nymphaeaceae	*m*-**GOG**	3
Ranunculales	Coriariaceae	*m*-**GOG**	4
		m-**DOG**	2
Hamamelidales	Hamamelidaceae	*m*-**DOG**	7
Fagales	Fagaceae	**CD**	6
		m-**DOG**	4
	Betulaceae	*m*-**DOG**	4
Casuarinales	Casuarinaceae	**CG**	1
		CFC	1
Theales	Theaceae	*m*-**DOG**	5
	Stachyuraceae	*m*-**DOG**	1
Rosales	Rosaceae	*m*-**GOG**	13
		m-**DOG**	6
		m-**GOD**	12
Myrtales	Lythraceae	*m*-**DOG**	13
	Trapaceae	*m*-**DOG**	2
	Onagraceae	*m*-**DOG**	3
	Melastomataceae	*m*-**DOG**	11
	Combretaceae	*m*-**DOG**	1
		CDF	3
		CFC	1
	Myrtaceae	*m*-**DOG**	1
		p-**DOG**	1
Cornales	Cornaceae	*m*-**DOG**	6
	Nyssaceae	*m*-**DOG**	2
Euphorbiales	Euphorbiaceae	*m*-**DOG**	18
		m-**GOG**	1
Lecythidales	Lecythidaceae	*m*-**DOG**	4
Violales	Tamaricaceae	*m*-**GOG**	11
		m-**GOGOG**	2

exclusively in a few genera (5). Thus, most *Rubus* species produce the tetramer lambertianin D (**157**) instead of sanguiin H-11 (**158**) (*141*). The hydrolyzable tannin monomers and oligomers of Rosaceae were found in herbaceous and frutescent species, but not in woody plants belonging to this family (5). However, both herbaceous and woody plants of Euphorbiaceae, and woody plants of Fagaceae, Betulaceae, Cornaceae, Combretaceae, Tamaricaceae, Nyssaceae, Hamamelidaceae, and Theaceae, *etc.*, have been found to yield oligomeric hydrolyzable tannins (6). Hydrolyzable tannin oligomers, presumably formed at a late stage of biosynthesis, would appear to be the better taxonomic markers than the monomers. Table 7 lists orders and families according to Cronquist's system (*175*) which produce oligomers.

4. Chemical Transformations of Ellagitannins

Chemical transformations of ellagitannins of unknown structure leading to the production of known compounds often give reliable structural information. However, there are some chemical transformations whose occurrence may be overlooked and which therefore introduce an element of confusion in the process of structure elucidation.

4.1. Isomerization by the Smiles Rearrangement

The Smiles rearrangement within an ellagitannin under mild condition was first observed in a dimer isolated from Tamaricaceae (*119*). Hirtellin C (**122**) (Scheme 6), a dimer from *Reaumuria hirtella* which contains both *m*-GOG and *p*-GOG groups as linking units, was isomerized to isohirtellin C (**173**) when its aqueous solution was kept at 95°C. This isomer has a symmetrical structure as revealed by the ^{1}H-NMR spectrum which looks like that of a monomer. The isomerization was attributed to a Smiles rearrangement of the isodehydrodigalloyl (*p*-GOG) group in hirtellin C (**122**) to a dehydrodigalloyl (*m*-GOG) group. The rearrangement was accelerated at higher pH (7.4 in 0.1 M phosphate buffer), almost complete isomerization of **122** to **173** occurring within 30 min at 40°C, although hirtellin C was fairly stable at room temperature in acidic solution below pH 3.0 or in aqueous acetone. This facile intramolecular nucleophilic aromatic substitution reaction under mild conditions may be due to the steric effect from adjacent hydroxyl groups rather than electronic stress (*176*). The rearrangement was successfully

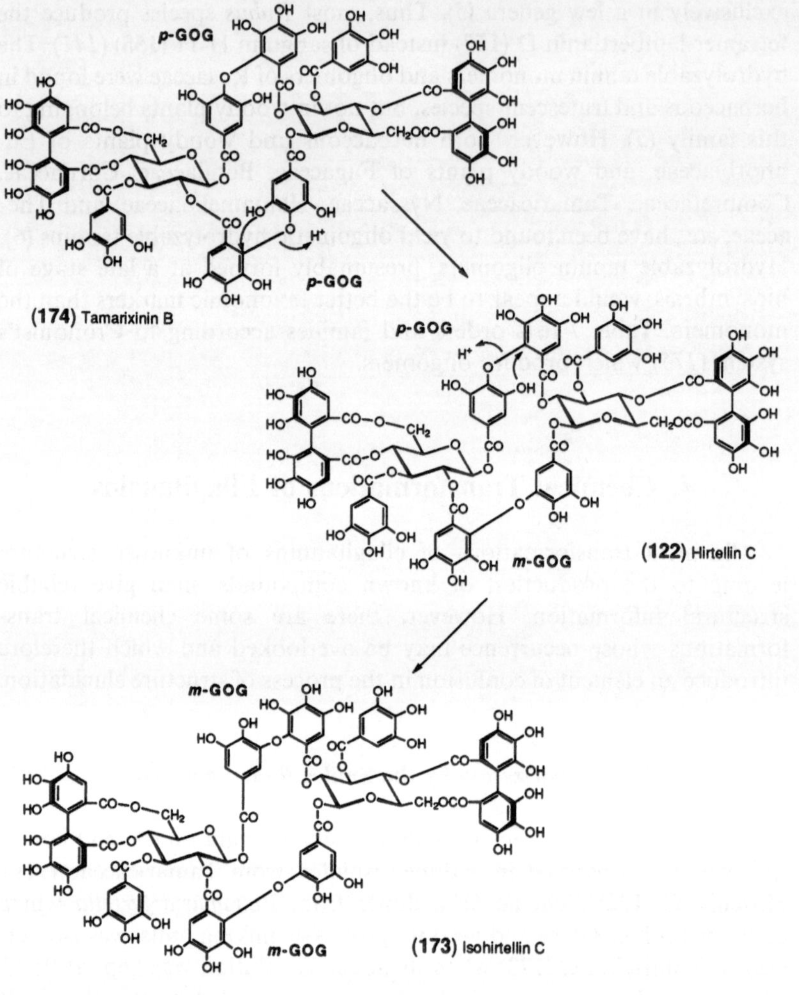

(174) Tamarixinin B

(122) Hirtellin C

(173) Isohirtellin C

Scheme 6

applied to the structure elucidation of tamarixinins B (174) and C (175) from *Tamarix pakistanica*, chemically correlating them with hirtellin C (122) and hirtellin A (121) of known structures (*154*) (Schemes 6 and 7).

The tergalloyl group in a tannin molecule can similarly be isomerized to a valoneoyl group through a Smiles rearrangement. For example, eucalbanin C (137), a *p*-DOG type dimer from *Eucalyptus alba*, was quantitatively converted into eucalbanin B (176) of the *m*-DOG type in a phosphate buffer (pH 7,4) at room temperature (*58*) (Scheme 8). The

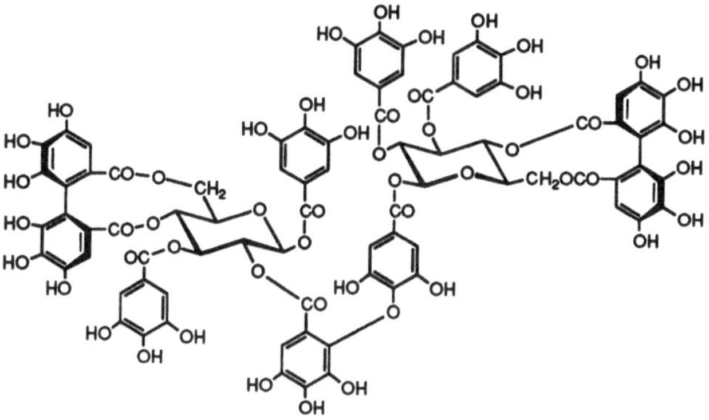

(175) Tamarixinin C

(121) Hirtellin A

Scheme 7

(137) Eucalbanin C

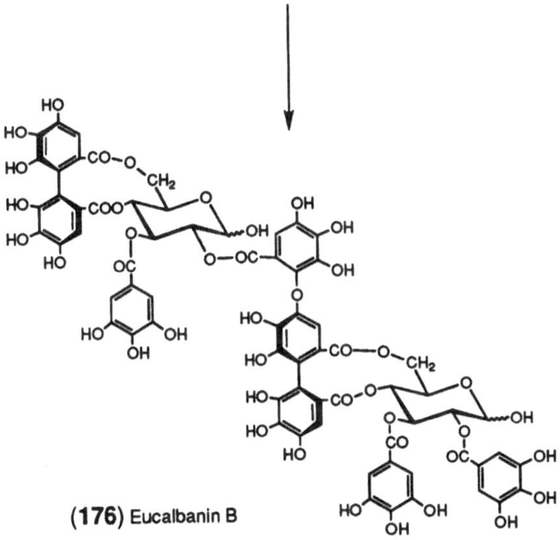

(176) Eucalbanin B

Scheme 8

dilactonized tergalloyl group in eucalbanin A (**41**) was also isomerized to the dilactonized valoneoyl group in cornusiin B (**39**) (*58*).

4.2. Chemical Transformation of Ellagitannins to *C*-Glucosidic and Complextannins

Biomimetic chemical transformations of ellagitannins to *C*-glucosidic tannins were achieved as follows: Casuarinin (**101**), a *C*-glucosidic ellagitannin, was prepared through acid catalyzed phenol-aldehyde coupling from liquidambin (**177**), an aldehydic ellagitannin which is presumed to be a biosynthetic precursor of **101** (Scheme 9) (*177*).

A series of complextannins (camelliatannins A–F) isolated from *Camellia japonica* leaves are grouped into two types according to their structures, a *C*-glucosidic type [camelliatannins A (**104**), B (**105**) and F (**107**)] and a type having a glucitol core [camelliatannins C (**178**) and E (**179**)]. In both types (–)-epicatechin is bound to C-1 of the sugar or polyalcohol core (*104, 105*). Structures of **178** and **179** were established by their conversion to camelliatannins B (**105**) and A (**140**), respectively, on treatment with polyphosphoric acid. Acid-catalyzed condensation of casuariin (**103**) with (–)-epicatechin afforded **104** and **105**, thus confirming the structures of these two compounds (Scheme 10) (*104*). The conversion of camelliatannin A (**104**) to camelliatannin F (**107**) which contains a highly functionalized cyclopentenone ring attached to C-1 and C-2 of the glucose core (*105*) on heating in ethanol-acetic acid is also noteworthy.

4.3. Hydrolysis of Ellagitannins

Partial hydrolysis of oligomers in boiling water or weak acidic media, under monitoring with HPLC, yields monomeric hydrolyzates of simpler or known structures. These hydrolyzates offer valuable information for structure elucidation of the oligomers. The susceptibility of the ester linkages to such cleavage depends largely on the location of the ester linkages on the glucose core(s). Generally, the dimers in which two carboxyl functions on the linking unit are involved in ester formation with one glucose O-1 and an oxygen other than O-1 of a second glucose are hydrolyzed preferentially at the former. The valoneoyl groups linking two monomers in rugosins D (**126**), E (**127**), F (**128**) (dimers) and G (**129**) (trimer) which belong to the class of *m*-DOG(4, 6-1′) type oligomers were thus hydrolyzed at the galloyl part prior to the HHDP part. As a result,

References, pp. 102–117

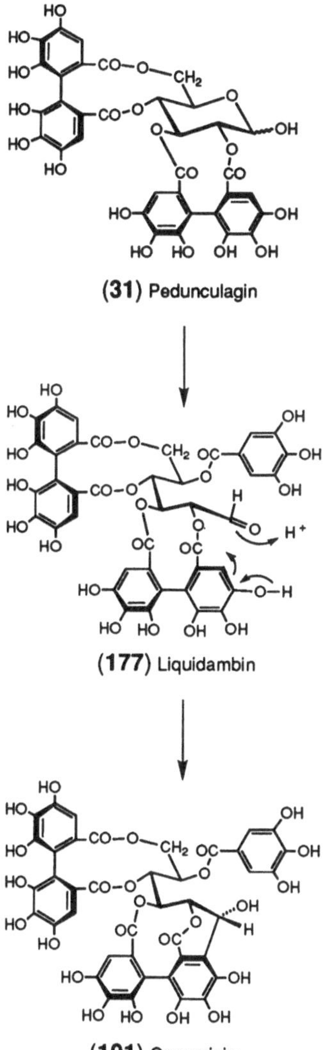

(31) Pedunculagin

(177) Liquidambin

(101) Casuarinin

Scheme 9

the constituent monomeric units, tellimagrandin I (**34**), rugosin A (**36**) and/or rugosin B (**37**), were liberated in the ratio representing that of each monomer in the oligomer molecule (*121*) (Scheme 11). An intermediary depsidone, such as euprostin C (**47**), was formed at an early stage of the reaction (*174*). Its formation is attributable to transesterification in the

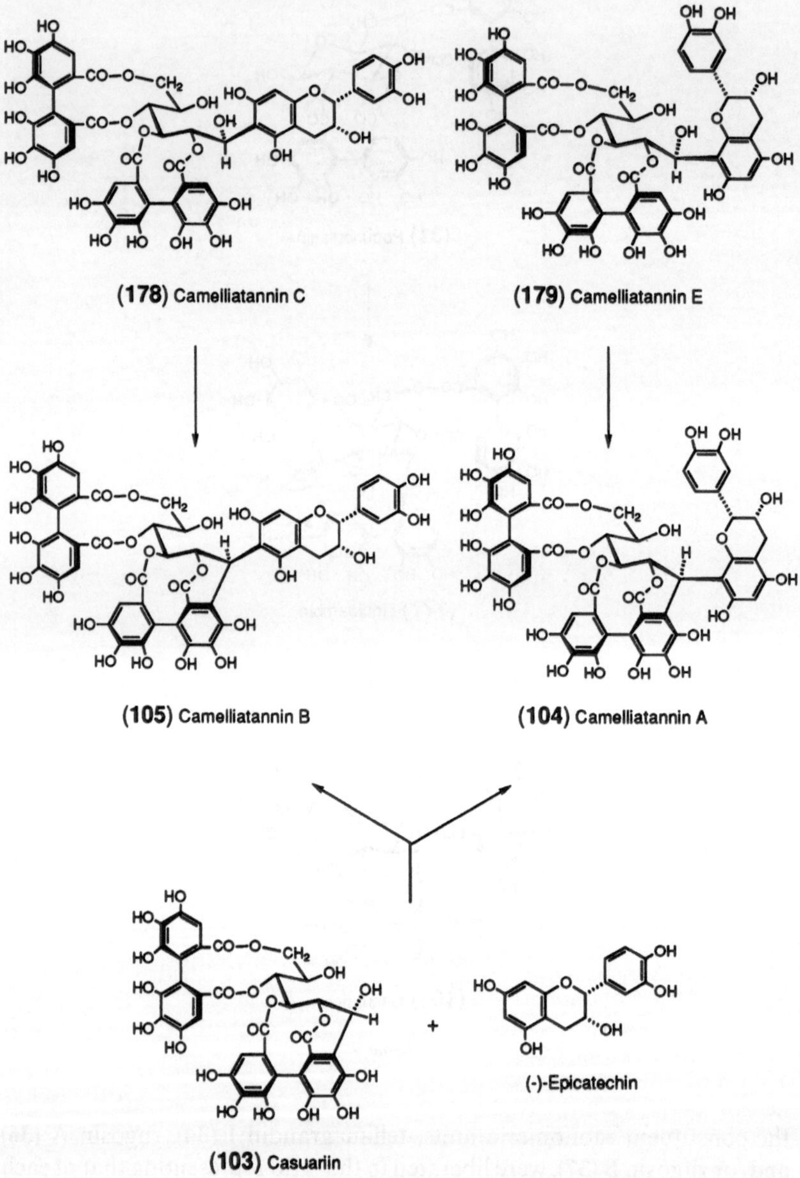

(178) Camelliatannin C

(179) Camelliatannin E

(105) Camelliatannin B

(104) Camelliatannin A

(103) Casuariin

+

(-)-Epicatechin

Scheme 10

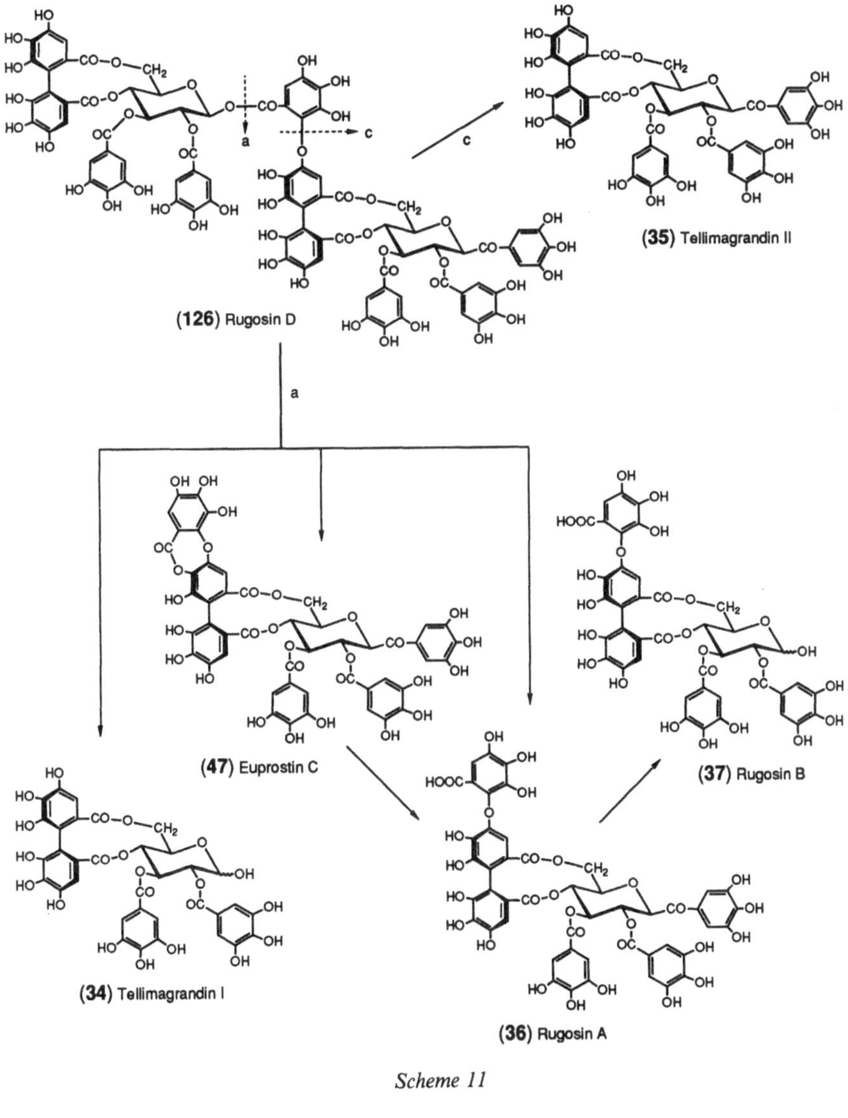

(126) Rugosin D

(35) Tellimagrandin II

(47) Euprostin C

(34) Tellimagrandin I

(36) Rugosin A

(37) Rugosin B

Scheme 11

valoneoyl group under participation of neighboring phenolic hydroxyl. The reaction rate and the product structure were pH-dependent (*174*).

The high susceptibility at the anomeric center to hydrolysis in hot water was independent of the type of linking unit, as exemplified by hydrolyses of the dimers hirtellin A (**121**) of the *m*-**GOG** (1-1′) type and sanguiin H-6 (**153**) of the *m*-**GOD**(6, 4-1′) type. In addition to tellima-grandin I (**34**), their hydrolysis liberated in the first instance a monomer with the *m*-**GOG** group at O-2 and, in the second instance, a monomer with the *p*-**GOD** group at O-4/O-6, respectively (*65, 112*). On the other hand, when the galloyl part of the valoneoyl group (*m*-**DOG**) or its analogs [*m*-**GOD** and *m*, *p*-**D(OG)**$_2$] in the oligomers is attached to a position other than the anomeric center of a glucose unit, the two ester linkages on the HHDP part are hydrolyzed preferentially yielding a monomer having a dilactonized valoneoyl group. Thus, *m*-**DOG** type dimers, such as cornusiin A (**130**) (Scheme 12), nobotanin F (**135**) (Scheme 13) and nobotanin B (**136**) (Scheme 14), yield on hydrolysis in boiling water a monomer retaining the ester bond on the galloyl moiety (**39** and

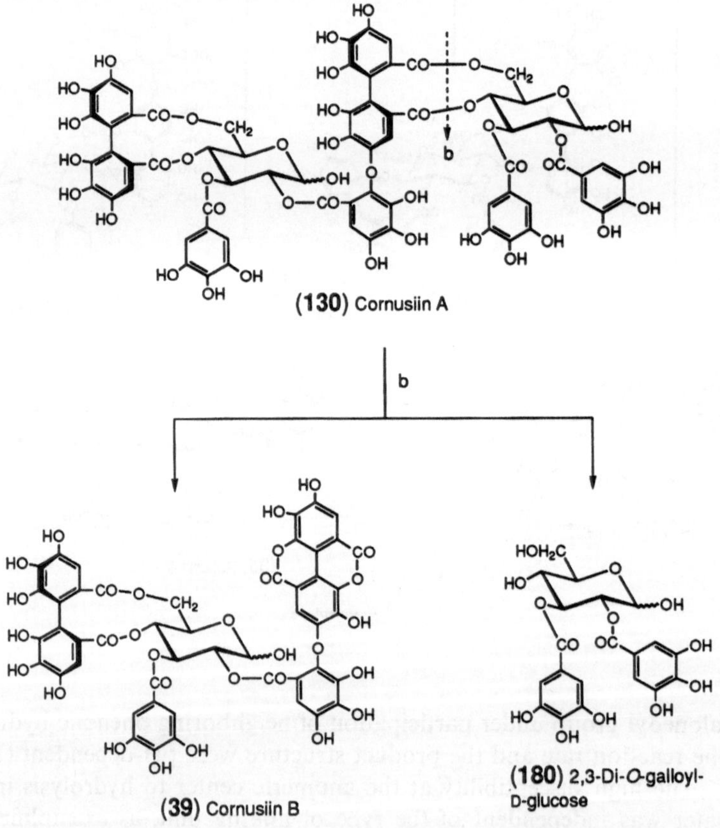

(**130**) Cornusiin A

(**39**) Cornusiin B

(**180**) 2,3-Di-*O*-galloyl-D-glucose

Scheme 12

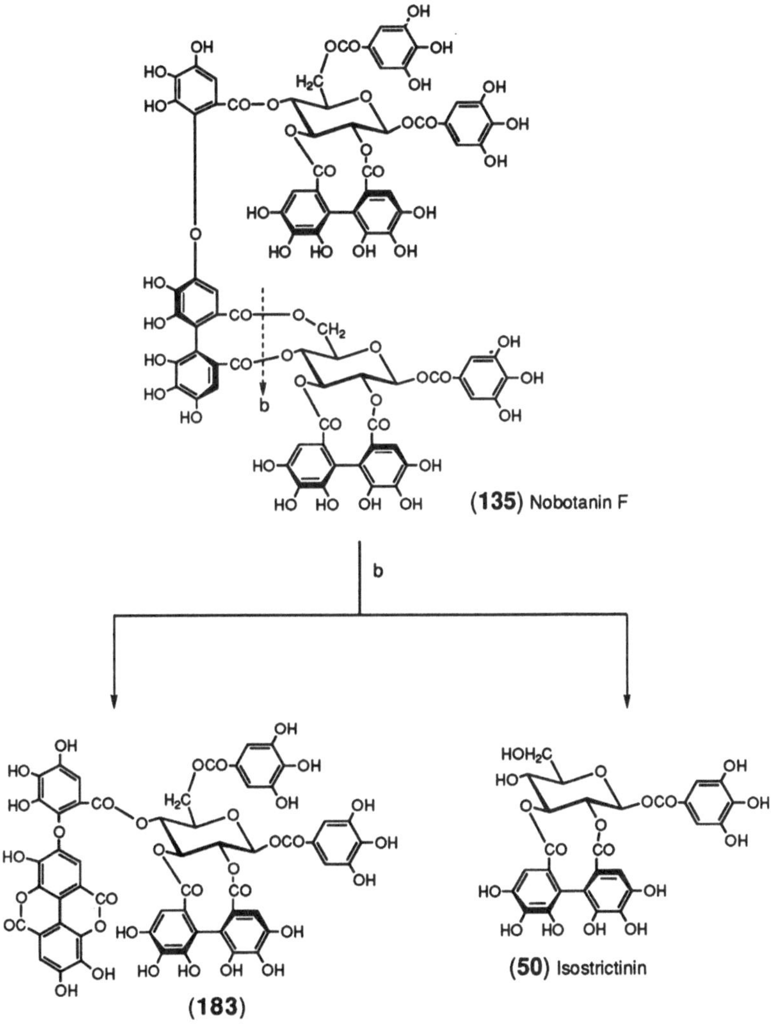

(135) Nobotanin F

(183)

(50) Isostrictinin

Scheme 13

183), along with the counterpart monomer (*123, 127, 128*). A similar hydrolysis was also observed in the case of oenothein B (**138**) as shown in Scheme 15.

It should be noted that the ether bond of the valoneoyl group in a tannin is sometimes cleaved under mild conditions as, for example, in the hydrolysis of isorugosin D (**169**) and the euphorbins A (**149**) and F (**150**) (Scheme 16) (*174*). This type of cleavage occurs in boiling water or weak

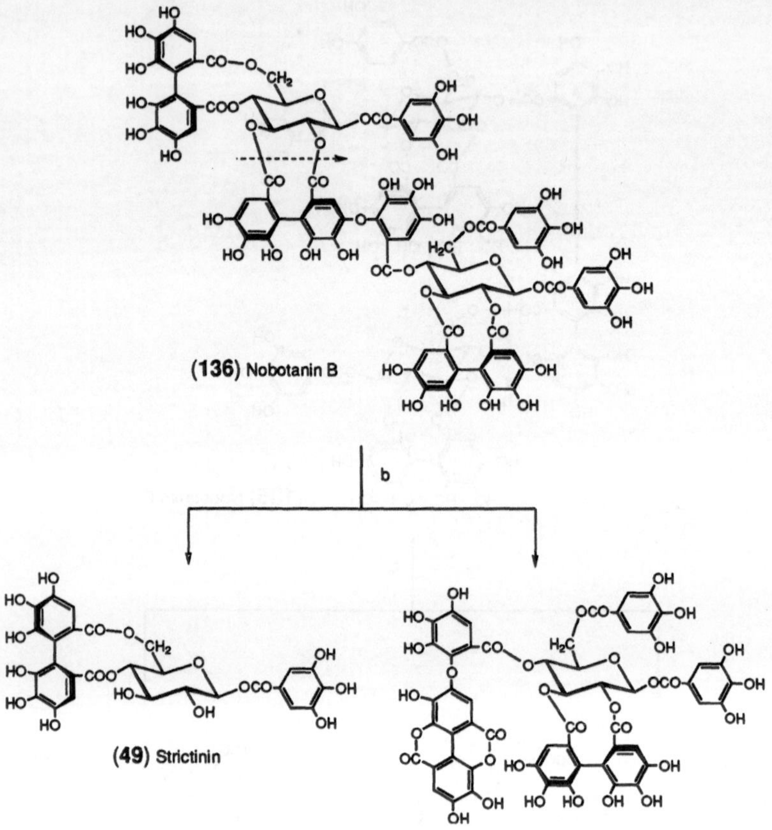

(136) Nobotanin B

(49) Strictinin

(183)

Scheme 14

acid, and more efficiently in weak alkaline medium. Treatment of rugosin B (37), with a phosphate buffer (pH 7, 4) at 60°C for 2.5 h, gave gallic acid and tellimagrandin I (34) as the major products. The reaction, which is analogous to the ether cleavage of dehydrodigallic acid (2), is ascribable to β-elimination followed by a disproportionation reaction (Scheme 17).

This ether bond cleavage under mild conditions may sometimes complicate structure determination of the oligomers, but may also be useful for structure elucidation as, for example, in the case of macrocyclic trimers oenothein A (159) and woodfordin D (160). These were hydro-lyzed in boiling water to give oenothein B (138) and gemin D (51) as the major primary hydrolyzates (Scheme 18) (142).

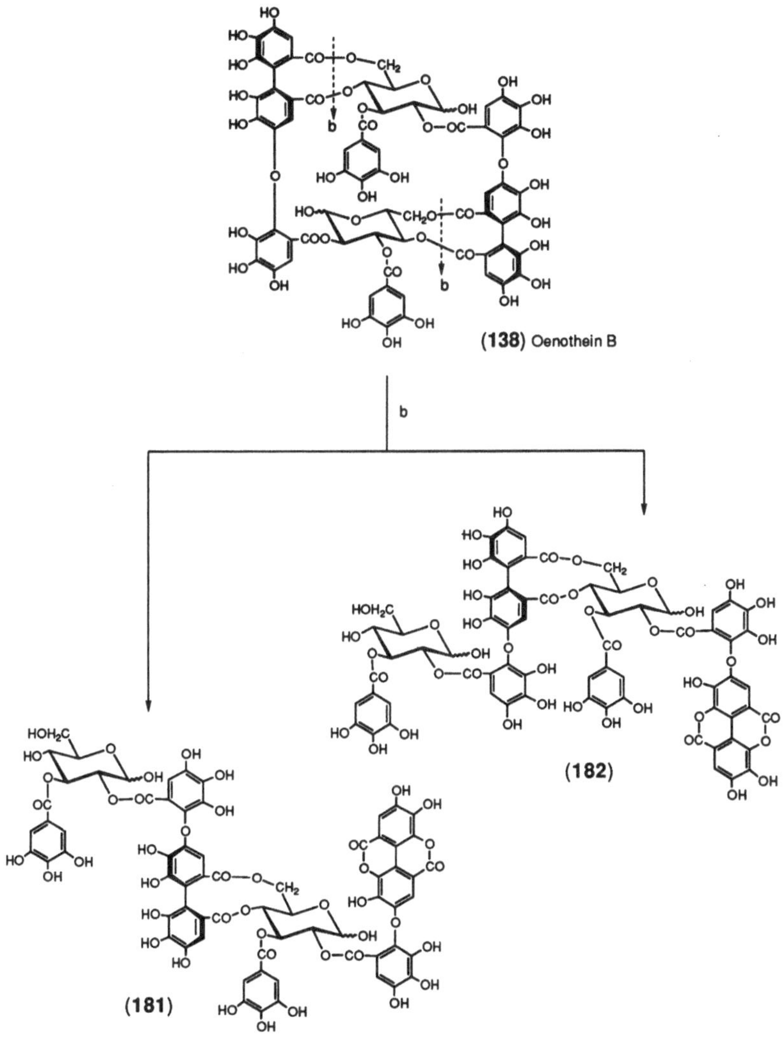

(138) Oenothein B

(182)

(181)

Scheme 15

5. Caffeic Acid Esters and Derivatives

Caffeic acid derivatives isolated from several plants have been regarded as the "tannin" of the species containing them. Some of them have pharmacological activities similar to those of gallo- and ellagitannins.

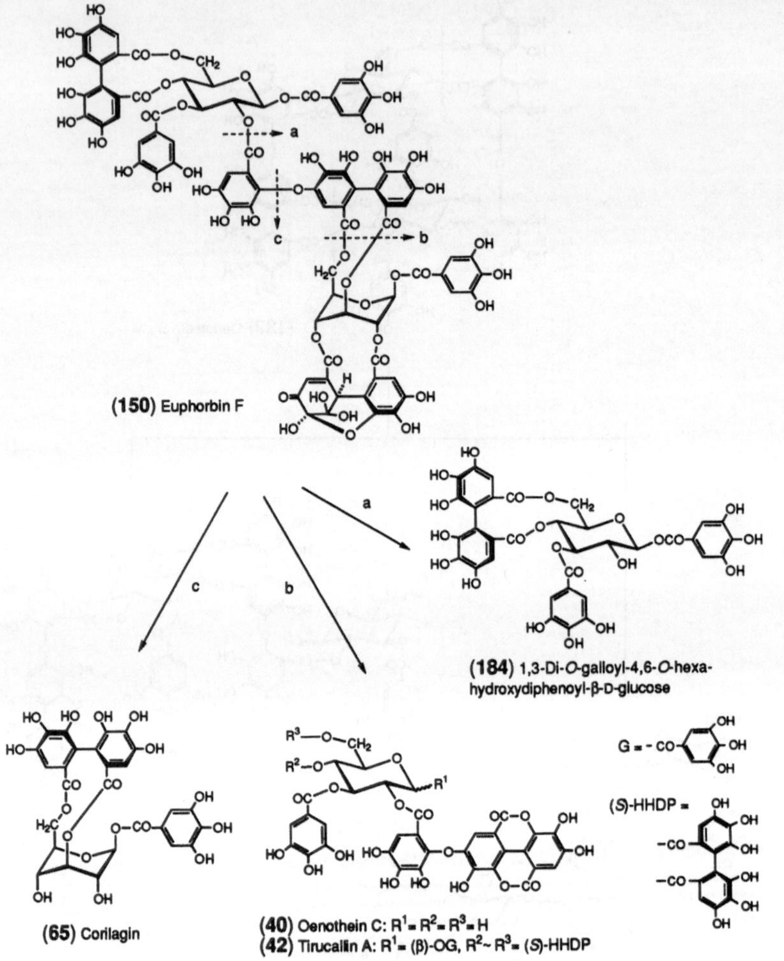

(150) Euphorbin F

(184) 1,3-Di-O-galloyl-4,6-O-hexa-
hydroxydiphenoyl-β-D-glucose

(65) Corilagin

(40) Oenothein C: R¹= R²= R³= H
(42) Tirucallin A: R¹= (β)-OG, R²~ R³= (S)-HHDP

G = - CO

(S)-HHDP =

Scheme 16

The name "caffetannin" has been used for chlorogenic acid (185) and its congeners in coffee beans (178). However, the protein-binding activity of chlorogenic acid, which is the main polyphenol in coffee, is almost negligible in spite of the name "caffetannin". Investigation of the polyphenolic components of several *Artemisia* species which are used medicinally in east Asia and regarded as rich in tannin revealed that the binding activity of the extracts of these plants is attributable to that of 3,5-di-O-caffeoylquinic acid (186) (178) and its isomers which have binding

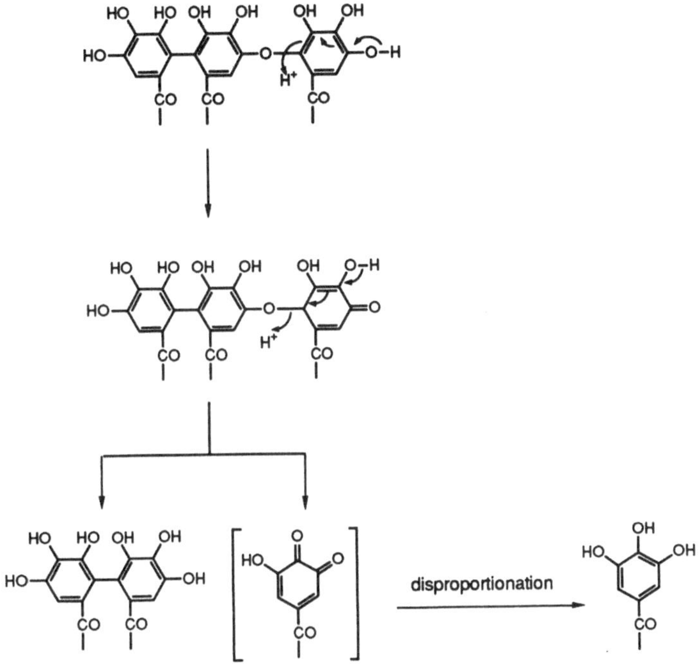

Scheme 17

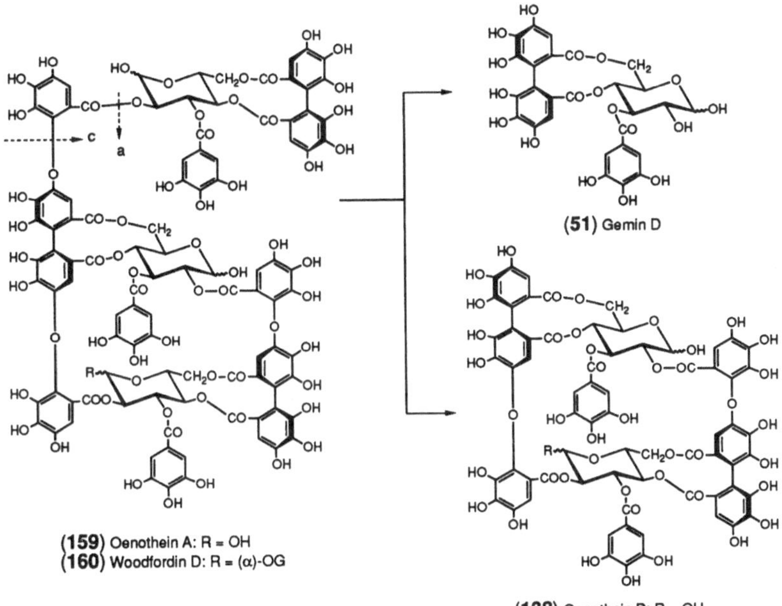

(159) Oenothein A: R = OH
(160) Woodfordin D: R = (α)-OG

(51) Gemin D

(138) Oenothein B: R = OH
(139) Woodfordin C: R = (α)-OG

Scheme 18

activities significantly stronger than chlorogenic acid. Di-*O*-caffeoyl-quinic acids are widely distributed in the families Compositae, Solanaceae and Caprifoliaceae. Recently, tricaffeoylquinic acids have been isolated from members of the Compositae, *e.g.*, 3,4,5-tri-*O*-caffeoylquinic acid (187) from *Chrysothamnus paniculatus* (*179*), 1,4,5-tri-*O*-caffeoyl-quinic acid (188) from *Arnica montana* (*180*) and 1,3,5-tri-*O*-caffeoylquinic acid (189) from *Xanthium strumarium* (*181*).

Another example of a caffeic acid derivative labeled a "tannin" is rosmarinic acid (190), which is widely distributed in Labiatae and called "labiataetannin" (*182, 183*). This compound, a caffeoyl ester of 3,4-dihydroxyphenyllactic acid, may be regarded as a dimer of caffeic acid. Lithospermic acid (191) (a trimer) (*184*), lithospermic acid B (or salvianolic acid C) (192) (*185, 186*) melitric acids A (193) and B (194) (*187*), rabdosiin (195) (*188*) and an isomer (196) of rabdosiin (tetramer) have also been found in some species of Labiatae and Boraginaceae (*189, 190*). The name "labiataetannin" may be applicable to these compounds, too.

Although a large number of caffeoyl esters of different types, such as those containing phenylpropanoid glycosides, has been found in various plants, these compounds, which have not been labeled "tannins", are not reviewed here.

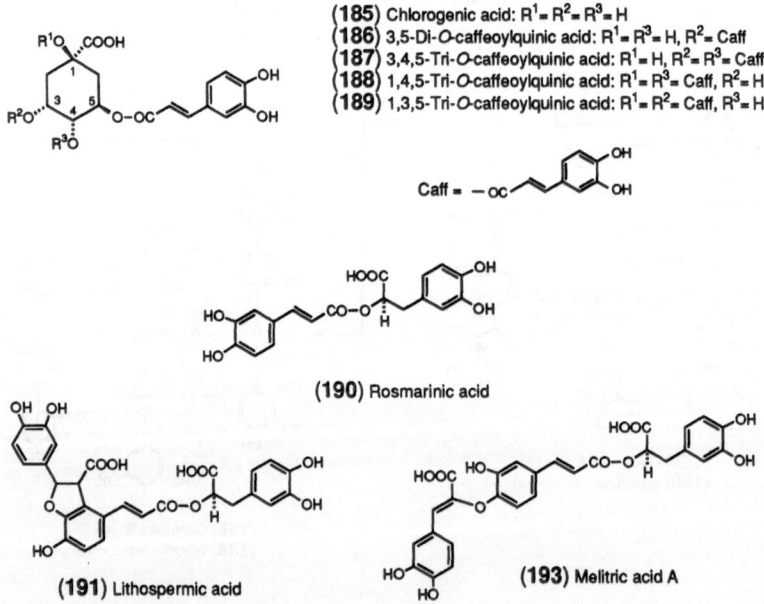

(185) Chlorogenic acid: $R^1 = R^2 = R^3 = H$
(186) 3,5-Di-*O*-caffeoylquinic acid: $R^1 = R^3 = H$, $R^2 = $ Caff
(187) 3,4,5-Tri-*O*-caffeoylquinic acid: $R^1 = H$, $R^2 = R^3 = $ Caff
(188) 1,4,5-Tri-*O*-caffeoylquinic acid: $R^1 = R^3 = $ Caff, $R^2 = H$
(189) 1,3,5-Tri-*O*-caffeoylquinic acid: $R^1 = R^2 = $ Caff, $R^3 = H$

(190) Rosmarinic acid

(191) Lithospermic acid

(193) Melitric acid A

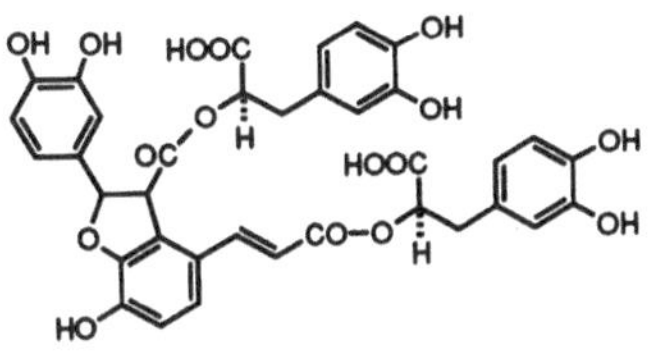

(192) Lithospermic acid B

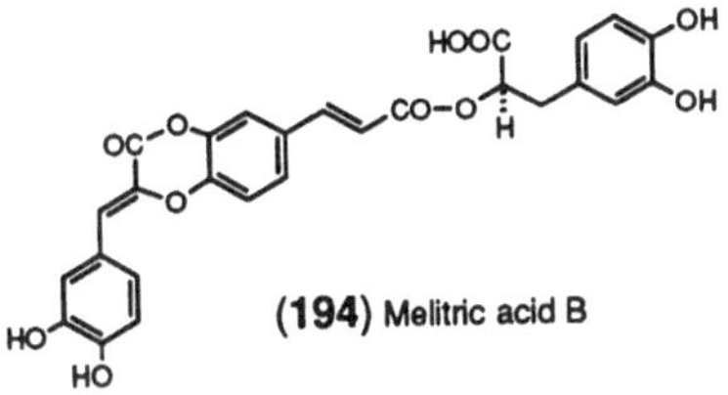

(194) Melitric acid B

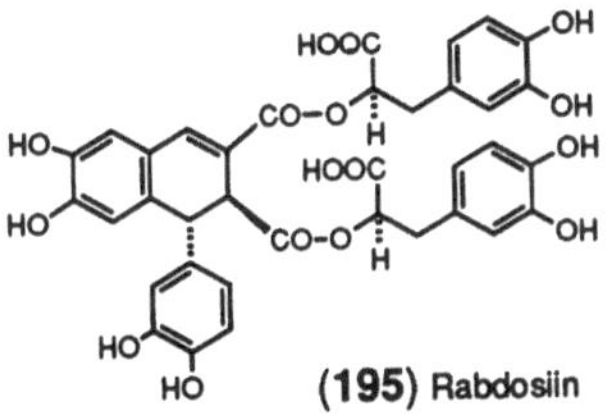

(195) Rabdosiin

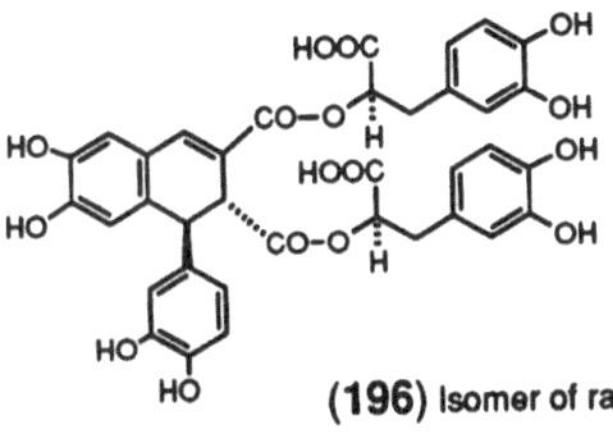

(196) Isomer of rabdosiin

6. Seasonal Change of Tannin Structures and Their Biogenesis

The C-glucosidic ellagitannins casuarinin (**101**), stachyurin (**102**) and casuariin (**103**), first isolated from *Casuarina stricta* (*52*), have also been found in the leaves of many Myrtaceae and Hamamelidaceae. They are presumed to be biosynthesized from pedunculagin (**31**) because of their almost invariable co-occurrence with pedunculagin and/or tellimagrandin I (**34**), the latter being a plausible precursor of pedunculagin, and also because of the seasonal variation in the content of these compounds in leaves of *Liquidambar formosana* (Hamamelidaceae) (*191*) (Fig. 5). The production of casuarinin (**101**) in the leaves of this plant in autumn was accompanied by a decrease and then by complete disappearance of the gallotannins pentagalloylglucose and trigalloylglucose and two ellagi-tannins, tellimagrandins I (**34**) and II (**35**), which are abundant in the young leaf in spring (*191*) (Scheme 19). This seasonal variation is in accord with the plausible hypothesis that these C-glucosidic tannins are bio-synthesized from the gallotannins and the ellagitannins found in the young leaf. An aldehyde liquidambin (5-O-galloyl-2,3;4,6-di-O-(S)-hexahydroxydiphenoyl-D-glucose) (**177**) mentioned in Sect. 4.2, is the most probable intermediate between pedunculagin (**31**) and casuarinin (**101**) (*177*).

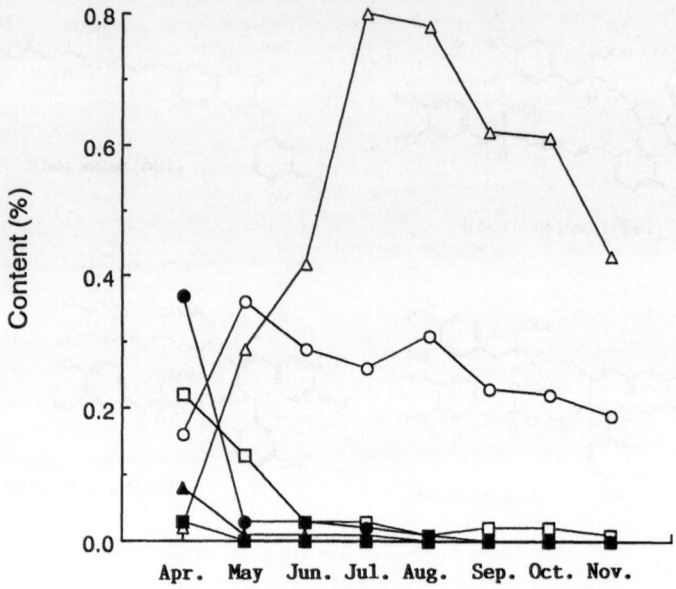

Fig. 5. Seasonal change of hydrolyzable tannins in *Liquidambar formosana*: △, casuarinin (**101**); ○, pedunculagin (**31**); □, casuarictin (**32**), ▲, 1,2,6-tri-*O*-galloyl-β-D-glucose (**3**); ■, 1,2,3,4,6-penta-*O*-galloyl-β-D-glucose (**6**); ●, tellimagrandin II (**35**)

7. Distribution of Hydrolyzable Tannins in Plants

The distribution of hydrolyzable tannins which are esters of gallic acid and its metabolites is limited to members of the Dicotyledoneae (mostly those within Choripetalae) of the Angiospermae except for *Spirogyra*. They are absent in most orders of Sympetalae which rank highest among the Dicotyledoneae in systems of plant evolution (*6*). This is unlike the wider distribution of condensed tannins and caffeic acid esters which are also found in Monocotyledoneae and Gymnospermae. More specific correlations with plant systematics have been found for oligomeric hydrolyzable tannins as described in Sects. 3.2–3.5. Fig. 6 shows the orders of plants producing hydrolyzable tannins in Cronquist's system of plant evolution (*175*).

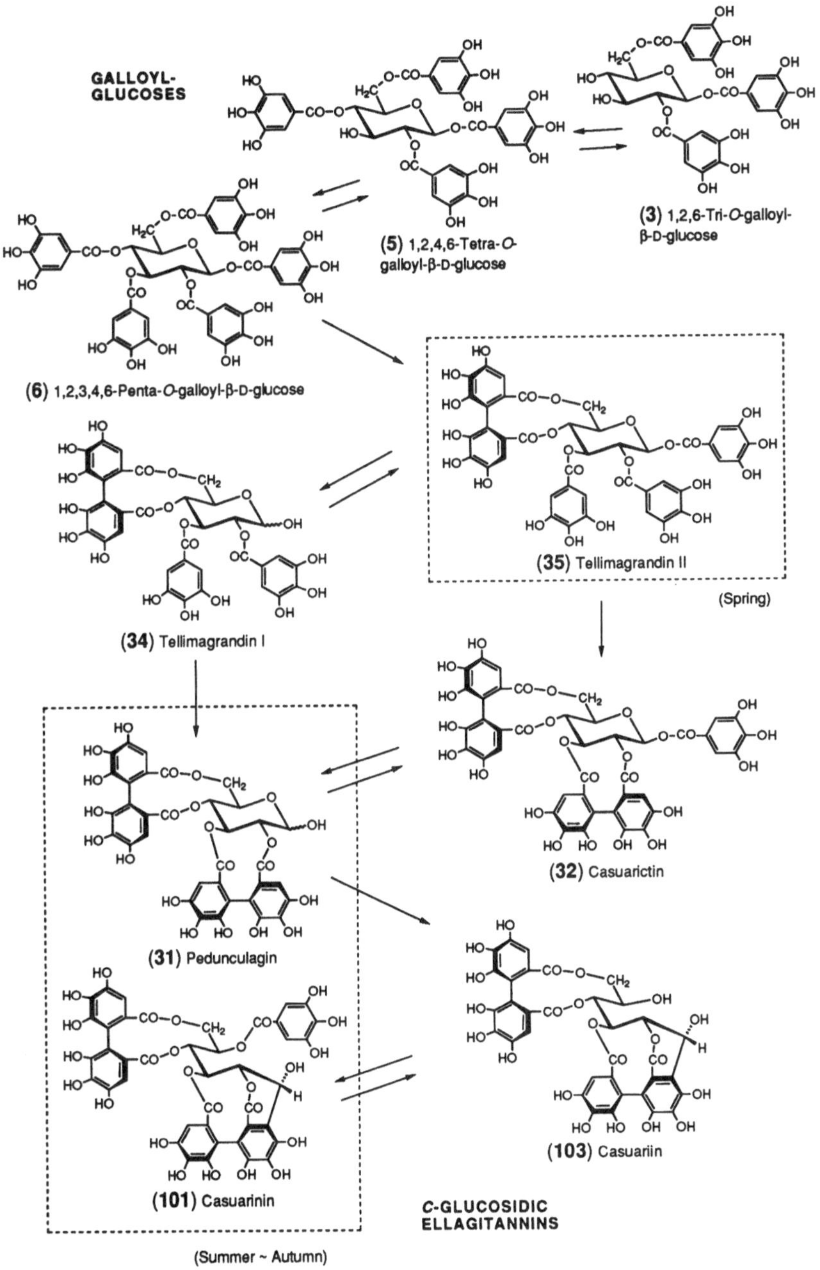

GALLOYL-GLUCOSES

(3) 1,2,6-Tri-O-galloyl-β-D-glucose

(5) 1,2,4,6-Tetra-O-galloyl-β-D-glucose

(6) 1,2,3,4,6-Penta-O-galloyl-β-D-glucose

(35) Tellimagrandin II

(Spring)

(34) Tellimagrandin I

(32) Casuarictin

(31) Pedunculagin

(103) Casuariin

(101) Casuarinin

C-GLUCOSIDIC ELLAGITANNINS

(Summer ~ Autumn)

Scheme 19

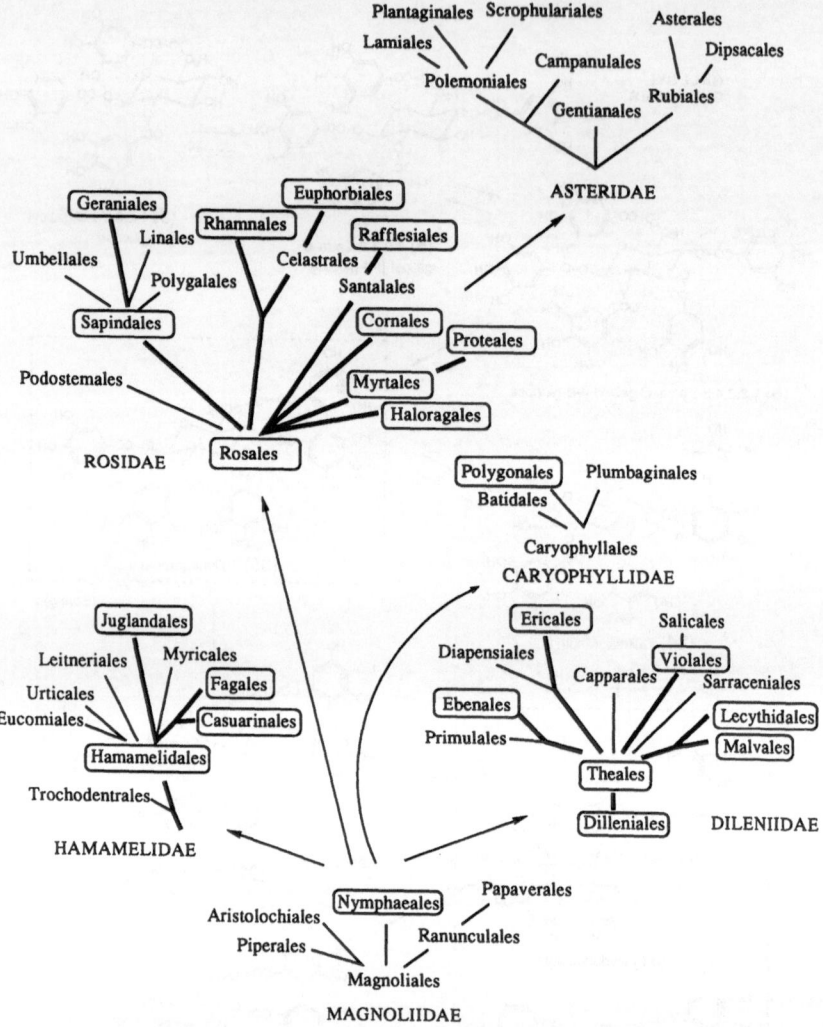

Fig. 6. Distribution of hydrolyzable tannins in Dicotyledoneae of Cronquist's system.
━━━ Evolutionary route of hydrolyzable tannin-producing plants. ⬭Order to
which hydrolyzable tannin-producing plants belong

8. Production of Hydrolyzable Tannins in Tissue Cultures

Although plant tissue cultures have been extensively studied in recent
years, only a limited number of reports on the production of tannins and
related polyphenols by tissue culture has appeared in spite of the

importance of these compounds. Examples of the production of procyanidins by cell cultures are: (+)-Catechin, (−)-epicatechin and its condensates, procyanidin B2 (197) and procyanidin B2-3′-O-gallate (198), etc.,
were produced by selected anthocyanin-rich red calli of buckwheat
(Fagopyrum esculentum) (192, 193). The production of galloylated procyanidin by these calli was stimulated by gallic acid in the medium (192).
Oligomeric procyanidins [procyanidin C1 (199) and cinnamtannin A2
(200)], along with (−)-epicatechin, hyperin and quercitrin, were produced in some cultured cell lines induced from seeds of Hypericum
erectum (194). Large amounts of procyanidins B2 (197), B4 and C1 (199)
were produced in callus and suspension cultures of Cinnamomum cassia
which is used as a spice and a crude drug in oriental medical prescriptions
(195).

The isolation of many hydrolyzable tannins from medicinal plants,
assisted by recent advances in analytical methods (1) and also the
discovery of their remarkable biological activities described in Sect. 10

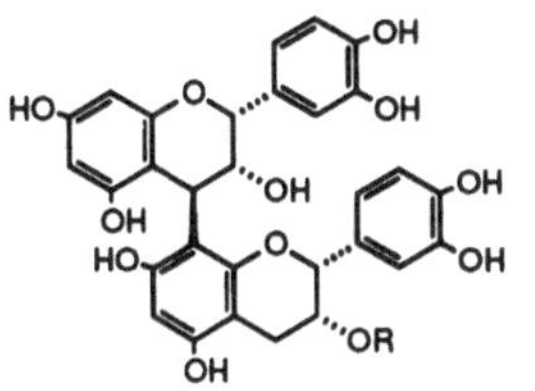

(197) Procyanidin B2: R = H
(198) Procyanidin B2-3′-O-gallate: R = G

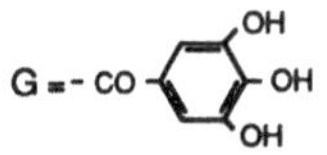

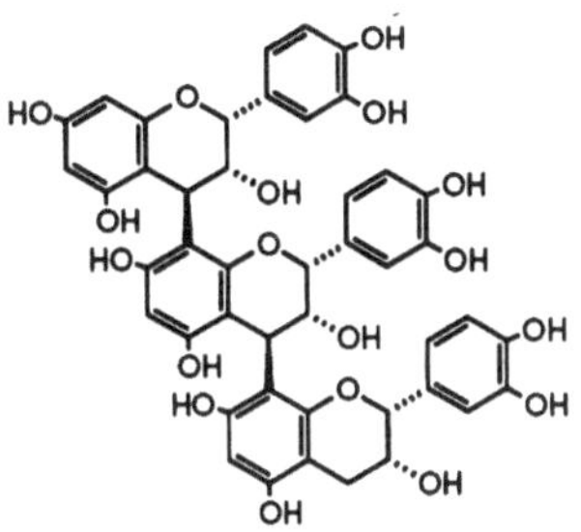

(199) Procyanidin C1

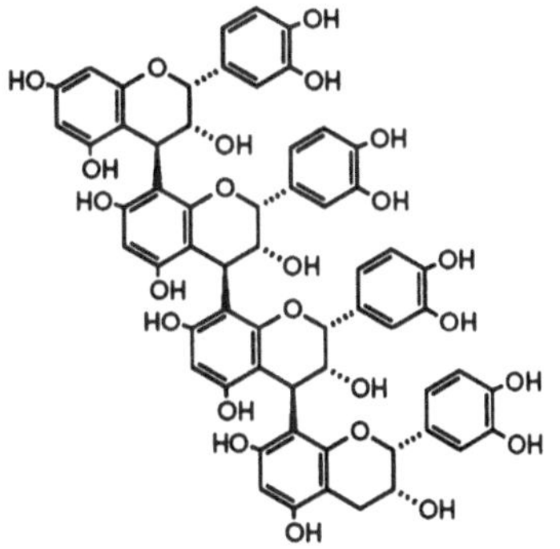

G = - CO

(200) Cinnamtannin A2

stimulated studies dealing with their production by cell culture. Galloyl-glucoses (1,2,6-tri- (**3**), 1,2,3,6-tetra- (**4**), 1,2,3,4,6-penta- (**6**) and 6-digalloyl-1,2,3-trigalloyl-β-D-glucose) were produced by calli and suspension cultures induced from flower buds of *Cornus officinalis*, the fruits of which have been used as an astringent, tonic and hemostatic in traditional Chinese medical prescriptions (*196, 197*). The content of the main tannin, 1,2,3,6-tetragalloyl-β-D-glucose (**4**) produced by the cultured cells, was shown to be 36 times that in the intact fruit. However, the cultured cells of *C. officinalis* did not produce ellagitannins which are main constituents of the intact fruit and are hence considered to lack the enzyme which catalyzes the oxidative coupling of galloyl groups leading to ellagitannins.

The production of ellagitannins by plant cell cultures has been carried out in several ways (*198, 199, 200, 201*). For example, two dehydroellagitannins [geraniin (**63**) and phyllanthusiin D (**98**)] were produced along with procyanidin B2 (**197**) and pentagalloylglucose by cell strains of *Geranium thunbergii* cultured in a medium containing suitably controlled hormones (*100*). A large amount of casuarictin (**32**), an ellagitannin monomer, accompanied by a dimer, nobotanin B (**136**), was produced by green callus tissues of *Heterocentron roseum* (a shrub native to Mexico), under illumination with fluorescent lamps (*198*). It is noteworthy that an ellagitannin trimer (nobotanin J) was also produced in this callus. Light was requisite for both ellagitannin production and cell growth in these plant cell cultures, since no ellagitannin was detected by HPLC in callus tissues grown in the dark (*198*). Hairy root cultures induced by infection with *Agrobacterium rhizogenes* strains generally showed more reproducible and faster growth than normal cell cultures and often produced the secondary metabolites in higher yield (*200*). Production of large amounts of geraniin (**63**) and sanguiin H-11 (**158**), an ellagitannin tetramer, in the hairy roots of *Geranium thunbergii* and *Sanguisorba officinalis*, respectively, was achieved by infection with *A. rhizogenes* A4 strain (*199, 200*). Thus, cell culture techniques may be useful for biochemical and biosynthetic studies of tannins.

9. Biosynthesis of Hydrolyzable Tannins

In spite of the discovery of numerous hydrolyzable tannins, their biosynthesis remains almost unsolved except for the biosynthesis of gallotannins. Even the biosynthesis of gallic acid itself, a fundamental constituent of hydrolyzable tannins, is still in dispute. As summarized by

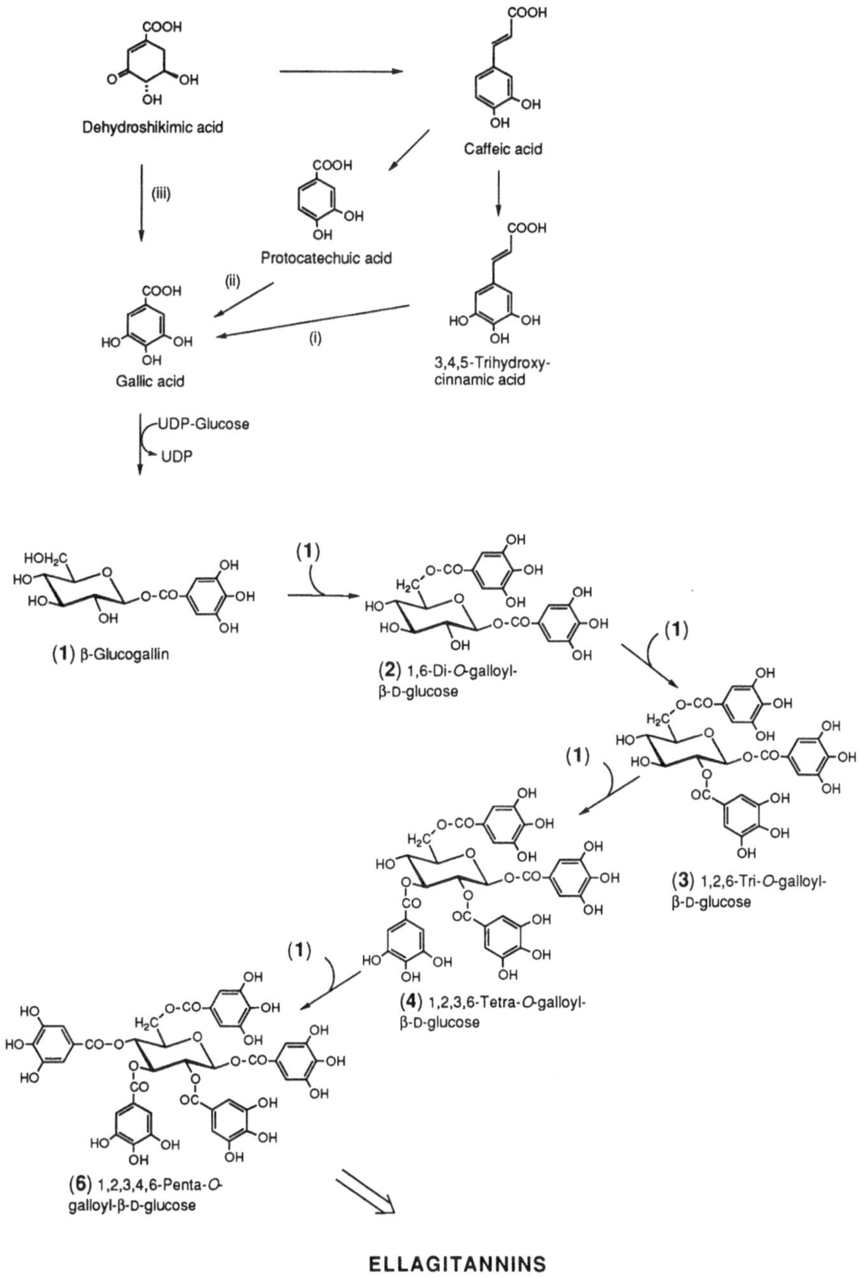

Dehydroshikimic acid

Caffeic acid

(iii)

Protocatechuic acid

(ii)

Gallic acid

(i)

3,4,5-Trihydroxy-
cinnamic acid

UDP-Glucose

UDP

(**1**) β-Glucogallin

(**2**) 1,6-Di-*O*-galloyl-
β-D-glucose

(**3**) 1,2,6-Tri-*O*-galloyl-
β-D-glucose

(**4**) 1,2,3,6-Tetra-*O*-galloyl-
β-D-glucose

(**6**) 1,2,3,4,6-Penta-*O*-
galloyl-β-D-glucose

ELLAGITANNINS

Scheme 20

HASLAM (*3*) and GROSS (*10*), the conflicting proposals made for the pathway of gallic acid biosynthesis are as follows (see Scheme 20): (i) β-oxidation of 3,4,5-trihydroxycinnamic acid derived from caffeic acid, (ii) hydroxylation of protocatechuic acid derived from caffeic acid, and (iii) direct aromatization of (dehydro) shikimic acid. The evidence for each pathway has been provided by investigations using several different plant species. It is now generally accepted that gallic acid is bio-synthesized by routes (ii) and (iii), and that the preferred route depends on the species and its stage of development (*3, 10*).

Recently the biosynthesis of gallotannins has been investigated in a series of experiments using cell-free extracts or partially purified enzymes from young leaves of *Quercus* and *Rhus* species. The primary metabolite, β-glucogallin (**1**), which is formed by reaction of UDP-glucose with free gallic acid, was shown to serve as the galloyl donor and also as the acceptor (*10*). The enzymatic disproportionation reaction of two moles of β-glucogallin (**1**) thus yields 1,6-digalloylglucose (**2**), which can, as well as β-glucogallin, act as galloyl donor in the next step, the formation of 1,2,6-trigalloylglucose (**3**). A series of gallotannins (di- ~ hexagalloylglucoses) are thus produced by β-glucogallin-dependent galloyltransferases in the presence of β-glucogallin (**1**) (*10*) (Scheme 20).

Ellagitannins have been regarded as metabolites produced through intramolecular oxidative C-C coupling between galloyl groups suitably located in a galloylglucose molecule (*2*). Although no experimental evidence for the formation of ellagitannins themselves *via* oxidative coupling of galloylglucoses has so far been obtained both in chemical and enzymatic studies, production of ellagic acid and alkyl hexahydroxydi-phenate by polyphenol oxidases and radical coupling reactions has been demonstrated (*202, 203*). An (*S*)-hexahydroxydiphenoyl ester has recently been synthesized by diastereoselective oxidative phenolic coupling of a digallate of deoxyglucose derivative, thus supporting the anticipated biogenesis of ellagitannins (*204*).

10. Biological and Pharmacological Activities of Hydrolyzable Tannins

Although for many years leather making has been the most important utilization of tannins for chemists, there have been many other uses of tannins produced by various plants. Examples are in dyeing (particularly the use of tannins as fixers), removal of cloudiness from products of brewing such as wine, beer and sake, and production of permanent-type inks. Tannins have also been used as anticorrosives, in the manufacturing of adhesives and in the production of gallic acid and pyrogallol, *etc.* As for their utilization in medicine, however, most medical scientists know only the name of what is marketed as "tannic acid" which consists of mixtures, in various proportions, of undefined gallotannins obtained from several kinds of tannin-rich galls such as Chinese galls, Turkish galls, *etc.* Some products made by complexation of tannin with alkaloids, for example, berberine and atropine-containing extracts, and proteins, for example, albumin, have been used in medicine, but the tannin used for this complexation was only the "tannic acid" described above.

It should be remembered that a large number of tannins of diverse chemical structures have been frequently consumed by humans and that these may have exerted various effects on human health. Among them are the tannins contained in many traditional medicinal plants. There are also many kinds of food, food additives and beverages which are rich in tannins and related polyphenols whose effects on human health have attracted strong interest in recent years (*205–209*).

One of the effects of tannins most widely concerned with health may be their ability to suppress "active oxygen", such as the inhibition of lipid peroxidation based on their radical-scavenging activity (*205, 209, 210*).

10.1. Antioxidant Activities

The antioxidant activity of tannins in foods and beverages was first demonstrated when their influence on suppressing the oxidation of ascorbic acid came to light (*210*). Later, the effect of tannins on Cu(II)-catalyzed autoxidation of ascorbic acid was studied by kinetic and ESR measurements. In these studies, geraniin showed a marked inhibitory effect attributable to its radical scavenging activity rather than to Cu(II)-blocking, while the inhibitory effect of polyphenols of lower molecular weight, such as gallic acid and pyrogallol, *etc.*, was low. The radical scavenging mechanism proposed for this inhibition was substantiated by ESR measurements of the inhibitory radicals generated from these polyphenols. The inhibitors with potent effects exhibited stable ESR signals, while polyphenols of lower molecular weight with low inhibitory effects gave only unstable or no ESR signals (*211*).

Tannins exhibited remarkable inhibitory effects on lipid peroxidation induced by adenine 5′-diphosphate (ADP) and ascorbic acid in rat liver mitochondria, and on that induced by ADP and NADPH in rat liver microsomes. Of 25 polyphenolic compounds all except some polyphenols of low molecular weight and methylated polyphenols showed significant inhibition in these two systems at a concentration of 1 µg/ml. However, marked differences in the inhibitory effects were observed, depending on their structures and experimental systems. Almost complete inhibition of the lipid peroxidation was exhibited in these two systems by some ellagitannins such as pedunculagin (31), isoterchebin (69), at a dose of 5 µg/ml. In both systems the inhibitory effects of most hydrolyzable tannins were stronger than those of condensed tannins. Some low molecular weight monomeric catechin analogs with a galloyl group represented by (−)-epigallocatechin gallate (27), the main polyphenol in green tea, showed inhibitory effects stronger than those produced by condensed tannins of larger molecular weight in these systems, particularly in the system with ADP and NADPH (212). Inhibitory effects in these systems were also exhibited by dicaffeoylquinic acids isolated from popular oriental medicinal plants of the genus Artemisia (213).

Oral administration of geraniin (63) and a geraniin-rich extract of Geranium thunbergii, one of the most popular medicinal plants in Japan, significantly lowered the lipid peroxide level in rat serum and rat liver, in which liver injury was produced by feeding peroxidized oil. The levels of serum cholesterol, glutamic oxaloacetic transaminase (GOT) and glutamic pyruvic transaminase (GPT), in the rats treated with peroxidized oil were also reduced (214). Oral administration of caffeoyl ester-rich extracts from several medicinal Artemisia species and caffeic acid derivatives also gave similar results (215).

The radical scavenging mechanism of tannins was confirmed by studies of the inhibitory effects of 25 tannins and related polyphenols on the autoxidation of methyl linoleate initiated by photo-irradiated AIBN. The effect was analyzed by kinetic studies and by in situ ESR measurements (216).

The radical scavenging ability of various polyphenols was also manifested by their ability to scavenge the 1,1-diphenyl-2-picrylhydrazyl (DPPH) radical, as measured by decolorization of the violet solution of the radical. The effects of licorice phenolics (217) and those of a number of polyphenols of various structures and origins were generally stronger than those of α-tocopherol and ascorbic acid (218). That a stable free radical was generated from these polyphenols upon scavenging the DPPH radical was demonstrated by an experiment employing alkyl gallates as the scavengers. In this experiment formation of the gallate radical was proved by ESR measurements and by isolation in high yield of dialkyl hexahydroxydiphenates produced by the mutual coupling of transient C-centered galloyl radicals (203).

Effects attributable to the radical scavenging activities of tannins were also observed when it was found that tannins exerted a significant inhibitory action on the carbon tetrachloride and galactosamine induced cytotoxicity in primary cultured rat hepatocyte (*219*).

In a study of the effects of tannins on arachidonic acid metabolism, geraniin (**63**) and corilagin (**65**) inhibited the formation of 5-HETE (lipoxygenase product) in rat peritoneal polymorphonuclear leukocytes, at concentrations of 10^{-6}–10^{-3} M. The effect was dose dependent. However, the formation of cyclooxygenase products (HHT, thromboxane B_2 and 6-keto-PGF$_1\alpha$) was not generally inhibited at these concentrations (*220*). Dicaffeoylquinic acids, rosmarinic acid (**190**) and other caffeic acid derivatives (*221*), and also licorice phenolics (*222*), displayed an analogous specific inhibition of the formation of the lipoxygenase products.

The inhibitory effects on the lipoxygenase dependent peroxidation of linoleic acid by caffeoylquinic acids and analogs were studied by kinetic and ESR measurements. The order of inhibitory potency in this study coincided with that obtained from the measurement of radical scavenging activities of these compounds against DPPH. The inhibition profiles of these compounds on lipid peroxidation in the lipoxygenase system were similar to those obtained in the biological systems of rat liver mitochondria and microsomes (*223*). In the ESR measurements in alkaline dimethyl sulfoxide solution, all caffeoylquinic acids exhibited relatively stable ESR signals assignable to the radical derived from the one electron oxidation of dihydroxyphenyl group. These results show that the radical scavenging mechanism is commonly operative in both chemical and biological peroxidation systems (*223*).

In an investigation of the protective effects exerted by tannins against oxidative damage induced in mouse ocular lenses by incubation with xanthine-xanthine oxidase, ADP and Fe^{3+} (X-XOD system), geraniin and pentagalloylglucose markedly decreased lipid peroxide concentration in the lens, while the effects by low molecular polyphenols were smaller (*224, 225*).

The effects of 25 tannins including low-molecular polyphenols on concentration of the superoxide anion radical generated in the hypoxanthine-xanthine oxidase system were estimated by ESR measurement of the adducts formed by 5,5-dimethyl-1-pyrroline-N-oxide (DMPO) and the radicals. Addition of each polyphenol to an ethanol solution of DPPH radical reduced the intensity of the DPPH radical signal with increase of the polyphenol concentration and finally elicited a signal assignable to a polyphenol radical, in a way similar to the appearance of the signal of *dl*-α-tocopherol radical in the same system. The radical scavenging mechanism of polyphenols on the superoxide anion radical was thus substantiated (*218*).

Inhibitory effects on the activity of xanthine oxidase were also demonstrated by some licorice polyphenols (226), by tannins of various types (227) and also by some gallates of flavonoid glycosides (33). Several licorice polyphenols were potent inhibitors of monoamine oxidase (MAO) (228), which is known to play a role in oxidative stress of the nervous system (229).

10.2. Antitumor Activities

A large amount of information about the inhibitory activities of tannins and related polyphenols on tumor incidence and propagation has been gathered in the last decade, although there were some proposals that the crude extract of some tannin-containing plants induced cancer (230) and that tumors were induced by extraordinarily high doses of low molecular weight polyphenols (231).

10.2.1. Inhibition of Mutagenicity of Carcinogens

Significant inhibition of mutagenicity of Trp-P-1 and MNNG by EGCG from green tea and several ellagitannins from medicinal plants, such as geraniin, mallotusinic acid, pedunculagin and agrimoniin, was reported (232). Remarkably strong inhibitions of N-OH-Trp-P-2, a direct-acting mutagen, by these polyphenols were also found (233). The strong inhibition by ellagic acid of the mutagenicity of 7β,8α-dihydroxy-9α,10α-epoxy-7,8,9,10-tetrahydrobenzo-[a] pyrene (B[a] p diol epoxide) was attributed to formation of a collision complex between ellagic acid and the carcinogen (234). Subsequent investigations of the anticarcinogenic activity of ellagic acid took various paths.

Because ellagic acid in plant extracts is produced mostly by hydrolysis of the ellagitannins originally present in the plant, the antimutagenic activity of extracts of Geranium species against some carcinogens such as B[a] p diol epoxide, a direct acting mutagen, and Trp-P-1, was studied (233). Whether the inhibitory activity correlates with the extent of hydrolysis of geraniin in the extract was also investigated. Marked upward or downward modification of the inhibitory activity depending on the extent of hydrolysis was observed in different ways according to the type of carcinogen (235). Effects of pH variation and of rat intestinal contents on the liberation of ellagic acid from ellagitannins were also reported (236). Tannic acid and some other phenolics inhibited the mutagenicity of bay-region diol epoxides of polycyclic aromatic hydrocarbons (*237*). Cytotoxic activities of tannins were also reported (*238*).

10.2.2. Inhibition of Tumor Promotion

Tumor promotion is a longer process than initiation in two stage chemical carcinogenesis and participation of active oxygen in the tumor promoting process has been proposed. Considerable efforts have been made in recent years to find agents inhibiting tumor promotion. (−)-Epigallocatechin gallate (**27**) (EGCG), which is the main polyphenolic component of "green tea tannin" and is a potent inhibitor of lipid peroxidation as mentioned above, is one of the most extensively investigated inhibitors of tumor promotion. Tumor promotion on mouse skin by teleocidin (*239*) and by okadaic acid (*51, 240*) after initiation with DMBA was significantly inhibited by EGCG. In an investigation of the inhibitory effects of EGCG on tumor promotion in the gastrointestinal tract, duodenal carcinogenesis produced by N-ethyl-N'-nitro-N-nitrosoguanidine in mice was significantly inhibited (to less than one third of the control) by oral administration of EGCG at the promotion stage (*241*). Inhibitory effects of EGCG have also been observed in several different systems of tumor promotion. Pentagalloylglucose (**6**) inhibited tumor promotion as potently as EGCG (**27**) (*51*).

Many other studies concerning the inhibition of tumor promotion by tea polyphenols or tea extracts have been carried out very recently (*242, 243*). The inhibitory effect of orally administered green tea extract against skin tumor (*243*) is one of them.

10.2.3. Host-Mediated Antitumor Activity of Oligomeric Hydrolyzable Tannins

Inhibition of tumor growth by administration of some oligomeric hydrolyzable tannins, either before or after intraperitoneal inoculation of tumor cells, is one of the most conspicuous antitumor effects of oligomeric hydrolyzable tannins (*244, 245*). It is noteworthy that this effect, first found in case of the dimer agrimoniin (**109**), was subsequently specifically found for some oligomeric hydrolyzable tannins. Notable effects were exhibited by most macrocyclic oligomers and by some non-cyclic oligomers among over 100 tannins and related polyphenols to be screened (*245, 246*). Examples of these active macrocyclic oligomers are the dimers oenothein B (**138**) and woodfordin C (**139**), the trimers oenothein A (**159**) and woodfordin D (**160**), and the tetramer woodfordin F. This effect is attributable to enhancement of the immune response of host animals as shown by stimulation of IL-1 production from human peripheral macrophages by these oligomers (*247*).

10.3. Antiviral Activities

Significant inhibitory activities of tannins have been observed against several viral species, while the antibacterial activities of tannins are generally not comparable with those of antibiotics.

Potent anti-HIV activities were found for the dimeric hydrolyzable tannins oenothein B (**138**), coriariin A (**116**) and agrimoniin (**109**) (*248*), and also of licorice phenolics (*249*). In screening 87 tannins and related compounds, significant anti-HIV activities were observed for some hydrolyzable tannins, gemin D (**51**) (monomer), camelliin B (**140**) and nobotanin B (**136**) (dimer), and also trapanin B (**133**) (tetramer) (*250*). Monomeric hydrolyzable tannins strongly inhibited herpes simplex virus infection (*50*) and exhibited related inhibitory activities (*251*). Inhibitory effects of tannins and related polyphenols on reverse transcriptase from RNA tumor virus (*252*), and on reverse transcriptase and replication of HIV (*248, 253, 254*), were reported. A mechanistic study using CD spectrometry showed strong interactions between polyphenols and DNA-RNA duplexes (*255*).

10.4. Mutually Reverse Effects on Enzyme

Sometimes the activities of tannins are inverted when the concentrations or the combination of tannin and substrate is varied. Tannins usually inhibit the activities of enzymes at relatively high concentrations. However, at low concentrations they often stimulate enzymatic activity. The correlation between inhibition and stimulation varies depending on the enzyme and the structure and concentration of tannin. Galloylglucoses smaller than hexagalloylglucose inhibited glucosyltransferase from *Streptococcus mutans*, and pentagalloylglucose and hexagalloylglucose inhibited this enzyme more potently than chlorhexidine. However, heptagalloylglucose and octagalloylglucose stimulated it at a concentration of $10^{-5} M$ although inhibition was observed at higher concentrations (*256*).

Several polyphenols of the tannin class inhibited adrenaline and ACTH-induced lipolysis and insulin-induced lipogenesis from glucose in fat cells of the rat (257, 258). However, ACTH-induced lipolysis was strongly promoted by polyphenolic compounds such as chebulinic acid (85) and tellimagrandin I (34) at concentrations between 5–100 μg/ml, although adrenaline-induced lipolysis was inhibited by these polyphenols regardless of the concentration (259).

Several caffeic acid derivatives such as 3,5-di-O-caffeoylquinic acid (186), rosmarinic acid (190) and caffeoylmalic acid increased the amount of the products from cyclooxygenase, such as prostaglandin E2, in arachidonic acid metabolism in human polymorphonuclear leukocytes (260), although the product formation caused by lipoxygenase were invariably inhibited as mentioned earlier. Enhancement of the production of several specific compounds in arachidonic acid metabolism was also observed on administration of licorice polyphenols (222), geraniin (63) and corilagin (65) (220).

Such relationships between polyphenols and enzyme activity should be further investigated in detail, along with the specific occurrence of each polyphenol in certain plant species, as carried out, for instance, in the case of the oligomeric hydrolyzable tannins in the Rosaceae (5).

10.5. Other Activities

Tannins reduce metallic ions, Cu^{2+}, Fe^{3+}, and Cr^{6+} to Cu^+, Fe^{2+}, and Cr^{3+}, at room temperature. Presumably this is accompanied by oxidation of the tannins to quinoids (261).

Mixing of tannins with alkaloids and heavy metal ions generally produces a precipitate regardless of tannin structure, when the concentration of tannin is not high ($< 1-2 \times 10^{-3}$ M). However, higher concentration of some tannins such as tannic acid (from Chinese gall) and punicalin caused resolution of the precipitates, presumably due to formation of water soluble complexes (262).

Histamine release from rat peritoneal mast cells induced by compound 48/80 and that induced by concanavalin A and phosphatidylserine was potentially inhibited by dicaffeoylquinic acids, and was moderately inhibited by monocaffeoylquinic acids and caffeic acid (263).

10.6. Absorption and Metabolism of Tannins in Animals

Studies on the absorption and metabolism of gallotannins in animals were published in 1950's (264). The major metabolite found in urine of rats receiving tannic acid (a mixture of gallotannins) by stomach tube was

4-O-methylgallic acid. This compound was also found to be the major metabolite in urine of rats or rabbits ingesting gallic acid or alkyl gallates and in a medium in which gallic acid was incubated with rat and rabbit liver slices. Pyrogallol and 2-O-methylpyrogallol were also found in urine of rats receiving gallic acid by intraperitoneal injection or of rabbits receiving gallic acid in the diet. Certain differences in the metabolism of tannic acid in rats or rabbits from metabolism in humans were pointed out (*264*).

On the other hand, the absorption and metabolism of ellagitannins is not yet clear. Although no metabolic study using isolated ellagitannins has so far been reported, an experiment involving oral administration of an ellagitannin (geraniin) which affected lipid metabolism in the rats (*214*) suggested absorption of the ellagitannin or its metabolites formed in the gastrointestinal tract. The metabolic conversion of ellagic acid, which is produced on the hydrolysis of ellagitannins, to 3,8-dihydroxybenzo-[*b,d*]pyran-6-one (**201**) and related compounds in animals has also been reported (*265*).

Compound **201** and 3-hydroxy-6H-dibenzo [*b,d*] pyran-6-one (**204**) were detected in the urine and serum of sheep given *Terminalia oblongata* leaves which contain ellagitannins such as chebulagic acid, punicalagin and teroblongin (= 1-α-O-galloylpunicalagin) (*266*). In one experiment, the leaves were administered into the sheep stomach; the freeze dried urine and serum were subsequently purified by column chromatography to afford the 3-O-glucuronide (**205**) of **204** and a mixture of 3-O-glucuronide (**202**) and 8-O-glucuronide (**203**) of **201**. Compounds **201** and **204** were also detected by HPLC analysis after acid hydrolysis of the freeze-dried serum (*266*).

(**201**) 3,8-Dihydroxy-6*H*-dibenzo[*b,d*]pyrane-
6-one: R¹ = R² = H
(**202**) 3,8-Dihydroxy-6*H*-dibenzo[*b,d*]pyrane-
6-one 3-O-glucuronide: R¹ = GA, R² = H
(**203**) 3,8-Dihydroxy-6*H*-dibenzo[*b,d*]pyrane-
6-one 8-O-glucuronide: R¹ = H, R² = GA

(**204**) 3-Hydroxy-6*H*-dibenzo[*b,d*]pyrane-
6-one: R = H
(**205**) 3-Hydroxy-6*H*-dibenzo[*b,d*]pyrane-
6-one 3-O-glucuronide: R = GA

GA =

Addendum

The unit previously named *m, m′*-**D(OG)**₂ (6) has now been renamed as the *m,m″*-**D(OG)**₂ unit, because a dimer (eurobustin A) having a *m′*-**DOG** (macaranoyl) unit was recently isolated from *Eucalyptus robusta* (Myrtaceae); KONDO, S., T. HATANO, T. YOSHIDA, T. OKUDA, C.-F. LU, L.-L. YANG, and K.-Y. YEN: Tannins Isolated from *Eucalyptus robusta*.

Eurobustin A

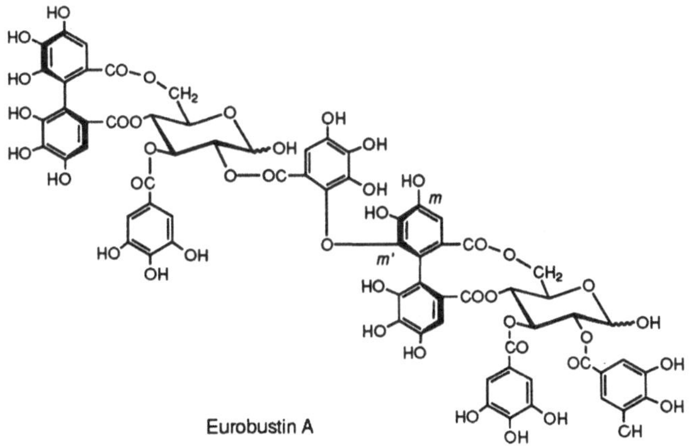

Chart 5

Abstract Papers of the 32nd Annual Meeting of the Chugoku-Shikoku Branch of the Pharmaceutical Society of Japan, p. 44 (1993).

Acknowledgements

We are grateful to Professor WERNER HERZ, of the Florida State University for contributions far beyond his editorial duties to this review.

List of Compounds

1 β-Glucogallin
2 1,6-Di-*O*-galloyl-β-D-glucose
3 1,2,6-Tri-*O*-galloyl-β-D-glucose
4 1,2,3,6-Tetra-*O*-galloyl-β-D-glucose
5 1,2,4,6-Tetra-*O*-galloyl-β-D-glucose
6 1,2,3,4,6-Penta-*O*-galloyl-β-D-glucose
7 Trigallic acid
8 Acertannin

9 Trigalloyl-1,5-anhydro-D-glucitols
10 Tetragalloyl-1,5-anhydro-D-glucitols
11 5-*O*-Polygalloylquinic acids
12 2-*O*-Galloylarbutin
13 6-*O*-Galloylarbutin
14 4,6-Di-*O*-galloylarbutin
15 Salidroside-6″-gallate
16 Cornuside

17 Bergenin
18 Norbergenin
19 4-*O*-Galloylbergenin
20 4-*O*-Galloylnorbergenin
21 11-*O*-Galloylbergenin
22 11-*O*-Galloylnorbergenin
23 3,4-Di-*O*-galloylbergenin
24 4,11-Di-*O*-galloylbergenin
25 3,4,11-Tri-*O*-galloylbergenin
26 3-*O*-Galloylepicatechin-(4β → 8)-3-*O*-
    galloyl-epicatechin-(4β → 8)-3-*O*-
    galloylepicatechin-(4β → 8)-3-*O*-
    galloylepicatechin
27 (−)-Epigallocatechin gallate
28 (−)-Epicatechin gallate
29 3,3′-Di-*O*-galloyl-(−)-epicatechin
30 3,4′-Di-*O*-galloyl-(−)-epicatechin
31 Pedunculagin
32 Casuarictin
33 Potentillin
34 Tellimagrandin I
35 Tellimagrandin II
36 Rugosin A
37 Rugosin B
38 Tergallagin
39 Cornusiin B
40 Oenothein C
41 Eucalbanin A
42 Tirucallin A
43 Alnusiin
44 Bicornin
45 Praecoxin C
46 Praecoxin D
47 Euprostin C
48 Rugosin C
49 Strictinin
50 Isostrictinin
51 Gemin D
52 Praecoxin A
53 Praecoxin B
54 Pterocaryanin C
55 Agrimonic acid A
56 Agrimonic acid B
57 Coriariin B
58 Coriariin J
59 Remurin A
60 Remurin B
61 Nupharin A
62 Nupharin B
63 Geraniin

64 Geraniin phenazine
65 Corilagin
66 Furosinin
67 Didehydrogeraniin
68 Terchebin
69 Isoterchebin
70 Furosin
71 Mallotusinic acid
72 Granatin A
73 Helioscopinin A
74 Carpinusin
75 Supinanin
76 Euphorscopin
77 Tanarinin
78 Macarinin A
79 Macaranin C
80 Dimethyl hexamethoxydiphenate
81 Schizandrin
82 Methylated phenylphenazine
83 4,6-Dehydrohexahydroxydiphenoyl-
    D-glucose
84 Chebulinic acid
85 Chebulagic acid
86 Macarinin B
87 Macarinin C
88 Phyllanthusiin A
89 Phyllanthusiin B
90 Phyllanthusiin C
91 Repandusinic acid
92 Geraniinic acid B
93 Geraniinic acid C
94 Heterophylliin E
95 Repandusinin
96 Ascorgeraniin (Elaeocarpusin)
97 Putranjivain A
98 Phyllanthusiin D
99 Castalagin
100 Vescalagin
101 Casuarinin
102 Stachyurin
103 Casuariin
104 Camelliatannin A
105 Camelliatannin B
106 Malabathrin A
107 Camelliatannin F
108 Malabathrin E
109 Agrimoniin
110 Laevigatin B
111 Laevigatin C
112 Laevigatin D

113 Laevigatin E
114 Davuriciin D2
115 Gemin A
116 Coriariin A
117 Nupharin C
118 Nupharin D
119 Nupharin E
120 Nupharin F
121 Hirtellin A
122 Hirtellin C
123 Hirtellin F
124 Hirtellin B
125 Tamarixinin A
126 Rugosin D
127 Rugosin E
128 Rugosin F
129 Rugosin G
130 Cornusiin A
131 Cornusiin C
132 Trapanin A
133 Trapanin B
134 Nobotanin A
135 Nobotanin F
136 Nobotanin B
137 Eucalbanin C
138 Oenothein B
139 Woodfordin C
140 Camelliin B
141 Camelliatannin D
142 Reginin A
143 Reginin B
144 Reginin C
145 Reginin D
146 Heterophylliin B
147 Heterophylliin C
148 Nobotanin K
149 Euphorbin A
150 Euphorbin F
151 Euphorbin B
152 Antidesmin A
153 Sanguiin H-6
154 Roshenin A
155 Roshenin B
156 Lambertianin C
157 Lambertianin D
158 Sanguiin H-11
159 Oenothein A
160 Woodfordin D
161 Euphorbin C
162 Euphorbin D
163 Euphorbin E

164 Alienanin A
165 Alienanin B
166 Anogeissinin
167 Anogeissusin A
168 Anogeissusin B
169 Isorugosin D
170 Isorugosin E
171 Isorugosin G
172 Camelliin A
173 Isohirtellin C
174 Tamarixinin B
175 Tamarixinin C
176 Eucalbanin B
177 Liquidambin
178 Camelliatannin C
179 Camelliatannin E
180 2,3-Di-O-galloyl-D-glucose
181 Hydrolyzate of oenothein B
182 Hydrolyzate of oenothein B
183 1,6-Di-O-galloyl-4-O-dilactonized-
valoneoyl-2,3-O-hexahydroxy-
diphenoyl-β-D-glucose
184 1,3-Di-O-galloyl-4,6-hexahydroxy-
diphenoyl-β-D-glucose
185 Chlorogenic acid
186 3,5-Di-O-caffeoylquinic acid
187 3,4,5-Tri-O-caffeoylquinic acid
188 1,4,5-Tri-O-caffeoylquinic acid
189 1,3,5-Tri-O-caffeoylquinic acid
190 Rosmarinic acid
191 Lithospermic acid
192 Lithospermic acid B
193 Melitric acid A
194 Melitric acid B
195 Rabdosiin
196 Isomer of rabdosiin
197 Procyanidin B2
198 Procyanidin B2 3'-O-gallate
199 Procyanidin C1
200 Cinnamtannin A2
201 3,8-Dihydroxy-6H-benzo[b,d]pyran-
6-one
202 3,8-Dihydroxy-6H-benzo[b,d]pyran-
6-one 3-O-glucuronide
203 3,8-Dihydroxy-6H-benzo[b,d]pyran-
6-one 8-O-glucuronide
204 3-Hydroxy-6H-dibenzo[b,d]pyran-6-
one
205 3-Hydroxy-6H-dibenzo[b,d]pyran-6-
one 3-O-glucuronide

References

1. Okuda, T., T. Yoshida, and T. Hatano: New Methods of Analyzing Tannins. J. Nat. Prod. **52**, 1 (1989).
2. Haslam, E.: The Metabolism of Gallic Acid and Hexahydroxydiphenic Acid in Higher Plants. Fortschritte Chem. Org. Naturst. **41**, 1 (1982).
3. Haslam, E.: Plant Polyphenols, Vegetable Tannins Revisited. Cambridge: Cambridge University Press. 1989.
4. Okuda, T., T. Yoshida, and T. Hatano: Oligomeric Hydrolyzable Tannins, a New Class of Plant Polyphenols. Heterocycles **30**, 1995 (1990).
5. Okuda, T., T. Yoshida, T. Hatano, M. Kubo, T. Orime, M. Yoshizaki, and N. Naruhashi: Hydrolysable Tannins as Chemotaxonomic Markers in the Rosaceae. Phytochem. **31**, 3091 (1992).
6. Okuda, T., T. Yoshida, and T. Hatano: Classification of Oligomeric Hydrolyzable Tannins and Specificity of Their Occurrence in Plants. Phytochem. **32**, 507 (1993).
7. Okuda, T., T. Yoshida, and T. Hatano: Chemistry and Biological Activities of Tannins in Medicinal Plants. In: Economic and Medicinal Plant Research, Vol. V (H. Wagner and N.R. Farnsworth, eds.), p. 129. London, New York: Academic Press. 1991.
8. Okuda, T., T. Yoshida, and T. Hatano: Ellagitannins as Active Constituents of Medicinal Plants. Planta Med. **55**, 117 (1989).
9. Okuda, T., T. Yoshida, and T. Hatano: Pharmacologically Active Tannins Isolated from Medicinal Plants. In: Plant Polyphenols, Synthesis, Properties, Significance (R.W. Hemingway and P.E. Laks, eds.), p. 539. New York: Plenum Press. 1992.
10. Gross, G.G.: Enzymatic Synthesis of Gallotannins and Related Compounds. In: Phenolic Metabolism in Plants (H.A. Stafford and R.K. Ibrahim, eds.), p. 297. New York: Plenum Press. 1992.
11. Schmidt, O.Th., and G. Klinger: Über natürliche Gerbstoffe, XXVIII: Synthese der 1,3,6-Trigalloylglucose. Liebigs Ann. Chem. **609**, 199 (1957).
12. Haddock, E.A., R.K. Gupta, S.M.K. Al-Shafi, K. Layden, E. Haslam, and D. Magnolato: The Metabolism of Gallic Acid and Hexahydroxydiphenic Acid in Plants: Biogenetic and Molecular Taxonomic Considerations. Phytochem. **21**, 1049 (1982).
13. Armitage, R., G.S. Bayliss, J.W. Gramshaw, E. Haslam, H.J.R.D. Haworth, K. Jones, H.J. Rogers, and T. Searle: Gallotannins, Part III: The Constitution of Chinese, Turkish, Sumach, and Tara Tannins. J. Chem. Soc. **1961**, 1842.
14. Delahaye, P., A. de Bruyn, F. van Damme, and M. Verzele: Trigallic Acid or Trimeric Gallic Acid. Bull. Soc. Chim. Belg. **92**, 469 (1983).
15. Mayer, W., N. Kunz, and F. Loebich: Die Struktur Hamamelitannins. Liebigs Ann. Chem. **688**, 232 (1965).
16. Ozawa, T., S. Kobayashi, R. Seki, and H. Imagawa: A New Gallotannin from Bark of Chestnut Tree, *Castanea crenata* Sieb. et Zucc. Agric. Biol. Chem. **48**, 1411 (1984).
17. Nonaka, G., K. Ishimaru, T. Tanaka, and I. Nishioka: Tannins and Related Compounds, XVII: Galloylhamameloses from *Castanea crenata* and *Sanguisorba officinalis* L. Chem. Pharm. Bull. (Japan) **32**, 483 (1984).
18. Ishimaru, K., G. Nonaka, and I. Nishioka: Gallic Acid Esters of *Proto*-quercitol, Quinic Acid and (−)-Shikimic Acid from *Quercus mongolica* and *Q. myrsinaefolia*. Phytochem. **26**, 1501 (1987).
19. Nishimura, H., G. Nonaka, and I. Nishioka: *Scyllo*-quercitol Gallates and Hexahydroxydiphenoates from *Quercus stenophylla*. Phytochem. **25**, 2599 (1986).
20. Nonaka, G., and I. Nishioka: Tannins and Related Compounds X: Rhubarb (2):

Isolation and Structures of a Glycerol Gallate, Gallic Acid Glucosides, Galloylglucoses and Isolindleyin. Chem. Pharm. Bull. (Japan) **31**, 1652 (1983).

21. PERKIN, A.G., and Y. UYEDA: Occurrence of a Crystalline Tannin in the Leaves of the *Acer ginnala*. J. Chem. Soc. **1922**, 66.

22. BOCK, K., F.N. LACOUR, R.S. JENSEN, and B.J. NIELSEN: The Structure of Acertannin. Phytochem. **19**, 2033 (1980).

23. HADDOCK, E.A., R.K. GUPTA, S.M.K. AL-SHAFI, E. HASLAM, and D. MAGNOLATO: The Metabolism of Gallic Acid and Hexahydroxydiphenic Acid in Plants, Part 1: Introduction, Naturally Occurring Galloyl Esters. J. Chem. Soc. Perkin Trans. 1 **1982**, 2515.

24. HATANO, T., S. HATTORI, Y. IKEDA, T. SHINGU, and T. OKUDA: Gallotannins Having 1,5-Anhydro-D-glucitol Core and Some Ellagitannins from *Acer* Species. Chem. Pharm. Bull. (Japan) **38**, 1902 (1990).

25. OKUDA, T., K. MORI, and T. HATANO: Relationship of the Structure of Tannins to the Binding Activities with Hemoglobin and Methylene Blue. Chem. Pharm. Bull. (Japan) **33**, 1424 (1985).

26. HASLAM, E., R.D. HAWORTH, and D.A. LAWTON: Gallotannins, Part VIII: The Preparation and Properties of Some Galloyl Esters of Quinic Acid. J. Chem. Soc. **1963**, 2173.

27. HATANO, T., Y. IKEGAMI, and T. OKUDA: Studies on the Tannins of Anacardiaceous Plants, Part 2: Tannins of *Pistacia chinensis*. Abstract Papers of the 33rd Annual Meeting of the Japanese Society of Pharmacognosy, p. 36 (1986).

28. NONAKA, G., M. AGETA, and I. NISHIOKA: Tannins and Related Compounds, XXV: A New Class of Gallotannins Possessing a (−)-Shikimic Acid Core from *Castanopsis cuspidata* var. *sieboldii* Nakai. (1). Chem. Pharm. Bull. (Japan) **33**, 96 (1985).

29. KASHIWADA, Y., G. NONAKA, and I. NISHIOKA: Galloylsucroses from Rhubarb. Phytochem. **27**, 1469 (1988).

30. STRACK, D., V. WRAY, J.W. METZGER, and W. GROSSE: Two Anthocyanins Acylated with Gallic Acid from the Leaves of *Victoria amazonica*. Phytochem. **31**, 989 (1992).

31. JI, S.-H., N. SAITO, M. YOKOI, A. SHIGIHARA, and T. HONDA: Galloylcyanidin Glycosides from *Acer*. Phytochem. **31**, 655 (1992).

32. KAMEDA, K., T. TAKAKU, H. OKUDA, Y. KIMURA, T. OKUDA, T. HATANO, I. AGATA, and S. ARICHI: Inhibitory Effects of Various Flavonoids Isolated from Leaves of Persimmon on Agniotensin-Converting Enzyme Activity. J. Nat. Prod. **50**, 680 (1987).

33. HATANO, T., T. YASUHARA, R. YOSHIHARA, Y. IKEGAMI, M. MATSUDA, K. YAZAKI, I. AGATA, S. NISHIBE, T. NORO, M. YOSHIZAKI, and T. OKUDA: Inhibitory Effects of Galloylated Flavonoids on Xanthine Oxidase. Planta Med. **57**, 83 (1991).

34. CHEN, X.-M., T. YOSHIDA, T. HATANO, M. FUKUSHIMA, and T. OKUDA: Galloylarbutin and Other Polyphenols from *Bergenia purpurascens*. Phytochem. **26**, 515 (1987).

35. YAZAKI, K., S. SHIDA, and T. OKUDA: Galloylhomoarbutin and Related Polyphenols from *Pyrola incarnata*. Phytochem. **28**, 607 (1989).

36. NONAKA, G., I. NISHIOKA, T. NAGASAWA, and H. OURA: Tannins and Related Compounds, I: Rhubarb (1). Chem. Pharm. Bull. (Japan) **29**, 2862 (1981).

37. ISHIMARU, K., G. NONAKA, and I. NISHIOKA: Phenolic Glucoside Gallates from *Quercus mongolica* and *Q. acutissima*. Phytochem. **26**, 1147 (1987).

38. AGETA, M., K. ISHIMARU, G. NONAKA, and I. NISHIOKA: Tannins and Related Compounds, LXIV: Six New Phenol Glucoside Gallates from *Castanopsis cuspidata* var. *sieboldii* Nakai (2). Chem. Pharm. Bull. (Japan) **36**, 870 (1988).

39. BRITTON, G., and E. HASLAM: Gallotannins, Part XII: Phenolic Constituents of *Arctostaphyllos uva-ursi* L. Spreng. J. Chem. Soc. **1965**, 7312.

40. NONAKA, G., H. NISHIMURA, and I. NISHIOKA: Tannins and Related Compounds, IV:

Seven New Phenol Glucoside Gallates from *Quercus stenophylla* Makino (1). Chem. Pharm. Bull. (Japan) **30**, 2061 (1982).

41. YOSHIKAWA, M., E. UCHIDA, A. KAWAGUCHI, I. KITAGAWA, and J. YAMAHARA: Galloyl-oxypaeoniflorin, Suffruticosides A, B, C and D, Five Antioxidative Glycosides, and Suffruticoside E, A Paeonol Glycoside, from Chinese Moutan Cortex. Chem. Pharm. Bull. (Japan) **40**, 2248 (1992).

42. HATANO, T., T. YASUHARA, R. ABE, and T. OKUDA: Galloylated Monoterpene Glucoside and a Dimeric Hydrolyzable Tannin from *Cornus officinalis*. Phytochem. **29**, 2975 (1990).

43. YOSHIDA, T., K. SENO, Y. TAKAMA, and T. OKUDA: Bergenin Derivatives from *Mallotus japonicus*. Phytochem. **21**, 1180 (1982).

44. SAIJO, R., G. NONAKA, and I. NISHIOKA: Gallic Acid Esters of Bergenin and Norbergenin from *Mallotus japonicus*. Phytochem. **29**, 267 (1990).

45. TANAKA, T., T. SUEYASU, G. NONAKA, and I. NISHIOKA: Tannins and Related Compounds, XXI: Isolation and Characterization of Galloyl and *p*-Hydroxybenzoyl Esters of Benzophenone and Xanthone *C*-Glucosides from *Mangifera indica* L. Chem. Pharm. Bull. (Japan) **32**, 2676 (1984).

46. MATSUO, T., and S. ITO: The Chemical Structure of Kaki-tannin from Immature Fruit of the Persimmon (*Diospyros kaki* L.). Agric. Biol. Chem. **42**, 1637 (1978).

47. OKUDA, T., T. YOSHIDA, T. HATANO, M. KUWAHARA, and S. IIDA: Inhibitory Effects of Crude Drugs on Proteases: Tannins and Related Polyphenols. Proc. Symp. Wakan-Yaku **15**, 111 (1982).

48. HATANO, T., K. URITA, and T. OKUDA: Tannins from *Saxifraga stronifera*. J. Med. Pharm. Soc. Wakan-Yaku **3**, 434 (1986).

49. SAKAGAMI, H., K. ASANO, S. TANUMA, T. HATANO, T. YOSHIDA, and T. OKUDA: Stimulation of Monocyte Iodination and IL-1 Production by Tannins and Related Compounds. Anticancer Res. **12**, 377 (1992).

50. FUKUCHI, K., H. SAKAGAMI, T. OKUDA, T. HATANO, S. TANUMA, K. KITAJIMA, Y. INOUE, S. INOUE, S. ICHIKAWA, M. NONOYAMA, and K. KONNO: Inhibition of Herpes Simplex Virus Infection by Tannins and Related Compounds. Antiviral Res. **11**, 285 (1989).

51. YOSHIZAWA, S., T. HORIUCHI, M. SUGANUMA, S. NISHIWAKI, J. YATSUNAMI, S. OKABE, T. OKUDA, Y. MUTO, K. FRENKEL, W. TROL, and H. FUJIKI: Penta-*O*-galloyl-β-D-glucose and (−)-Epigallocatechin Gallate. Cancer Preventive Agents. In: ACS Symposium Series 507. Phenolic Compounds in Food and Their Effects on Health II: Antioxidants and Cancer Prevention (M.-T. HUANG, C.-T. HO, C.Y. LEE, eds.), p. 316. Washington, DC: American Chemical Society. 1992.

52. OKUDA, T., T. YOSHIDA, M. ASHIDA, and K. YAZAKI: Structures of Pedunculagin, Casuarictin, Strictinin, Casuarinin, Casuariin and Stachyurin. J. Chem. Soc. Perkin Trans. 1 **1983**, 1765.

53. WILKINS, C.K., and B.A. BOHM: Ellagitannins from *Tellima grandiflora*. Phytochem. **15**, 211 (1976).

54. HATANO, T., N. OGAWA, T. YASUHARA, and T. OKUDA: Tannins of Rosaceous Plants, VIII: Hydrolyzable Tannin Monomers Having a Valoneoyl Group from Flower Petals of *Rosa rugosa* Thunb. Chem. Pharm. Bull. (Japan) **38**, 3308 (1990).

55. TANAKA, T., G. NONAKA, and I. NISHIOKA: Tannins and Related Compounds, XLII: Isolation and Characterization of Four New Hydrolyzable Tannins, Terflavins A and B, Tergallagin and Tercatain from the Leaves of *Terminalia catappa* L. Chem. Pharm. Bull. (Japan) **34**, 1039 (1986).

56. OKUDA, T., T. HATANO, N. OGAWA, R. KIRA, and M. MATSUDA: Cornusiin A, A

Dimeric Ellagitannin Forming Tautomers, and Accompanying New Tannins in *Cornus officinalis.* Chem. Pharm. Bull. (Japan) **32**, 4662 (1984).

57. YOSHIDA, T., K. YOKOYAMA, O. NAMBA, and T. OKUDA: Tannins and Related Polyphenols of Euphorbiaceous Plants, VIII: Tirucallins A, B and Euphorbin F, Monomeric and Dimeric Ellagitannins from *Euphorbia tirucalli* L. Chem. Pharm. Bull. (Japan) **39**, 1137 (1991).

58. YOSHIDA, T., T. MARUYAMA, A. NITTA, and T. OKUDA: Eucalbanins A, B and C, Monomeric and Dimeric Hydrolyzable Tannins from *Eucalyptus alba* Reinw. Chem. Pharm. Bull. (Japan) **40**, 1750 (1992).

59. YOSHIDA, T., T. HATANO, T. KUWAJIMA, and T. OKUDA: Oligomeric Hydrolyzable Tannins and Their ¹H-NMR Spectra and Partial Degradation. Heterocycles **33**, 463 (1992).

60. HATANO, T., K. YAZAKI, A. OKONOGI, and T. OKUDA: Tannins of *Stachyurus* Species, II: Praecoxins A, B, C and D, Four New Hydrolyzable Tannins from *Stachyurus praecox* Leaves. Chem. Pharm. Bull. (Japan) **39**, 1689 (1991).

61. YOSHIDA, T., O. NAMBA, L. CHEN, Y. LIU, and T. OKUDA: Ellagitannin Monomers and Oligomers from *Euphorbia prostrata* Ait., and Oligomers from *Loropetalum chinense* Oliv. Chem. Pharm. Bull. (Japan) **38**, 3296 (1990). Prostratins A, B and C in this paper were renamed as euprostins A, B and C.

62. OKUDA, T., T. YOSHIDA, M. KUWAHARA, M.U. MEMON, and T. SHINGU: Structures of Potentillin, Agrimonic Acids A and B, and Agrimoniin, a Dimeric Ellagitannin. Chem. Pharm. Bull. (Japan) **32**, 2165 (1984).

63. HATANO, T., S. HATTORI, and T. OKUDA: Tannins of *Coriaria japonica* A. Gray, I: Coriariins A and B, New Dimeric and Monomeric Hydrolyzable Tannins. Chem. Pharm. Bull. (Japan) **34**, 4092 (1986).

64. HATANO, T., R. YOSHIHARA, S. HATTORI, M. YOSHIZAKI, T. SHINGU, and T. OKUDA: Tannins of *Coriaria japonica* A. Gray, III: Structures of Coriariins G, H, I and J. Chem. Pharm. Bull. (Japan) **40**, 1703 (1992).

65. YOSHIDA, T., A.F. AHMED, M.U. MEMON, and T. OKUDA: New Monomeric and Dimeric Hydrolyzable Tannins from *Reaumuria hirtella* and *Tamarix pakistanica.* Chem. Pharm. Bull. (Japan) **39**, 2849 (1991).

66. SAIJO, R., G. NONAKA, and I. NISHIOKA: Tannins and Related Compounds, LXXXIV: Isolation and Characterization of Five New Hydrolyzable Tannins from the Bark of *Mallotus japonicus.* Chem. Pharm. Bull. (Japan) **37**, 2063 (1989).

67. ISHIMATSU, M., T. TANAKA, G. NONAKA, I. NISHIOKA, M. NISHIZAWA, and T. YAMAGISHI: Tannins and Related Compounds, LXXV: Isolation and Characterization of Novel Diastereomeric Ellagitannins, Nupharins A and B, and Their Homologues from *Nuphar Japonicum* DC. Chem. Pharm. Bull. (Japan) **37**, 129 (1989).

68. OKUDA, T., T. YOSHIDA, T. HATANO, K. YAZAKI, and M. ASHIDA: Ellagitannins of Casuarinaceae, Stachyuraceae and Myrtaceae. Phytochem. **21**, 2871 (1982).

69. YOSHIDA, T., Y. MARUYAMA, M.U. MEMON, T. SHINGU, and T. OKUDA: Gemins D, E and F, Ellagitannins from *Geum japonicum.* Phytochem. **24**, 1041 (1985).

70. OKUDA, T., T. YOSHIDA, and T. HATANO: Hydrated Stereostructure and Equilibration of Geraniin. J. Chem. Soc. Perkin Trans. 1 **1982**, 9.

71. OKUDA, T., K. MORI, and T. HATANO: The Distribution of Geraniin and Mallotusinic Acid in the Order Geraniales. Phytochem. **19**, 547 (1980).

72. YOSHIDA, T., K. HABA, F. NAKATA, Y. OKANO, T. SHINGU, and T. OKUDA: Tannins and Related Polyphenols of Melastomataceous Plants, III: Nobotanins G, H and I, Dimeric Hydrolyzable Tannins from *Heterocentron roseum.* Chem. Pharm. Bull. (Japan) **40**, 66 (1992).

73. YAZAKI, K., T. HATANO, and T. OKUDA: Structures of Didehydrogeraniin, Furosinin and Furosin. J. Chem. Soc. Perkin Trans. 1 **1989**, 2289.
74. SCHMIDT, O. TH., J. SCHULZ, and R. WURMB: Terchebin. Liebigs Ann. Chem. **706**, 169 (1967).
75. OKUDA, T., T. HATANO, H. NITTA, and R. FUJII: Hydrolyzable Tannins Having Enantiomeric Dehydrohexahydroxydiphenoyl Group: Revised Structure of Terchebin and Structure of Granatin B. Tetrahedron Lett. **21**, 4361 (1980).
76. FÜRSTENWERTH, H., and H. SCHILDKNECHT: Isoterchebin, der gelbe Farbstoff des Zistrosenwürgers *Cytinus hypocistis* (Rafflesiaceae, Schmarotzerblumengewächse). Liebigs Ann. Chem. **1976**, 112.
77. OKUDA, T., T. HATANO, and T. YASUI: Revised Structure of Isoterchebin, Isolated from *Cornus officinalis*. Heterocycles **16**, 1681 (1981).
78. OKUDA, T., T. YOSHIDA, K. MORI, and T. HATANO: Tannins of Medicinal Plants and Drugs. Heterocycles **15**, 1323 (1981).
79. OKUDA, T., and K. SENO: Mallotusinic Acid and Mallotinic Acid, New Hydrolyzable Tannins from *Mallotus japonicus*. Tetrahedron Lett. **1978**, 139.
80. LEE, S.-H., T. TANAKA, G. NONAKA, and I. NISHIOKA: Tannins and Related Compounds, XCV: Isolation and Characterization of Helioscopinins and Helioscopins. Four New Hydrolyzable Tannins from *Euphorbia helioscopia* L. Chem. Pharm. Bull. (Japan) **38**, 1518 (1990).
81. AKAZAWA, M., Y. KASHIWADA, G. NONAKA, and I. NISHIOKA: Tannins and Related Compounds of *Carpinus laxiflora*. Abstract Papers (III) of the 109th Annual Meeting of the Pharmaceutical Society of Japan, p. 158 (1989).
82. LEE, S.-H., T. TANAKA, G. NONAKA, and I. NISHIOKA: Tannins and Related Compounds, CV: Monomeric and Dimeric Hydrolyzable Tannins Having a Dehydrohexahydroxydiphenoyl Group, Supinanin, Euphorscopin, Euphorhelin and Jolkianin, from *Euphorbia* Species. Chem. Pharm. Bull. (Japan) **39**, 630 (1991).
83. LIN, J.-H., G. NONAKA, and I. NISHIOKA: Tannins and Related Compounds, XCIV: Isolation and Characterization of Seven New Hydrolyzable Tannins from the Leaves of *Macaranga tanarius*. Chem. Pharm. Bull. (Japan) **38**, 1218 (1990).
84. LIN, J.-H., M. ISHIMATSU, T. TANAKA, G. NONAKA, and I. NISHIOKA: Structures of Macaranins and Macarinins, New Hydrolyzable Tannins Possessing Macaranoyl and Tergalloyl Ester Groups, from the Leaves of *Macaranga sinensis* (Baill.) Muell.-Arg. Chem. Pharm. Bull. (Japan) **38**, 1844 (1990).
85. IKEYA, Y., H. TAGUCHI, I. YOSHIOKA, and H. KOBAYASHI: The Constituents of *Schizandra chinensis* Baill. Isolation and Structure Determination of Five New Lignans, Gomisin A, B, C, F and G and the Absolute Structure of Schizandrin. Chem. Pharm. Bull. **27**, 1383 (1979).
86. OKUDA, T., T. YOSHIDA, T. HATANO, T. KOGA, N. TOH, and K. KURIYAMA: Circular Dichroism of Hydrolyzable Tannins, I: Ellagitannins and Gallotannins. Tetrahedron Lett. **23**, 3937 (1982).
87. HARADA, N., and K. NAKANISHI: The Exciton Chirality Method and Its Application to Configurational and Conformational Studies of Natural Products. Acc. Chem. Res. **5**, 257 (1972).
88. OKUDA, T., T. YOSHIDA, T. HATANO, T. KOGA, N. TOH, and K. KURIYAMA: Circular Dichroism of Hydrolyzable Tannins, II: Dehydroellagitannins. Tetrahedron Lett. **23**, 3941 (1982).
89. YOSHIDA, T., R. FUJII, and T. OKUDA: Revised Structures of Chebulinic Acid and Chebulagic Acid. Chem. Pharm. Bull. (Japan) **28**, 3713 (1980).
90. UDDIN, M., and E. HASLAM: Gallotannins, Part 15: Some Observations on the Structure of Chebulinic Acid and Its Derivatives. J. Chem. Soc. (C) **1967**, 2381.

91. YOSHIDA, T., H. ITOH, S. MATSUNAGA, R. TANAKA, and T. OKUDA: Hydrolyzable Tannins with $^1C_4$ Glucose Core from *Phyllanthus flexuosus* Muell. Arg. Chem. Pharm. Bull. (Japan) **40**, 53 (1992).

92. NAMBA, O., T. HATANO, T. YOSHIDA, and T. OKUDA: Studies on the Constituents of *Geranium thunbergii*, Part 17: New Ellagitannins. Abstract Papers (III) of the 108th Annual Meeting of the Pharmaceutical Society of Japan, p. 339 (1988).

93. YOSHIDA, T., Z.-X. JIN, and T. OKUDA: Heterophylliins A, B, C, D, and E, Ellagitannin Monomers and Dimers from *Corylus heterophylla*. Chem. Pharm. Bull. (Japan) **39**, 876 (1991).

94. SAIJO, R., G. NONAKA, and I. NISHIOKA: Isolation and Characterization of Four New Hydrolyzable Tannins from the Leaves of *Mallotus repandus*. Chem. Pharm. Bull. (Japan) **37**, 2624 (1989).

95. OKUDA, T., T. YOSHIDA, T. HATANO, K. YAZAKI, R. KIRA, and Y. IKEDA: Chromatography of Tannins, II: Preparative Fractionation of Hydrolyzable Tannins by Centrifugal Partition Chromatography. J. Chromatogr. **362**, 375 (1986).

96. OKUDA, T., T. YOSHIDA, T. HATANO, Y. IKEDA, T. SHINGU, and T. INOUE: Constituents of *Geranium thunbergii* Sieb. et Zucc., XIII: Isolation of Water-soluble Tannins by Centrifugal Partition Chromatography, and Biomimetic Synthesis of Elaeocarpusin. Chem. Pharm. Bull. (Japan) **34**, 4075 (1986).

97. TANAKA, T., G. NONAKA, I. NISHIOKA, K. MIYAHARA, and T. KAWASAKI: Isolation and Structure Elucidation of Elaeocarpusin, a Novel Ellagitannin from *Elaeocarpus sylvestris* var. *ellipticus*. J. Chem. Soc. Perkin Trans. 1 **1986**, 369.

98. OKUDA, T., T. YOSHIDA, and T. HATANO: Biomimetic Synthesis of Elaeocarpusin. Heterocycles **24**, 1841 (1986).

99. FOO, L.Y., and H. WONG: Phyllanthusiin D, an Unusual Hydrolysable Tannin from *Phyllanthus amarus*. Phytochem. **31**, 711 (1992).

100. YAZAKI, K., T. YOSHIDA, and T. OKUDA: Tannin Production in Cell Suspension Cultures of *Geranium thunbergii*. Phytochem. **30**, 501 (1991).

101. MAYER, W.: Über die Gerbstoffe aus dem Holz der Edelkastanie und Eiche. Das Leder **22**, 277 (1971).

102. NONAKA, G., T. SAKAI, T. TANAKA, K. MIHASHI, and I. NISHIOKA: Tannins and Related Compounds, XCVII: Structure Revision of C-glycosidic Ellagitannins, Castalagin, Vescalagin, Casuarinin and Stachyurin, and Related Hydrolyzable Tannins. Chem. Pharm. Bull. (Japan) **38**, 2151 (1990).

103. OKUDA, T., T. YOSHIDA, T. HATANO, K. YAZAKI, Y. IKEGAMI, and T. SHINGU: Guavins A, C and D, Complex Tannins from *Psidium guajava*. Chem. Pharm. Bull. (Japan) **35**, 443 (1987). The structure of guavin A in this paper has been revised to be an isomer of malabathrin E (**108**) (*106*), in which the (−)-epicatechin moiety of **108** is replaced by (+)-catechin unit.

104. HATANO, T., S, SHIDA, L. HAN, and T. OKUDA: Camelliatannins A and B, Two New Complex Tannins from *Camellia japonica* L. Chem. Pharm. Bull. (Japan) **39**, 876 (1991).

105. HATANO, T., L. HAN, S. TANIGUCHI, T. CHOU, T. SHINGU, H. SAKAGAMI, M. TAKEDA, H. NAKASHIMA, T. MURAYAMA, N. YAMAMOTO, T. YOSHIDA, and T. OKUDA: Anti-HIV Tannins from *Camellia japonica* and Related Plant Species. Symposium Papers of the 34th Symposium on the Chemistry of Natural Products, p. 510 (1992).

106. YOSHIDA, T., F. NAKATA, K. HOSOTANI, A. NITTA, and T. OKUDA: Three New Complex Tannins from *Melastoma malabathricum* L. Chem. Pharm. Bull. (Japan) **40**, 1727 (1992).

107. OKUDA, T., T. YOSHIDA, M. KUWAHARA, M.U. MEMON, and T. SHINGU: Agrimoniin and Potentillin, Ellagitannin Dimer and Monomer Having α-Glucose Core. J. Chem. Soc. Chem. Commun. **1982**, 163.

108. YOSHIDA, T., L. CHEN, T. SHINGU, and T. OKUDA: Euphorbins A and B, Novel Dimeric Dehydroellagitannins from *Euphorbia hirta* L. Chem. Pharm. Bull. (Japan) **36**, 2940 (1988).
109. YOSHIDA, T., O. NAMBA, L. CHEN, and T. OKUDA: Euphorbin C, an Equilibrated Dimeric Dehydroellagitannin Having a New Tetrameric Galloyl Group. Chem. Pharm. Bull. (Japan) **38**, 86 (1990).
110. YOSHIDA, T., O. NAMBA, C.-F. LU, L.-L. YANG, K.-Y. YEN, and T. OKUDA: Antidesmin A, a New Dimeric Hydrolyzable Tannin from *Antidesma pentandrum* var. *barbathum*. Chem. Pharm. Bull. (Japan) **40**, 338 (1992).
111. TANAKA, T., G. NONAKA, I. NISHIOKA, I. KOUNO, and F.-C. HO: Dehydroellagitannin from *Bishofia javanica*. Phytochem. **38**, 509 (1995).
112. TANAKA, T., F. NONAKA, and I. NISHIOKA: Tannins and Related Compounds, Part 28: Revision of the Structures of Sanguiins H-6, H-2, and H-3, and Isolation and Characterization of Sanguiin H-11, a Novel Tetrameric Hydrolysable Tannin, and Seven Related Tannins, from *Sanguisorba officinalis*. J. Chem. Res. (M) **1985**, 2001.
113. YOSHIDA, T., Y. MARUYAMA, M.U. MEMON, T. SHINGU, and T. OKUDA: Gemins A, B, and C, New Dimeric Ellagitannins from *Geum japonicum*. J. Chem. Soc. Perkin Trans. 1 **1985**, 315.
114. YOSHIDA, T., K. TANAKA, X.-M. CHEN, and T. OKUDA: Hydrolyzable Tannins with Dehydrodigalloyl Group from *Rosa laevigata* Mischx. Chem. Pharm. Bull. (Japan) **37**, 920 (1989).
115. YOSHIDA, T., K. TANAKA, X.-M. CHEN, and T. OKUDA: Dimeric Ellagitannins, Laevigatins E, F and G, from *Rosa laevigata*. Phytochem. **28**, 2451 (1989).
116. YOSHIDA, T., Z.-X. JIN, and T. OKUDA: Hydrolyzable Tannin Oligomers from *Rosa davurica*. Phytochem. **30**, 2747 (1991).
117. HATANO, T., T. HATTORI, S., and T. OKUDA: Coriariins C, D, E and F, New Dimeric and Monomeric Hydrolyzable Tannins. Chem. Pharm. Bull. (Japan) **34**, 4533 (1986).
118. ISHIMATSU, M., T. TANAKA, G. NONAKA, I. NISHIOKA, M. NISHIZAWA, and T. YAMAGISHI: Tannins and Related Compounds, LXXIX: Isolation and Characterization of Novel Dimeric and Trimeric Hydrolyzable Tannins, Nupharins C, D, E and F, from *Nuphar japonicum* DC. Chem. Pharm. Bull. (Japan) **37**, 1735 (1989).
119. YOSHIDA, T., A.F. AHMED, and T. OKUDA: Tannins of Tamaricaceous Plants, III: New Dimeric Hydrolyzable Tannins from *Reaumuria hirtella*. Chem. Pharm. Bull. (Japan) **41**, 672 (1993).
120. YOSHIDA, T., T. HATANO, A.F. AHMED, A. OKONOGI, and T. OKUDA: Structures of Isorugosin E and Hirtellin B, Dimeric Hydrolyzable Tannins Having a Trisgalloyl Group. Tetrahedron **47**, 3575 (1991).
121. HATANO, T., N. OGAWA, T. SHINGU, and T. OKUDA: Rugosins D, E, F and G, Dimeric and Trimeric Hydrolyzable Tannins with Valoneoyl Group from Flower Petals of *Rosa rugosa* Thunb. Chem. Pharm. Bull. (Japan) **38**, 3341 (1990).
122. YOSHIDA, T., Z.-X. JIN, and T. OKUDA: Unpublished data.
123. HATANO, T., N. OGAWA, R. KIRA, T. YASUHARA, and T. OKUDA: Tannins of Cornaceous Plants, I: Cornusiins A, B and C, Dimeric, Monomeric and Trimeric Hydrolyzable Tannins from *Cornus officinalis*, and Orientation of Valoneoyl Group in Related Tannins. Chem. Pharm. Bull. (Japan) **37**, 2083 (1989).
124. HATANO, T., T. YASUHARA, and T. OKUDA: Tannins of Cornaceous Plants, II: Cornusiins D, E and F, New Dimeric and Trimeric Hydrolyzable Tannins from *Cornus officinalis*. Chem. Pharm. Bull. (Japan) **37**, 2665 (1989).
125. HATANO, T., Y. IKEGAMI, T. SHINGU, and T. OKUDA: Camptothins A and B, New

Dimeric Hydrolyzable Tannins from *Camptotheca acuminata* Decne. Chem. Pharm. Bull. (Japan) **36**, 2017 (1988).

126. HATANO, T., A. OKONOGI, K. YAZAKI, and T. OKUDA: Trapanins A and B, Oligomeric Hydrolyzable Tannins from *Trapa japonica* Flerov. Chem. Pharm. Bull. (Japan) **38**, 2707 (1990).

127. YOSHIDA, T., H. OHBAYASHI, K. ISHIHARA, W. OHWASHI, K. HABA, Y. OKANO, T. SHINGU, and T. OKUDA: Hydrolyzable Tannins from *Tibouchina semidecandra* Cogn. Chem. Pharm. Bull. (Japan) **39**, 2233 (1991).

128. YOSHIDA, T., W. OHWASHI, K. HABA, H. OHBAYASHI, K. ISHIHARA, Y. OKANO, T. SHINGU, and T. OKUDA: Nobotanins B, C and E, Hydrolyzable Tannin Dimer and Trimers from *Tibouchina semidecandra* Cogn. Chem. Pharm. Bull. (Japan) **39**, 2264 (1991).

129. HATANO, T., T. YASUHARA, M. MATSUDA, K. YAZAKI, T. YOSHIDA, and T. OKUDA: Oenothein B, a Dimeric Hydrolysable Tannin with Macrocyclic Structure and Accompanying Tannins from *Oenothera erythrosepala*. J. Chem. Soc. Perkin Trans. 1 **1990**, 2735.

130. CHOU, T., T. YOSHIDA, and T. OKUDA: Three Hydrolyzable Tannin Oligomers Having Macrocyclic Structure from *Oenothera laciniata*. Abstract Papers of the XVth International Botanical Congress, Yokohama, p. 377 (1993).

131. OKUDA, T., T. YOSHIDA, T. HATANO, K. YAZAKI, R. KIRA, and Y. IKEDA: Preparative Fractionation of Hydrolyzable Tannins by Centrifugal Partition Chromatography. J. Chromatogr. **362**, 375 (1986).

132. YOSHIDA, T., T. CHOU, A. NITTA, K. MIYAMOTO, R. KOSHIURA, and T. OKUDA: Woodfordin C, a Macro-ring Hydrolyzable Tannin Dimer with Antitumor Activity and Accompanying Dimers from *Woodfordia fruticosa* Flowers. Chem. Pharm. Bull. (Japan) **38**, 1211 (1990).

133. KADOTA, S., Y. TAKAMORI, K.N. NYEIN, T. KIKUCHI, K. TANAKA, and H. EKIMOTO: Constituents of *Woodfordia fruticosa* Kurz, I: Isolation, Structure, and Proton and Carbon-13 Nuclear Magnetic Resonance Signal Assignments of Woodfruticosin (Woodfordin C), an Inhibitor of Deoxyribonucleic Acid Topoisomerase II. Chem. Pharm. Bull. (Japan) **38**, 2687 (1990).

134. YOSHIDA, T., T. CHOU, Y. MARUYAMA, and T. OKUDA: Camelliins A and B, Two New Dimeric Hydrolyzable Tannins from Flower Buds of *Camellia japonica* L. and *Camellia sasanqua* Thunb. Chem. Pharm. Bull. (Japan) **38**, 2681 (1990).

135. YOSHIDA, T., T. CHOU, A. NITTA, and T. OKUDA: Monomeric and Dimeric Hydrolyzable Tannins Having a Dilactonized Valoneoyl Group from *Schima wallichii* Koeth. Chem. Pharm. Bull. (Japan) **39**, 2247 (1991).

136. XU, Y.-M., T. SAKAI, T. TANAKA, G. NONAKA, and I. NISHIOKA: Tannins and Related Compounds, CVI: Preparation of Aminoalditol Derivatives of Hydrolyzable Tannins Having α- and β-Glucopyranose Cores, and Its Application to the Structure Elucidation of New Tannins, Reginins A and B and Flosin A, Isolated from *Lagerstroemia flos-reginae* Retz. Chem. Pharm. Bull. (Japan) **39**, 639 (1991).

137. XU, Y.-M., T. TANAKA, G. NONAKA, and I. NISHIOKA: Tannins and Related Compounds, CVII: Structure Elucidation of Three New Monomeric and Dimeric Ellagitannins, Flosin B and Reginins C and D, Isolated from *Lagerstroemia flos-regina* Retz. Chem. Pharm. Bull. (Japan) **39**, 647 (1991).

138. YOSHIDA, T., Z.-X. JIN, and T. OKUDA: Heterophylliins A, B, C, D and E, Ellagitannin Monomers from *Corylus heterophylla* Fisch. Chem. Pharm. Bull. (Japan) **39**, 49 (1991).

139. YOSHIDA, T., T. HATANO, and T. OKUDA: Chromatography of Tannins, IV: Separation

of Labile Oligomeric Hydrolysable Tannins and Related Polyphenols by Centrifugal Partition Chromatography. J. Chromatogr. **467**, 139 (1989).

140. YOSHIDA, T., W.-S. FENG, and T. OKUDA: Tannins and Related Polyphenols of Rosaceous Medicinal Plants, XII: Roshenins A–E, Dimeric Hydrolyzable Tannins from *Rosa henryi* Boul. Chem. Pharm. Bull. (Japan) **40**, 1997 (1992).

141. TANAKA, T., H. TACHIBANA, G. NONAKA, I. NISHIOKA, and F.-L. HSU: Tannins and Related Compounds, CXXII: New Dimeric Trimeric and Tetrameric Ellagitannins, Lambertianins A–D, from *Rubus lambertianus* Seringe. Chem. Pharm. Bull. (Japan) **41**, 1214 (1993).

142. YOSHIDA, T., T. CHOU, M. MATSUDA, T. YASUHARA, K. YAZAKI, T. HATANO, A. NITTA, and T. OKUDA: Woodfordin D and Oenothein A, Trimeric Hydrolyzable Tannins of Macro-ring Structure with Antitumor Activity. Chem. Pharm. Bull. (Japan) **39**, 1157 (1991).

143. YOSHIDA, T., T. CHOU, A. NITTA, and T. OKUDA: Hydrolyzable Tannin Oligomers with Macrocyclic Structures, and Accompanying Tannins from *Woodfordia fruticosa* Kurz. Chem. Pharm. Bull. (Japan) **40**, 2023 (1992).

144. YOSHIDA, T., O. NAMBA, L. CHEN, and T. OKUDA: Euphorbin E, a Hydrolyzable Tannin Dimer of Highly Oxidized Structure from *Euphorbia hirta*. Chem. Pharm. Bull. (Japan) **38**, 1113 (1990).

145. NONAKA, G., T. SAKAI, K. MIHASHI, and I. NISHIOKA: Tannins and Related Compounds, CIX: Isolation of Alienanins A and B, Novel C, C-Linked Ellagitannin Dimers from *Quercus aliena* Blume. Chem. Pharm. Bull. (Japan) **39**, 884 (1991).

146. PENHOAT, C.L.M.H., V.M.F. MICHON, S. PENG, C. VIRIOT, A. SCALBERT, and D. GAGE: Structural Elucidation of New Ellagitannins from *Quercus robur* L. Roburins A–E. J. Chem. Soc. Perkin Trans. 1 **1991**, 1653.

147. LIN, T.-C., T. TANAKA, G. NONAKA, I. NISHIOKA, and T.-J. YOUNG: Tannins and Related Compounds, CVIII: Isolation and Characterization of Novel Complex Tannins (Flavano-ellagitannins), Anogeissinin and Anogeissusins A and B, from *Anogeissus acuminata* (Roxb ex DC.) Guill. et Perr. var. *lanceolata* Wall. ex Clarke. Chem. Pharm. Bull. (Japan) **39**, 1144 (1991).

148. YOSHIDA, T., F. NAKATA, K. HOSOTANI, A. NITTA, and T. OKUDA: Tannins from *Melastoma malabathricum*. Phytochem. **31**, 2829 (1992).

149. YOSHIDA, T., X.-M. CHEN, T. HATANO, M. FUKUSHIMA, and T. OKUDA: Tannins and Related Polyphenols of Rosaceous Plants, IV: Roxbins A and B from *Rosa roxburghii* Fruits. Chem. Pharm. Bull. (Japan) **35**, 1817 (1987).

150. TANAKA, T., A. MORITA, G. NONAKA, T.-C. LIN, I. NISHIOKA, and F.-C. HO: Tannins and Related Compounds, CIII: Isolation and Characterization of New Monomeric, Dimeric and Trimeric Ellagitannins, Calamansanin and Calamanins A, B and C, from *Terminalia calamansanai* (Blanco) Rolfe. Chem. Pharm. Bull. (Japan) **39**, 60 (1991).

151. NAKATA, F., T. YOSHIDA, and T. OKUDA: Ellagitannins from *Tibouchina semidecandra*. Abstract Papers (2) of the 112th Annual Meeting of the Pharmaceutical Society of Japan, p. 206 (1992).

152. KOBAYASHI, S., T. HATANO, T. YOSHIDA, T. OKUDA, C.-F. LU, L.-L. YANG, K.-Y. YEN: Tannins from *Barringtonia asiatica*. Abstract Papers (2) of the 112th Annual Meeting of the Pharmaceutical Society of Japan, p. 208 (1992).

153. ATALLAH, A.F., T. YOSHIDA, and T. OKUDA: Tannins of Tamaricaceous Plants, V: New Dimeric, Trimeric and Tetrameric Ellagitannins from *Reaumuria hirtella*. Chem. Pharm. Bull. (Japan) **42**, 246 (1994).

154. YOSHIDA, T., A.F. AHMED, and T. OKUDA: Tamarixinins B and C, Dimeric Hydrolyzable Tannins from *Tamarix pakistanica*. Phytochem. **33**, 197 (1993).

155. AGATA, I., T. HATANO, Y. NAKAYA, T. SUGAYA, S. NISHIBE, T. YOSHIDA, and T. OKUDA: Tannins and Related Polyphenols of Euphorbiaceous Plants, VIII: Eumaculin A and Eusupinin A, and Accompanying Polyphenols from *Euphorbia maculata* L. and *E. supina* Rafin. Chem. Pharm. Bull. (Japan) **39**, 881 (1991).

156. YOSHIDA, T., S. TANEI, Y.-Z. LIU, K. YUAN, C.-R. JI, and T. OKUDA: Hydrolyzable Tannins from *Loropetalum chinense*. Phytochem. **32**, 1287 (1993).

157. HATANO, T., R. KIRA, T. YASUHARA, and T. OKUDA: Tannins of Hamamelidaceous Plants, III: Isorugosins A, B and D, New Ellagitannins from *Liquidambar formosana*. Chem. Pharm. Bull. (Japan) **36**, 3920 (1988).

158. HATANO, T., A. OKONOGI, and T. OKUDA: Oligomeric Hydrolyzable Tannins from *Liquidambar formosana* and Spectral Analysis of the Orientation of Valoneoyl Groups in Their Molecules. In: Plant Polyphenols. Synthesis, Properties, Significance (R.W. HEMINGWAY and P.E. LAKS, eds.), p. 195. New York: Plenum Press. 1992.

159. NONAKA, G., S. NAKAYAMA, and I. NISHIOKA: Tannins and Related Compounds, LXXXIII: Isolation and Structure of Hydrolyzable Tannins, Phillyraeoidins A–E from *Quercus phillyraeoides*. Chem. Pharm. Bull. (Japan) **37**, 2030 (1989).

160. TANAKA, T., M. UEDA, G. NONAKA, and I. NISHIOKA: Tannins from *Lagerstroemia indica* L. Abstract Papers of the 108th Annual Meeting of the Pharmaceutical Society of Japan, p. 278 (1988).

161. YOSHIDA, T., O. NAMBA, K. YOKOYAMA, and T. OKUDA: Hydrolyzable Tannin Oligomers from Euphorbiaceous Plants. Symposium Papers of the 31st Symposium on the Chemistry of Natural Products, p. 601 (1989).

162. LIN, J.-H., T. TANAKA, G. NONAKA, I. NISHIOKA, and I.-S. CHEN: Tannins and Related Compounds, XCVIII: Structures of Three New Dimeric Ellagitannins, Excoecarianin and Excoecarinins A and B, Isolated from the Leaves of *Excoecaria kawakamii* Hayata. Chem. Pharm. Bull. (Japan) **38**, 2162 (1990).

163. AGATA, I., T. SUGAYA, S. NISHIBE, T. HATANO, T. YOSHIDA, and T. OKUDA: Tannins of *Euphorbia maculata* and *E. supina*. Abstract Papers (2) of the 111th Annual Meeting of the Pharmaceutical Society of Japan, p. 177 (1991).

164. SAIJO, R., G. NONAKA, I. NISHIOKA, I.-S. CHEN, and T.-H. HWANG: Tannins and Related Compounds, LXXXVIII: Isolation and Characterization of Hydrolyzable Tannins from *Mallotus japonicus* (Thunb.) Muell.-Arg. Chem. Pharm. Bull. (Japan) **37**, 2940 (1989).

165. SHIMOKAWA, H., G. NONAKA, and I. NISHIOKA: Tannins of *Casuarina glauca*. Abstract Papers (2) of the 111th Annual Meeting of the Pharmaceutical Society of Japan, p. 147 (1991); Abstract Papers of the 38th Annual Meeting of the Japanese Society of Pharmacognosy, p. 104 (1991).

166. YOSHIDA, T., Y. IKEDA, H. OHBAYASHI, K. ISHIHARA, W. OHWASHI, T. SHINGU, and T. OKUDA: Dimeric Ellagitannins in Plants of Melastomataceae. Chem. Pharm. Bull. (Japan) **34**, 2676 (1986).

167. BATE-SMITH, E.C.: Detection and Determination of Ellagitannins. Phytochem. **11**, 1153 (1972).

168. OKUDA, T., K. MORI, and N. HAYASHI: Application of Determination Methods of Tannin Activity with Hemoglobin and of Ellagitannin with Nitrous Acid. Yakugaku Zasshi **96**, 1143 (1976).

169. OKUDA, T., T. YOSHIDA, and T. HATANO: Application of Centrifugal Partition Chromatography to Separation of Tannins and Related Polyphenols. J. Liq. Chromatogr. **11**, 2447 (1988).

170. OKUDA, T., T. YOSHIDA, T. HATANO, K. MORI, and T. FUKUDA: Fractionation of Pharmacologically Active Plant Polyphenols by Centrifugal Partition Chromatography. J. Liq. Chromatogr. **13**, 3637 (1990).

171. YOSHIDA, T., T. HATANO, and T. OKUDA: Two-Dimensional NMR Spectra of Hy-drolyzable Tannins Which Form Equilibrium Mixtures. Mag. Reson. Chem. **30**, 546 (1992).
172. YOSHIDA, T., T. HATANO, T. OKUDA, M.U. MEMON, and T. SHINGU: Spectral and Chromatographic Analyses of Tannins, I: ^{13}C Nuclear Magnetic Resonance Spectra of Hydrolyzable Tannins. Chem. Pharm. Bull. (Japan) **32**, 1790 (1984).
173. HATANO, T., T. YOSHIDA, T. SHINGU, and T. OKUDA: ^{13}C Nuclear Magnetic Reso-nance Spectra of Hydrolyzable Tannins, II: Tannins Forming Anomer Mixtures. Chem. Pharm. Bull. (Japan) **36**, 2925 (1988).
174. YOSHIDA, T., T. HATANO, T. KUWAJIMA, and T. OKUDA: Oligomeric Hydrolyzable Tannins and Their ^{1}H NMR Spectra and Partial Degradation. Heterocycles **33**, 463 (1992).
175. CRONQUIST, A.: The Evolution and Classification of Flowering Plants, 2nd Ed. New York: The New York Botanical Garden. 1988.
176. MAKI, Y., M. SUZUKI, O. TOYOTA, and M. TAKAYA: Studies on the Smiles Rearrange-ment, XII: Synthesis and Structural Assignment of Two Isomeric N-Phenyl-2,3-diazaphenothiazinones. Chem. Pharm. Bull. (Japan) **21**, 241 (1973).
177. OKUDA, T., T. HATANO, T. KANEDA, M. YOSHIZAKI, and T. SHINGU: Liquidambin, a New Ellagitannin from *Liquidambar formosana*. Phytochem. **26**, 2053 (1987).
178. OKUDA, T., T. HATANO, I. AGATA, S. NISHIBE, and K. KIMURA: Tannins in *Artemisia montana, A. princeps* and Related Species of Plant. Yakugaku Zasshi **106**, 894 (1986).
179. TIMMERMANN, B.N., J.J. HOFFMAN, S.D. JOLAD, K.H. SCHRAM, R.E. KLENCK, and R.B. BATES: Constituents of *Chrysothamnus paniculatus* 3: 3,4,5-Tricaffeoylquinic Acid (A New Shikimate Prearomatic) and 3,4-, 3,5- and 4,5-Dicaffeoylquinic Acids. J. Nat. Prod. **46**, 365 (1983).
180. MERFORT, I.: Caffeoylquinic Acids from Flowers of *Arnica montana* and *Arnica chamissonis*. Phytochem. **31**, 2111 (1992).
181. AGATA, I., S. GOTO, T. HATANO, S. NISHIBE, and T. OKUDA: 1,3,5-Tri-O-caffeoylquinic Acid from *Xanthium strumarium*. Phytochem. **33**, 508 (1993).
182. OKUDA, T., T. HATANO, I. AGATA, and S. NISHIBE: The Components of Tannic Activities in Labiatae Plants, I: Rosmarinic Acid from Labiatae Plants in Japan. Yakugaku Zasshi **106**, 1108 (1986).
183. HEGNAUER, R.: Chemotaxonomie der Pflanzen, Bd. 4, p. 327. Basel-Stuttgart: Birk-häuser. 1966.
184. KELLY, C.J., R.C. HARRUFF, and M. CARMACK: The Polyphenolic Acids of *Litho-spermum ruderale*, II: Carbon-13 Nuclear Magnetic Resonance of Lithospermic and Rosmarinic Acids. J. Org. Chem. **41**, 449 (1976).
185. WAGNER, H.: Phenolic Compounds in Plants of Pharmaceutical Interest. In: Bio-chemistry of Plant Phenolics (T. SWAIN, J.B. HARBORNE, and C.F. VAN SUMERE, eds.), p. 598. New York: Plenum Press. 1979.
186. AI, C.-B, and L.-N. LI: Stereostructure of Salvianolic Acid B and Isolation of Salvianolic Acid C from *Salvia miltiorrhiza*. J. Nat. Prod. **51**, 145 (1988).
187. AGATA, I., H. KUSAKABE, T. HATANO, S. NISHIBE, and T. OKUDA: Melitric Acids A and B, New Trimeric Caffeic Acid Derivatives from *Melissa officinalis*. Chem. Pharm. Bull. (Japan) **41**, 1608 (1993).
188. AGATA, T., T. HATANO, S. NISHIBE, and T. OKUDA: A Tetrameric Derivative of Caffeic Acid from *Rabdosia japonica*. Phytochem. **28**, 2447 (1989).
189. NISHIZAWA, M., M. TSUDA, and K. HAYASHI: Two Caffeic Acid Tetramers Having Enantiomeric Phenyldihydronaphthalene Moieties from *Macrotomia euchroma*. Phy-tochem. **29**, 2645 (1990).

190. AGATA, I., Y. NAKAYA, S. NISHIBE, T. HATANO, and T. OKUDA: Polyphenols of *Rabdosia trichocarpa*. Abstract Papers (II) of the 110th Annual Meeting of the Pharmaceutical Society of Japan, p. 222 (1990).

191. HATANO, T., R. KIRA, M. YOSHIZAKI, and T. OKUDA: Seasonal Changes in the Tannins of *Liquidambar formosana* Reflecting Their Biogenesis. Phytochem. **25**, 2787 (1986).

192. MOUMOU, Y., F. TROTIN, J. DUBOIS, J. VASSEUR, and E. EL-BOUSTANI: Influence of Culture Condition on Polyphenol Production by *Fagopyrum esculentum* Tissue Cultures. J. Nat. Prod. **55**, 33 (1992).

193. MOUMOU, Y., J. VASSEUR, F. TROTIN, and J. DUBOIS: Catechin Production by Callus Cultures of *Fagopyrum esculentum*. Phytochem. **31**, 1239 (1992).

194. YAZAKI, K., and T. OKUDA: Procyanidins in Callus and Multiple Shoot Cultures of *Hypericum erectum* Thunb. Planta Med. **56**, 490 (1990).

195. YAZAKI, K., and T. OKUDA: Condensed Tannin Production in Callus and Suspension Cultures of *Cinnamomum cassia* Blume. Phytochem. **29**, 1559 (1990).

196. YAZAKI, K., and T. OKUDA: Gallotannin Production in Cell Cultures of *Cornus officinalis* Sieb. et Zucc. Plant Cell Rep. **8**, 346 (1989).

197. YAZAKI, K., and T. OKUDA: *Cornus officinalis: In vitro* Culture and the Production of Gallotannins. In: Biotechnology in Agriculture and Biochemistry, Vol. 21: Medicinal and Aromatic Plants, IV (Y.P. BAJAJ ed.), p. 104. Berlin-Heidelberg-New York: Springer. 1993.

198. YAZAKI, K., and T. OKUDA: Ellagitannin Formation in Callus Cultures of *Heterocentron roseum*. Phytochem. **29**, 1127 (1990).

199. ISHIMARU, K., and K. SHIMOMURA: Tannin Production in Hairy Root Culture of *Geranium thunbergii*. Phytochem. **30**, 825 (1991).

200. HATANO, T., T. YOSHIDA, and T. OKUDA: Chromatography of Tannins, III: Multiple Peaks in High-performance Liquid Chromatography of Some Hydrolyzable Tannins. J. Chromatogr. **435**, 285 (1988).

201. ISHIMARU, K., M. HIROSE, K. TAKAHASHI, K. KOYAMA, and K. SHIMOMURA: Tannin Production in Root Culture of *Sanguisorba officinalis*. Phytochem. **29**, 3827 (1990).

202. MAYER, W., E.H. HOFFMANN, N. LÖSCH, H. WOLF, B. WOLTER, and G. SCHILLING: Dehydrierungsreaktionen mit Gallussäureestern. Liebigs Ann. Chem. **1984**, 929.

203. YOSHIDA, T., K. MORI, T. HATANO, T. OKUMURA, I. UEHARA, K. KOMAGOE, Y. FUJITA, and T. OKUDA: Studies on Inhibition Mechanism of Autoxidation by Tannins and Flavonoids, V: Radical-Scavenging Effects of Tannins and Related Polyphenols on 1,1-Diphenyl-2-picrylhydrazyl Radical. Chem. Pharm. Bull. (Japan) **37**, 1919 (1989).

204. FELDMAN, K.S., and S.M. ENSEL: Ellagitannin Chemistry. The First Example of Biomimetic Diastereoselective Oxidative Coupling of a Glucose-Derived Digalloyl Substrate. J. Am. Chem. Soc. **115**, 1162 (1993).

205. OKUDA, T., T. YOSHIDA, and T. HATANO: Antioxidant Effects of Tannins and Related Polyphenols. In: Phenolic Compounds in Food and Their Effects on Health, II (M.-T. HUANG, C.-T. HO, and C.Y. LEE, eds.), p. 87. Washington, DC: American Chemical Society. 1992.

206. OKUDA, T., T. YOSHIDA, and T. HATANO: Polyphenols from Asian Plants: Structural Diversity and Antitumor and Antiviral Activities. In: Phenolic Compounds in Food and Their Effects on Health, II (M.-T. HUANG, C.-T. HO, and C.Y. LEE, eds.), p. 160. Washington, DC: American Chemical Society. 1992.

207. OKUDA, T., T. YOSHIDA, and T. HATANO: Antioxidant Phenolics in Oriental Medicine. In: Active Oxygens, Lipid Peroxides, and Antioxidants (K. YAGI, ed.), p. 333. Tokyo: Japan Sci. Soc. Press/Boca Raton: CRC Press. 1993.

208. OKUDA, T., T. YOSHIDA, and T. HATANO: Chemistry and Antioxidant Effects of Phenolics from Licorice, Tea and Composite and Labiate Herbs. In: Food Phytochemicals for Cancer Prevention, II (C.-T. HO, T. OSAWA, M.-T. HUANG, and R.T. ROSEN, eds.), p. 133. Washington, DC: American Chemical Society. 1992.

209. OKUDA, T., T. YOSHIDA, and T. HATANO: Antioxidant Polyphenols in Oriental Medicine. In: Proceedings of 5th International Congress on Oxygen Radicals (K. YAGI, ed.). Tokyo: Publication Center for Academic Societies, Japan (1993).

210. YOSHIDA, T., S. KOYAMA, and T. OKUDA: Inhibitory Effects of Tannins on Cupric Ion-catalyzed Autoxidation of Ascorbic Acid. Yakugaku Zasshi 101, 695 (1981).

211. FUJITA, Y., K. KOMAGOE, Y. SASAKI, I. UEHARA, T. OKUDA, and T. YOSHIDA: Inhibition Mechanism of Tannins on Cu(II)-Catalyzed Oxidation of Ascorbic Acid. Yakugaku Zasshi 107, 17 (1987).

212. OKUDA, T., Y. KIMURA, T. YOSHIDA, T. HATANO, H. OKUDA, and S. ARICHI: Studies on the Activities of Tannins and Related Compounds, I: Inhibitory Effects on Lipid Peroxidation in Mitochondria and Microsomes of Liver. Chem. Pharm. Bull. (Japan) 31, 1625 (1983).

213. KIMURA, Y., H. OKUDA, T. OKUDA, T. HATANO, I. AGATA, and S. ARICHI: Inhibitory Effects on Lipid Peroxidation in Mitochondria and Microsomes of Liver. Planta Med. 50, 459 (1984).

214. KIMURA, Y., H. OKUDA, K. MORI, T. OKUDA, and S. ARICHI: Effects of Various Extracts of Geranii Herba and Geraniin on Liver Injury and Lipid Metabolism in Rats Fed Peroxidized Oil. Chem. Pharm. Bull. (Japan) 32, 1866 (1984).

215. KIMURA, Y., H. OKUDA, T. OKUDA, T. HATANO, I. AGATA, and S. ARICHI: Effects of Extracts of Leaves of Artemisia Species, and Caffeic Acid and Chlorogenic Acid on Lipid Metabolic Injury in Rats Fed Peroxidized Oil. Chem. Pharm. Bull. (Japan) 33, 2028 (1985).

216. FUJITA, Y., K. KOMAGOE, Y. NIWA, I. UEHARA, R. HARA, H. MORI, and T. OKUDA: Inhibition Mechanism of Tannins Isolated from Medicinal Plants and Related Compounds on Autoxidation of Methyl Linoleate. Yakugaku Zasshi 108, 528 (1988).

217. HATANO, T., H. KAGAWA, T. YASUHARA, and T. OKUDA: Two New Flavonoids and other Constituents in Licorice Root: Their Relative Astringency and Radical Scavenging Effects. Chem. Pharm. Bull. (Japan) 36, 2090 (1988).

218. HATANO, T., R. EDAMATSU, M. HIRAMATSU, A. MORI, Y. FUJITA, T. YASUHARA, T. YOSHIDA, and T. OKUDA: Effects of Tannins and Related Polyphenols on Superoxide Anion Radical, and on 1,1-Diphenyl-2-picrylhydrazyl Radical. Chem. Pharm. Bull. (Japan) 37, 2016 (1989).

219. HIKINO, H., Y. KISO, T. HATANO, T. YOSHIDA, and T. OKUDA: Antihepatotoxic Actions of Tannins. J. Ethnopharmacol. 14, 19 (1985).

220. KIMURA, Y., H. OKUDA, T. OKUDA, and S. ARICHI: Effects of Geraniin, Corilagin, and Ellagic Acid Isolated from Geranii Herba on Arachidonate Metabolism in Leukocytes. Planta Med. 52, 337 (1986).

221. KIMURA, Y., H. OKUDA, T. OKUDA, T. HATANO, and S. ARICHI: Effects of Caffetannins and Related Compounds on Arachidonate Metabolism in Human Polymorphonuclear Leukocytes. J. Nat. Prod. 50, 392 (1987).

222. KIMURA, Y., H. OKUDA, T. OKUDA, and S. ARICHI: Effects of Chalcones Isolated from Licorice Roots on Leukotriene Biosynthesis in Human Polymorphonuclear Neutrophils. Phytother. Res. 2, 140 (1988).

223. FUJITA, Y., I. UEHARA, Y. MORIMOTO, M. NAKASHIMA, T. HATANO, and T. OKUDA: Inhibition Mechanism of Caffetannins Isolated from Leaves of Artemisia Species on Lipoxygenase Dependent Lipid Peroxidation. Yakugaku Zasshi 108, 129 (1988).

224. IWATA, S., Y. FUKAYA, K. NAKAZAWA, and T. OKUDA: Effects of Tannins on the Oxidative Damage of Mouse Ocular Lens, I. J. Ocular Pharmacol. 3, 227 (1987).

225. FUKAYA Y., K. NAKAZAWA, T. OKUDA, and S. IWATA: Effect of Tannin on Oxidative Damage of Ocular Lens. Jpn. J. Ophthalmol. 32, 166 (1988).

226. HATANO, T., T. YASUHARA, T. FUKUDA, T. NORO, and T. OKUDA: Structures of Licopyranocoumarin, Licoarylcoumarin and Glisoflavone, and Inhibitory Effects of Licorice Phenolics and Xanthine Oxidase. Chem. Pharm. Bull. (Japan) 37, 3005 (1989).

227. HATANO, T., T. YASUHARA, R. YOSHIHARA, I. AGATA, T. NORO, and T. OKUDA: Inhibitory Effects of Tannins and Related Polyphenols on Xanthine Oxidase. Chem. Pharm. Bull. (Japan) 38, 1224 (1990).

228. HATANO, T., T. FUKUDA, T. MIYASE, T. NORO, and T. OKUDA: Structures of Glicoricone and Licofuranone, and Inhibitory Effects of Licorice Constituents on Monoamine Oxidase. Chem. Pharm. Bull. (Japan) 39, 1238 (1991).

229. CHEN, G.: The Pathobiology of Parkinson's Disease: Biochemical Aspects of Dopamine Neuron Senescence. J. Neural. Transm., Suppl. 19, 89 (1983).

230. MORTON, J.F.: Widespread Tannin Intake via Stimulants and Masticatories, Especially Guarana, Kola Nut, Betel Vine, and Accessories. In: Plant Polyphenols, Synthesis, Properties, Significance (R.W. HEMINGWAY and P.E. LAKS, eds.), p. 739. New York: Plenum Press. 1992.

231. HIROSE, M., S. FUKUSHIMA, H. SHIRAI, R. HASEGAWA, T. KATO, H. TANAKA, E. ASAKAWA, and N. ITO: Stomach Carcinogenicity of Caffeic Acid, Sesamol and Catechol in Rats and Mice. Jpn. J. Cancer Res. 81, 207 (1990).

232. OKUDA, T., K. MORI, and H. HAYATSU: Inhibitory Effects of Tannins and Related Compounds on Mutagenic Substances (2). Abstract Papers of the 102nd Annual Meeting of Pharmaceutical Society of Japan, p. 588 (1982).

233. OKUDA, T., K. MORI, and H. HAYATSU: Inhibitory Effect of Tannins on Direct-acting Mutagens. Chem. Pharm. Bull. (Japan) 32, 3755 (1984).

234. SAYER, J.M., H. YAGI, A.W. WOOD, A.H. CONNEY, and D.M. JERINA: Extremely Facile Reaction Between the Ultimate Carcinogen Benzo[a]pyrene 7,8-Diol-9,10-epoxide and Ellagic Acid. J. Am. Chem. Soc. 104, 5562 (1982).

235. OKUDA, T., K. MORI, and M. ISHINO: Studies on the Constituents of Geranium thunbergii, Part 8: Transformations of Geraniin upon Decoction. Yakugaku Zasshi 99, 505 (1979).

236. DANIEL, E.M., S. RATNAYAKE, T. KINSTLE, and G.D. STONER: The Effects of pH and Rat Intestinal Contents on the Liberation of Ellagic Acid from Purified and Crude Ellagitannins. J. Nat. Prod. 54, 946 (1991).

237. HUANG, M.-T., R.L. CHANG, A.W. WOOD, H.L. NEWMARK, J.M. SAYER, H. YAGI, D.M. JERINA, and A.H. CONNEY: Inhibition of the Mutagenicity of Bay-region Diol-epoxide of Polycyclic Aromatic Hydrocarbons by Tannic Acid, Hydroxylated Anthraquinones and Hydroxylated Cinnamic Acid Derivatives. Carcinogenesis 6, 237 (1985).

238. KASHIWADA, Y., G. NONAKA, I. NISHIOKA, J.-J. CHANG, and K.-H. LEE: Tannins and Related Compounds as Selective Cytotoxic Agents. J. Nat. Prod. 55, 1033 (1992).

239. YOSHIZAWA, S., T. HORIUCHI, H. FUJIKI, T. YOSHIDA, T. OKUDA, and T. SUGIMURA: Antitumor Promoting Activity of (−)-Epigallocatechin Gallate, the Main Constituent of "Tannin" in Green Tea. Phytother. Res. 1, 44 (1987).

240. FUJIKI, H., M. SUGANUMA, S. YOSHIZAWA, J. YATSUNAMI, S. NISHIWAKI, H. FURUYA, S. OKABE, R. NISHIWAKI-MATSUSHIMA, S. MATSUNAGA, Y. MUTO, T. OKUDA, and T. SUGIMURA: Sarcophytol A and (−)-Epigallocatechin Gallate (EGCG), Nontoxic Inhibitors of Cancer Development. In: Cancer Chemoprevention (L. WATTENBERG, M. LIPKIN, C.W. BOONE, and G.J. KELLOFF, eds.), p. 393. Boca Raton: CRC Press. 1992.

241. FUJITA, Y., T. YAMANE, M. TANAKA, K. KUWATA, J. OKUZUMI, T. TAKAHASHI, H. FUJIKI, and T. OKUDA: Inhibitory Effect of (−)-Epigallocatechin Gallate on Carcinogenesis with N-ethyl-N'-nitro-N-nitrosoguanidine in Mouse Duodenum. Jpn. J. Cancer Res. **80**, 503 (1989).

242. KOMORI, A., J. YATSUNAMI, S. OKABE, S. ABE, K. HARA, M. SUGANUMA, S. KIM, and H. FUJIKI: Anticarcinogenic Activity of Green Tea Polyphenols. Jpn. J. Clin. Oncol. **23**, 186 (1993).

243. WANG, Z.-Y., M.-T. HUANG, T. FERRARO, C.-Q. WONG, Y.R. LOU, K. REUHL, M. IATROPOULOS, C.S. YANG, and A.H. CONNEY: Inhibitory Effect of Green Tea in the Drinking Water on Tumorigenesis by Ultraviolet Light and 12-O-Tetradecanoylphorbol-13-acetate in the Skin of SKH-1 Mice. Cancer Res. **52**, 1162 (1992).

244. MIYAMOTO, K., N. KISHI, R. KOSHIURA, T. YOSHIDA, T. HATANO, and T. OKUDA: Relationship Between the Structures and the Antitumor Activities of Tannins. Chem. Pharm. Bull. (Japan) **35**, 814 (1987).

245. MIYAMOTO, K., M. SASAKURA, E. MATSUI, R. KOSHIURA, T. MURAYAMA, T. HATANO, T. YOSHIDA, and T. OKUDA: Antitumor Activity of Oenothein B, a Unique Macrocyclic Ellagitannin. Jpn. J. Cancer Res. **84**, 99 (1993).

246. MIYAMOTO, K., M. NOMURA, T. MURAYAMA, T. FURUKAWA, T. HATANO, T. YOSHIDA, R. KOSHIURA, and T. OKUDA: Antitumor Activities of Ellagitannins Against Sarcoma-180 in Mice. Biol. Pharm. Bull. **16**, 379 (1993).

247. MIYAMOTO, K., T. MURAYAMA, M. NOMURA, T. HATANO, T. YOSHIDA, T. FURUYA, R. KOSHIURA, and T. OKUDA: Antitumor Activity and Interleukin-1 Induction by Tannins. Anticancer Res. **31**, 37 (1993).

248. ASANAKA, M., T. KURIMURA, R. KOSHIURA, T. OKUDA, M. MORI, and H. YOKOI: Inhibitory Effect of Ellagitannins on the *in vitro* Replication of Human Immunodeficiency Virus (HIV). AIDS Research Newsletter **1**, 72 (1987).

249. HATANO, T., T. YASUHARA, K. MIYAMOTO, and T. OKUDA: Anti-Human Immunodeficiency Virus Phenolics from Licorice. Chem. Pharm. Bull. (Japan) **36**, 2286 (1988).

250. NAKASHIMA, H., T. MURAKAMI, N. YAMAMOTO, H. SAKAGAMI, S. TANUMA, T. HATANO, T. YOSHIDA, and T. OKUDA: Inhibition of Human Immunodeficiency Viral Replication by Tannins and Related Compounds. Antiviral Res. **18**, 91 (1992).

251. TAKECHI, M., Y. TANAKA, M. TAKEHARA, G. NONAKA, and I. NISHIOKA: Structure and Antiherpetic Activity Among the Tannins. Phytochem. **24**, 2245 (1985).

252. KAKIUCHI, N., M. HATTORI, T. NAMBA, M. NISHIZAWA, T. YAMAGISHI, and T. OKUDA: Inhibitory Effects of Tannins on Reverse Transcriptase from RNA Tumor Virus. J. Nat. Prod. **48**, 614 (1985).

253. NISHIZAWA, M., T. YAMAGISHI, G.E. DUTCHMAN, W.B. PARKER, A.J. BODNER, R.E. KILKUSKIE, Y.-C. CHENG, and K.-H. LEE: Isolation and Characterization of Four New Tetragalloylquinic Acids as a New Class of HIV Reverse Transcriptase Inhibitors from Tannic Acid. J. Nat. Prod. **52**, 762 (1989).

254. NONAKA, G., I. NISHIOKA, M. NISHIZAWA, T. YAMAGISHI, Y. KASHIWADA, G.E. DUTCHMAN, A.J. BODNER, R.E. KILKUSKIE, Y.-C. CHENG, and K.-H. LEE: Inhibitory Effects of Tannins on HIV Reverse Transcriptase and HIV Replication in H9 Lymphocyte Cells. J. Nat. Prod. **53**, 587 (1990).

255. KAKIUCHI, N., X. WANG, M. HATTORI, T. OKUDA, and T. NAMBA: Circular Dichroism Studies on the Ellagitannins-Nucleic Acids Interaction. Chem. Pharm. Bull. (Japan) **35**, 2875 (1987).

256. KAKIUCHI, N., M. HATTORI, M. NISHIZAWA, T. YAMAGISHI, T. OKUDA, and T. NAMBA: Inhibitory Effect of Various Tannins on Glucan Synthesis by Glucosyltransferase from *Streptococcus mutans*. Chem. Pharm. Bull. (Japan) **34**, 720 (1986).

257. KIMURA, Y., H. OKUDA, T. OKUDA, T. YOSHIDA, T. HATANO, and S. ARICHI: Effects of Various Tannins and Related Compounds on Adrenaline-induced Lipolysis in Fat Cells. Chem. Pharm. Bull. (Japan) 31, 2497 (1983).
258. KIMURA, Y., H. OKUDA, T. OKUDA, T. YOSHIDA, T. HATANO, and S. ARICHI: Effects of Various Tannins and Related Compounds on Adrenocorticotropic Hormone-induced Lipolysis and Insulin-induced Lipogenesis from Glucose in Fat Cells. Chem. Pharm. Bull. (Japan) 31, 2501 (1983).
259. HATANO, T., T. OKUDA, Y. FUJITA, Y. KIMURA, H. OKUDA, and S. ARICHI: Effects of Tannins Suppressing Lipid-peroxidation, and Their Effects on Fat Cells. J. Med. Pharm. Soc. Wakan-Yaku 1, 40 (1984).
260. HATANO, T., T. OKUDA, Y. FUJITA, Y. KIMURA, H. OKUDA, and S. ARICHI: Effects of Caffetannins of Arachidonate Metabolism. J. Med. Pharm. Soc. Wakan-Yaku 4, 350 (1987).
261. OKUDA, T., K. MORI, M. SHIOTA, and K. IDA: Reduction of Heavy Metal Ions and Solubilization of Precipitates. Yakugaku Zasshi 102, 735 (1982).
262. OKUDA, T., K. MORI, and M. SHIOTA: Effects of the Interaction of Tannins with Co-existing Substances, Part 3: Formation and Solubilization of Precipitates with Alkaloids. Yakugaku Zasshi 102, 854 (1982).
263. KIMURA, Y., H. OKUDA, T. OKUDA, T. HATANO, I. AGATA, and S. ARICHI: Studies on the Activities of Tannins and Related Compounds from Medicinal Plants and Drugs, VI: Inhibitory Effects of Caffeoylquinic Acids on Histamine Release from Rat Peritoneal Mast Cells. Chem. Pharm. Bull. (Japan) 33, 690 (1985).
264. BOOTH, A.N., M.S. MASRI, D.J. ROBBINS, O.H. EMERSON, F.T. JONES, and F. DEEDS: The Metabolic Fate of Gallic Acid and Related Compounds. J. Biol. Chem. 234, 3014 (1959).
265. DOYLE, B., and L.A. GRIFFITHS: The Metabolism of Ellagic Acid in the Rat. Xenobiotica 10, 247 (1980).
266. OKUDA, T., T. HATANO, M. MATSUDA, and P.B. ENGLISH: Studies on the Adsorption and the Metabolism of Tannins, I: A New Tannin and Related Compounds from Terminalia Species. Abstract Papers of the 32nd Annual Meeting of the Japanese Society of Pharmacognosy, p. 9 (1985).

(Received January 3, 1994)

Some Aspects of Guanidine Secondary Metabolites

ROBERTO GOMES DE SOUZA BERLINCK,
Departamento de Química e Física Molecular,
Instituto de Química de São Carlos,
Universidade de São Paulo, CP 369, 13560-970,
São Carlos, SP, Brasil

In memoriam Prof. Toshio Goto

Contents

1. Introduction

Guanidine natural products have been the subject of several reviews (*1–12*). More recently, MORI (*8, 9*) has discussed many aspects of natural guanidine derivatives, CHEVOLOT (*10*) has focused on guanidine secondary metabolites isolated from marine organisms, and MARESCAU (*11*) and also ROBIN and MARESCAU (*12*) have discussed the biological role of both primary and secondary guanidine metabolites. The selection of topics is justifiable: a wide variety of guanidine compounds has been isolated from living organisms and many of them have been synthesized. Apart from their academic interest, some of these compounds have direct implications in public health (for example the well-known tetrodotoxin and saxitoxin), in agriculture (blasticidin S and peramine), and in medicinal chemistry (the bleomycins and the argiopin-like spider toxins are some examples). Also, since guanidine biosynthesis is directly linked with arginine metabolism, many guanidine secondary metabolites are of interest in connection with studies of arginine degradation and transformation. Guanidines of synthetic origin have presented much interest as drugs (*13, 13a*), mainly as antihistamines (*14, 15*), as antihypertensives (*16*) and as potent sweeteners (*17*).

As a consequence of the basicity of guanidine, its derivatives are very polar compounds and this requires unusual techniques for isolation and chromatographic detection (*18*). The basicity of guanidine also confers pronounced biological activities. In this review, some of the more important chemical and physico-chemical features of guanidine natural products will be discussed. These will include spectroscopic properties, isolation methods, synthesis, biosynthesis and in some cases, biological activities. Only secondary metabolites will be considered including simple arginine derivatives directly involved in secondary metabolism as, for example, in bufotoxins and in some peptide derivatives. The immense literature dealing with such compounds as tetrodotoxin, saxitoxin, the bleomycins, streptomycin and related guanidine derivatives is largely beyond the scope of this review.

2. Physico-Chemical Properties of Guanidine

Guanidine (**1**) is the imine of urea. It is one of the strongest organic bases known (pKa = 13.5) (*19, 21*). PAULING (*19*) related its high basicity to resonance stabilisation of the guanidinium ion by about 8 kcal/mole as compared with guanidine and stated that N,N,N',N'-tetraalkylguanidines ought to be less strongly basic. Nevertheless mono- or poly-substituted guanidines are bases as strong as guanidine itself (*20, 21*). The pKa of *N,N*-dialkylguanidines has also been the subject of a small dispute many years ago (*21, 22*).

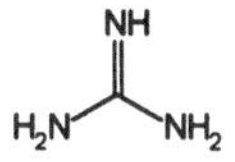

(1)

pKa measurements of guanidine derivatives have been utilised for identification of this functional group, since values above 12 are not usual for other organic functionalities. However, some guanidine derivatives such as leonurine (**2**) (*23*), tetrodotoxin (**3**) (*24*), monobromophakellin (**4**), dibromophakellin (**5**) (*25*) and saxitoxin (**6**) (*26*) have pKa values either unusually low or not accurately determinable.

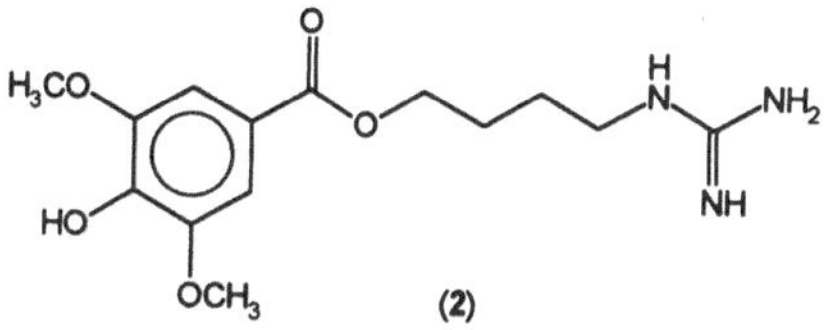

(2)

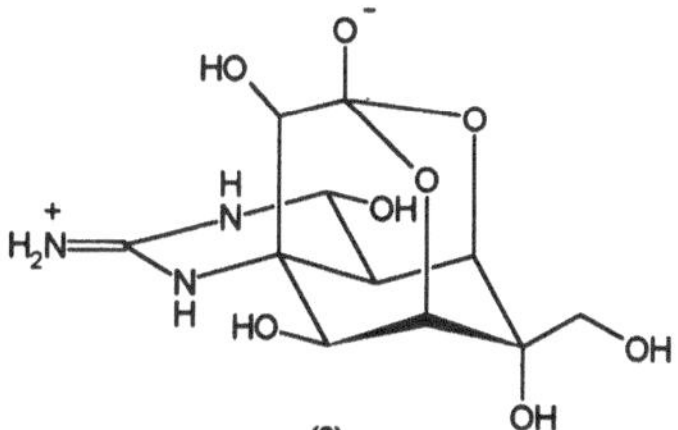

(3)

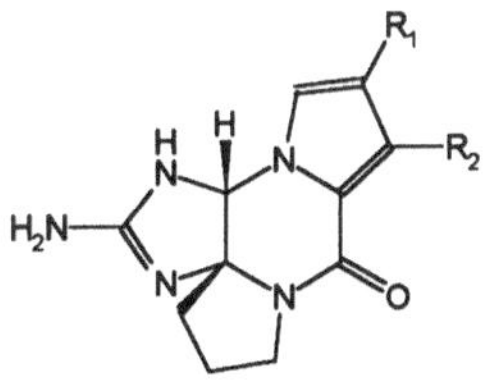

(4): R₁=H, R₂=Br
(5): R₁=R₂=Br

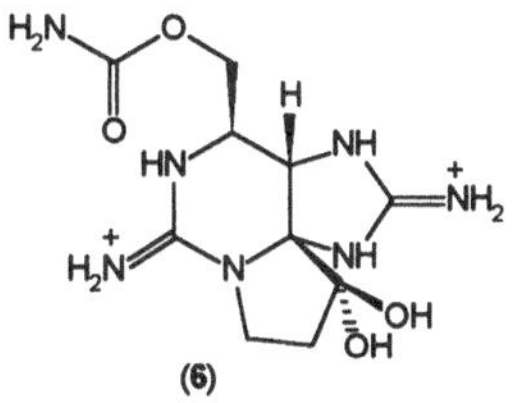

(6)

In the infrared region (*27–31*) guanidine and its derivatives show a strong C=N stretching band between 1550 and 1690 cm^{-1} (*31*), a region where absorption of other functional groups occurs such as the C=O stretching band of amides, ureas and carbamates, the C=N band of imines and oximes and the C=C band of enols and enamines (*32*).

In the ultraviolet absorption region (*33–36*) simple non-conjugated guanidine compounds exhibit a weak absorption at λ_{max} 265 nm ($\varepsilon = 15$, H_2O) and a stronger absorption below λ_{max} 220 nm ($5600 < \varepsilon < 39$).

Guanidines react with ninhydrin in strongly alkaline medium (pH > 10) to give green or blue fluorescent compounds (excitation peaks at nearly 305 nm and 390 nm; emission spectra with maxima at 495 nm) depending on the type of substitution (37, 38). Apparently, although highly sensitive, this method has never been used for the detection of a guanidine function in natural products.

Mass spectrometry of guanidine compounds has received some attention (39–45). A pioneer study (39) of guanidine and ten of its derivatives under electron impact interprets fragmentation patterns in terms of three ring intermediates, alkyl migration and decompositions involving expulsion of N$^{..}$ with release of kinetic energy. Also, the large number of metastable transitions indicates that the fragmentation processes involved proceed along many competing pathways. The presence of an abundant fragment ion of m/z 43 $(CN_2H_3)^+$ was described and the loss of a m/z 43 ion fragment is a characteristic feature of monosubstituted guanidine derivatives (40–42). Moreover, the utilisation of new mass spectrometry techniques such as field desorption and fast atom bombardment has much enhanced the detection of quasi-molecular ions of polar compounds. Some examples are studies dealing with the bleomycin family of antibiotics (41–43) and some antifungal guanidines (44, 45). Derivatisation of a monosubstituted guanidine with pentane-2,4-dione is widely employed; detection of a fragment ion at m/z 123 and its higher homologues, shows the presence of a 2-alkylamino-4,6-dimethylpyrimidyl ring, which confirms the presence of the monoalkylsubstituted guanidine function (46–49). Acetylguanidine derivatives also give informative mass spectra (41, 42, 50, 51).

Among ^{1}H-NMR studies on guanidine (52–56) that of CORRAL et al. (56) demonstrates that the chemical shift and the signal multiplicity of the proton on the nitrogen which bears an alkyl substituent can be used to identify the substitution pattern in various substituted guanidines.

More recently, ^{13}C-NMR spectroscopy has become a powerful method for identification of organic functional groups (57) but few systematic reports have appeared dealing with the analysis and the behaviour of the guanidine carbon chemical shift (30, 58–60). For example, the carbon atom of guanidine salts has been shown (58, 59) to be shielded when compared with the corresponding free bases and the guanidine carbon in the conjugated acylimino form (7) is deshielded compared with the acylamino form (8) of acylated guanidines (30, 60). By contrast, the guanidine carbon of aryl conjugated guanidines seems to be shielded (58, 59). However, the unambiguous assignment of the guanidine carbon signal is still difficult in complex compounds. When there are protons attached to the adjacent carbons, useful alternatives are ^{1}H–^{13}C

(7)   (8)

long range correlation NMR techniques such as COLOC (64), HMBC (65) or LR-HETCOR (66). Otherwise, chemical evidence is required.

The positive response to the SAKAGUCHI reagent (62, 63 and references therein) together with the synthesis of 2-amino-4,6-dimethylpyrimidyl derivatives mentioned above confirms the presence of a monosubstituted guanidine derivative. However, these methods are not useful for poly-substituted guanidines.

3. Guanidine Secondary Metabolites from Microorganisms

3.1. Streptothricins and Related Antibiotics

Microorganisms and particularly Actinomycetes comprise the main group of living organisms which produce guanidine secondary metab-olites. Actinomycetes probably have a complex arginine metabolism but few studies have been made to clarify these biosynthetic pathways.

The streptothricin family of antibiotics was isolated from Streptomy-cetes in the early 40's (67), and has shown antibacterial activity against a wide variety of gram-positive and gram-negative bacteria. Streptothricin F was isolated from *Streptomyces lavendulae* strains by various methods (68–70), and purified by successive crystallizations of different salts (68–72). Improved separation methods were early employed for the separation of the streptothricin mixtures, such as paper chromatography (114, 115), Sephadex LH-20 (116), carboxymethylcellulose (117), among others (118, 119). The structure of streptothricin F (9, n = 1) was estab-lished by extensive chemical degradation and analysis of spectral data (73–88) and BYCROFT and KING (89) determined the structure of the amino acid streptolidine (10) by X-ray analysis of its dihydrochloride.

Streptothricin-like antibiotics possess the general structure (9) and differ in the number of β-lysine residues [streptolin, for example, has two β-lysine residues (85)], the position of the carbamate function, the substituents on the lactam nitrogen and on the nitrogen involved in the amide linkage. In the case of streptothricin F the only remaining problem was the position of the carbamate, earlier postulated by VAN TAMELEN et al. (85) as attached to the 4-hydroxyl group of α-D-gulosamine. This

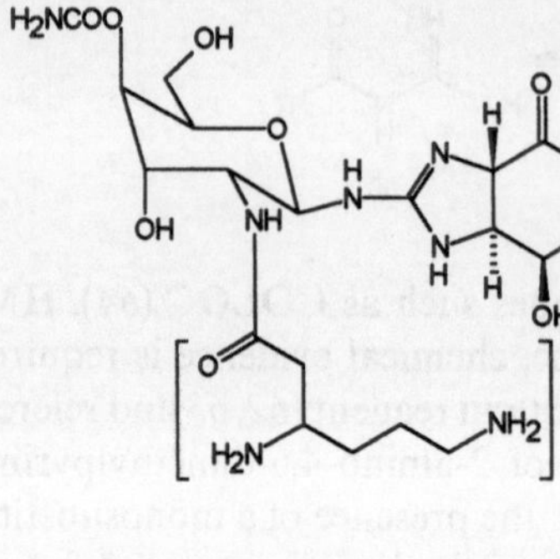

(9)

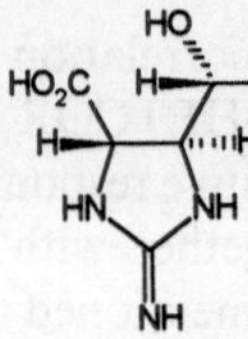

(10)

could be established only in 1981 by ^{1}H-NMR analyses of streptothricin F (9, n = 1) and the synthetic models (11) and (12) (91) and by ^{13}C-NMR analysis of the related antibiotics, LL-AC541 (13) and LL-AB664 (14) (90).

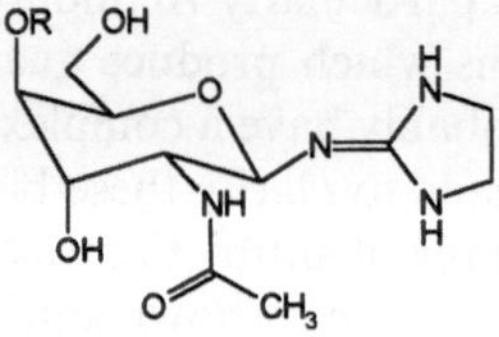

(11): R=H
(12): R=CONH$_2$

In addition to these last two compounds (90, 92) other streptothricin-like antibiotics isolated from *Streptomyces* are: glycinothricin (15), which is the deformimino derivative of LL-AB664 (93), N-methylstreptothricin F (16) (94), albothricin (17), which is the 4-desoxy analog of streptothricin (95), two inseparable isomers, A-269A (18) and A-269A' (19), which differ from streptothricin in the amino acid substituent on α-D-gulose and from each other in the position of carbamate function (96), and finally, a series of Nβ-acetylated derivatives of streptothricins E, D and F, named AN-201 I (20), AN-201 II (21) and AN-201 III (22), from *Streptomyces nojiriensis* (97). All these compounds were isolated from Streptomycetes and have a *trans* junction between the 2-aminoimidazolyl and the δ-lactam rings. They possess high antibacterial activity, but are nephrotoxic (77), and hence unavailable for clinical use.

References, pp. 248–295

(13): R_1=OCONH$_2$, R_2=OH, R_3=CH$_3$, R_4=OH, R_5=CH$_3$, R_6=CH$_2$NH(C=NH)H
(14): R_1=OCONH$_2$, R_2=OH, R_3=H, R_4=OH, R_5=CH$_3$, R_6=CH$_2$NH(C=NH)H
(15): R_1=OH, R_2=OCONH$_2$, R_3=CH$_3$, R_4=OH, R_5=CH$_3$, R_6=CH$_2$NH$_2$
(16): R_1=OCONH$_2$, R_2=OH, R_3=CH$_3$, R_4=β-OH, R_5=H, R_6=β-lysyl
(17): R_1=OCONH$_2$, R_2=OH, R_3=CH$_3$, R_4=R_5=H, R_6=β-lysyl
(18): R_1=OCONH$_2$, R_2=OH, R_3=H, R_4=β-OH, R_5=CH$_3$, R_6=CH$_2$NHCH$_3$
(19): R_1=OH, R_2=OCONH$_2$, R_3=H, R_4=β-OH, R_5=CH$_3$, R_6=CH$_2$NHCH$_3$
(20): R_1=OCONH$_2$, R_2=OH, R_3=H, R_4=OH, R_5=H, R_6=CH$_2$CH(NHAc)-(CH$_2$)$_3$-β-lysyl
(21): R_1=OCONH$_2$, R_2=OH, R_3=H, R_4=OH, R_5=H, R_6=CH$_2$CH(NHAc)-(CH$_2$)$_3$-β-lysyl-β-lysyl
(22): R_1=OCONH$_2$, R_2=OH, R_3=H, R_4=OH, R_5=H, R_6=CH$_2$CH(NHAc)-(CH$_2$)$_3$-NH$_2$

Biosynthesis of the streptolidine moiety of streptothricins was originally proposed by BYCROFT and KING (89) to follow Scheme 1 from arginine *via* dehydroarginine (23) (99) and the capreomycidin isomer (24) *via* an intramolecular rearrangement. Much subsequent work, not only on streptotricin biosynthesis, but also on viomycin (100) and capreomycin (110), has shown that in different guanidine derived antibiotics obtained from different Streptomyces strains somewhat distinct biosynthetic routes may be preferred as is depicted in Scheme 2. For example, GRÄFE et al. (98) have observed the incorporation of high levels of [U-^{14}C]arginine into streptolidine, in addition to smaller amounts of [U-^{14}C]proline and [U-^{14}C]glutamic acid. Biosynthetic studies on racemomycin D (25, n = 4) showed that L-proline, L-alanine, L-leucine and L-arginine greatly stimulate the production of this antibiotic by *Streptomyces lavendulae* (101). [U-^{14}C]arginine shows the best amino acid incorporation, and is the direct precursor of the streptolidine residue. On the other hand, [1-^{13}C]acetate is efficiently incorporated in the biosynthesis of racemomycin A (26) (102). GOULD et al. (103–110) suggest that the experiments of SAWADA et al. (102) imply the distribution of acetate into an array of intermediates interconnected by primary metabolic grids, before streptothricin biosynthesis begins. It has also been demonstrated that the D-gulosamine moiety of streptothricin F (9, n = 1) and racemomycin A (26) arises from D-glucosamine (108, 112, 113). HERBERT (111) recently reviewed these results.

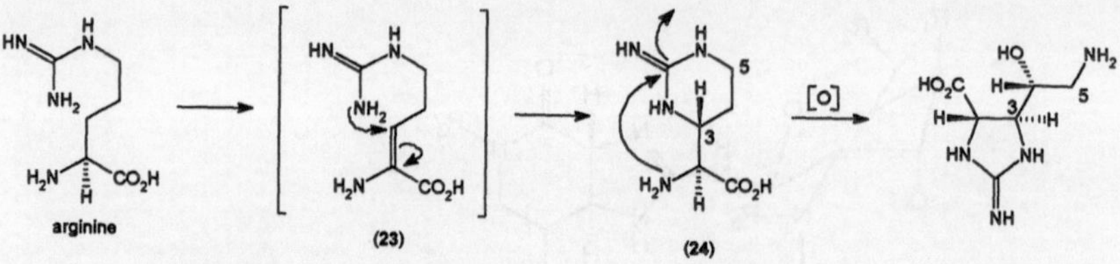

Scheme 1. Biogenetic mechanism of formation of streptolidine proposed by BYCROFT

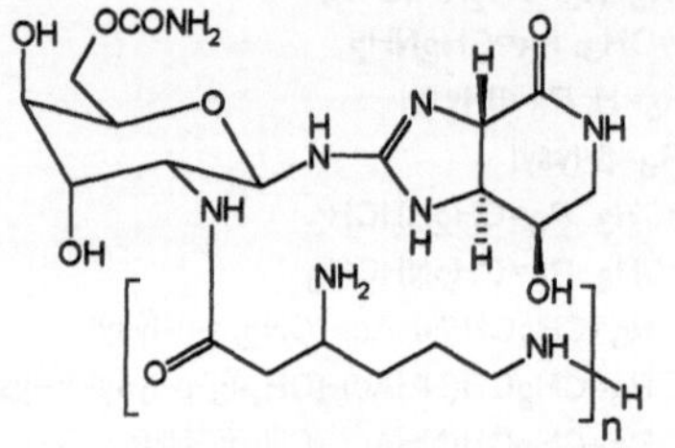

(25) 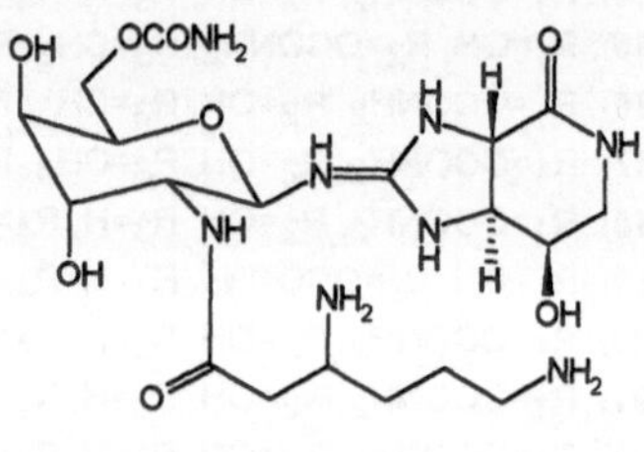   (26)

The synthesis of streptolidine (10) has been accomplished (Scheme 3) (*120*). The key intermediate, 3,4,6-triaminopentalactone (28), was prepared by stereoselective conversion of the hydroxyl groups of D-ribose to the corresponding amino groups. Treatment of (28) with 1N NaOH and cyanogen bromide, in order to protect the amino group on C-5 from guanidination gave lactam (35), with the streptolidine ring being formed selectively on amino groups 2 and 3. The bis-4-hydroxyazobenzene-4'-sulfonate dihydrate salt of streptolidine (10) was finally obtained by acid (6N HCl) hydrolysis of (35), followed by cation-exchange chromatography.

Further, in order to verify the structure proposal for streptothricin, its N^g-streptolidyl-2-amino-2-deoxy-β-D-gulopyranoside moiety was synthesized (Scheme 4) (*121*). A suitably protected precursor of the streptolidine moiety [2,5-diamino-3-(Boc-amino)-2,3,5-trideoxy-4-O-methoxyethoxymethyl-D-arabino-δ-lactam] (45), prepared through a multistep stereospecific conversion from D-xylose, was coupled with glycosyl isothiocyanate (46). The thiourea derivative (47) was subsequently subjected to a multi-step procedure, yielding the desired N^g-streptolidyl-2-amino-2-deoxy-β-D-gulopyranoside (48).

The total synthesis of streptothricin F (9, n = 1) has also been reported (Scheme 5) (*122*). Removal of the protecting glycosidic allyl group of (56) and treatment with p-nitrobenzoyl chloride in pyridine gave

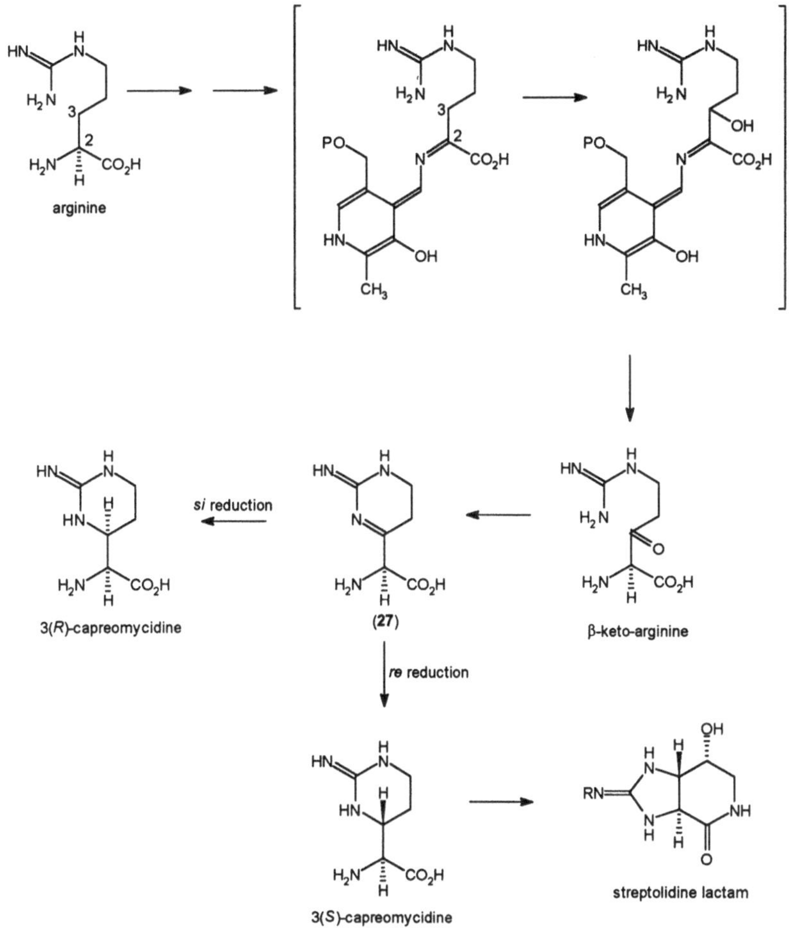

Scheme 2. Biogenetic mechanism of formation of streptolidine and capreomycidines proposed by GOULD

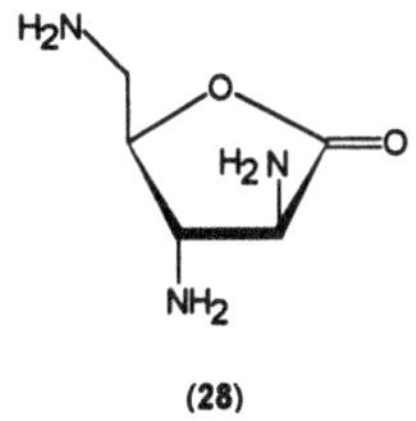

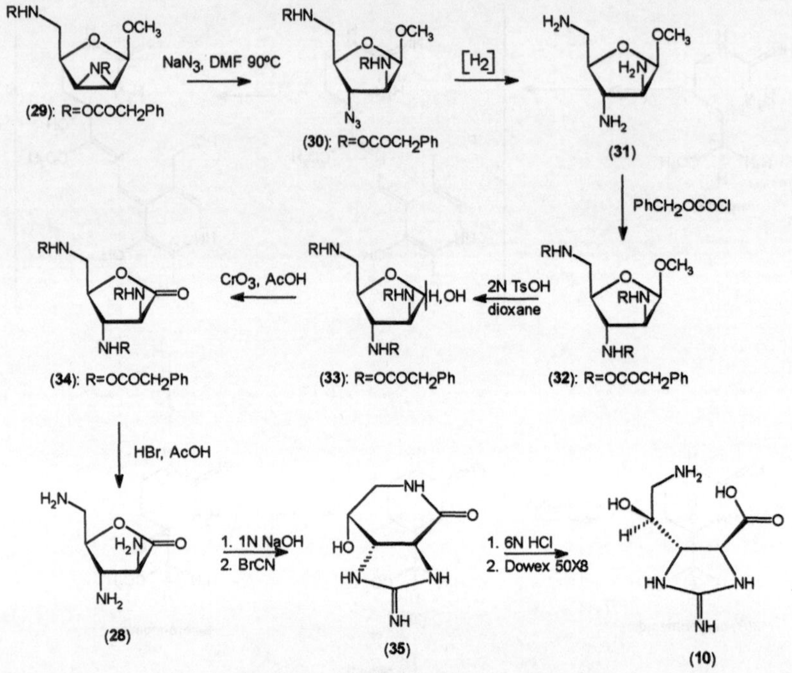

Scheme 3. Synthesis of streptolidine

(**57**), which was converted to the β-glycosyl isothiocyanate (**58**). This was coupled with the previously synthesized 2,5-diamino-3-*O*-benzyloxy-methyl-4-*t*-butoxycarbonylamino-2,3,5-trideoxy-D-arabino-1,5-lactam (*122*). Treatment of (**59**), as previously described for (**47**), yielded strepto-thricin F trihydrochloride (**9**, n = 1). Some semi-synthetic derivatives of racemomycins have also been prepared (*123*) and their biological activi-ties evaluated.

3.2. Streptomycin and Related Antibiotics

Almost simultaneously with the isolation of streptothricin from *Streptomyces lavendulae*, streptomycin was isolated from *Streptomyces griseus* (*124–128*). Streptomycin was also active against a wide variety of gram-positive and gram-negative bacteria, while its toxicity for animals was sufficiently low (*71*) to permit its utilization in the therapy of various infectious diseases.

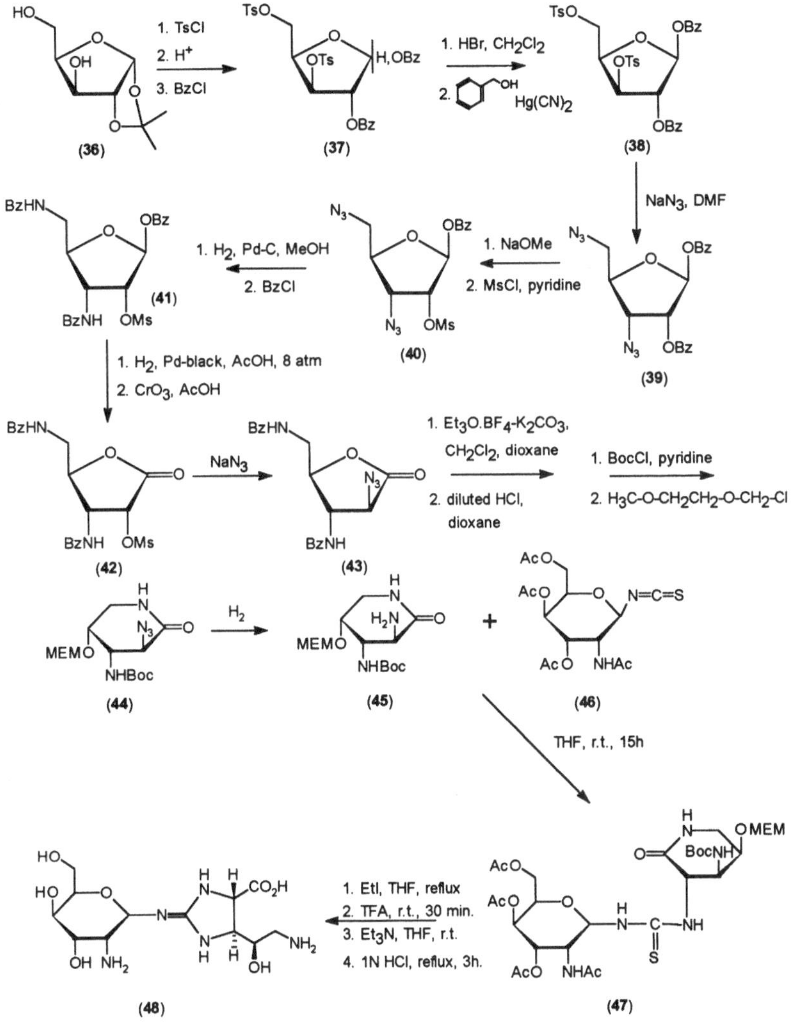

Scheme 4. Synthesis of N^g-streptolidyl-2-amino-2-deoxy-β-D-gulopyranoside moiety of streptothricin F

Streptomycin was isolated and purified by various methods (*71, 72, 129, 130, 206–208*) and its structure (**60**), as well as those of its components streptidine, streptose and N-methyl-L-glucosamine were established by a series of chemical degradation experiments (*129–156, 159, 161–170*). Moreover, the syntheses of streptomycin (**60**) and of streptidine (**61**) have

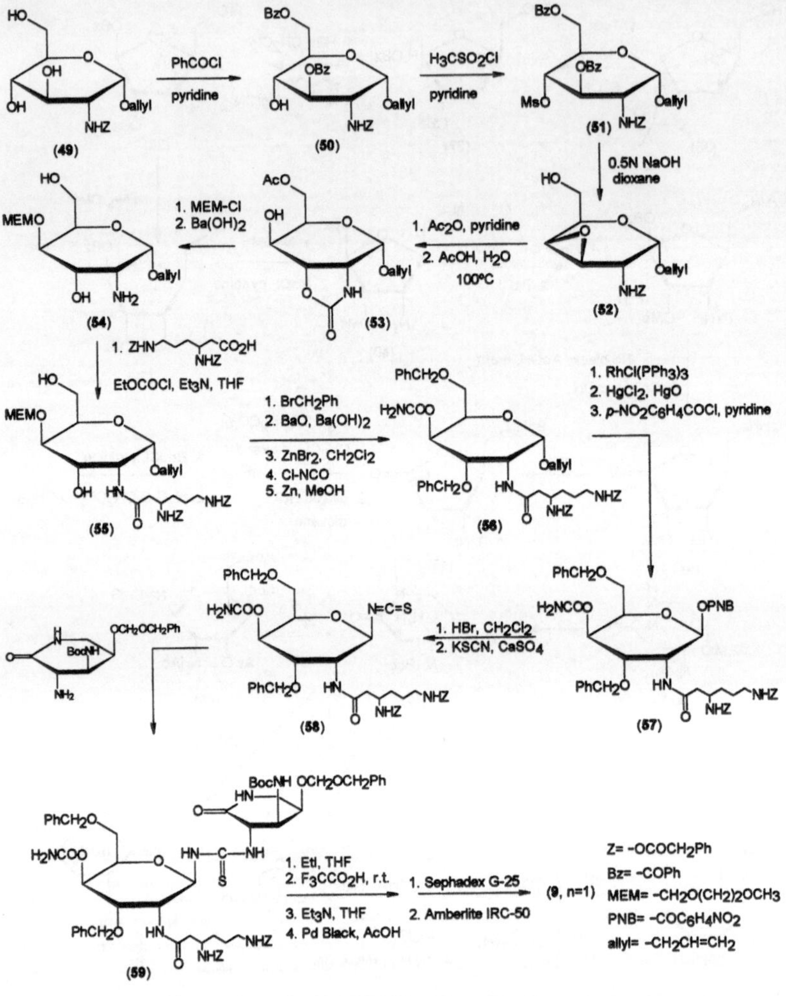

Scheme 5. Synthesis of streptothricin F

been accomplished. That of streptidine is depicted in Scheme 6 (*157, 158, 160*). The synthesis of streptomycin was achieved (*185*) from dihydro-streptomycin (DSM) synthesized earlier (*186*). Dihydrostreptomycin (DSM) was selectively benzyloxycarbonylated at the *N*-methyl group of the L-glucosamine portion, giving 2″-*N*-benzyloxycarbonyldihydrostrep-tomycin dihydrochloride (**72**). Treatment of (**72**) with 2,2-dimethoxypro-

(60)   (61)

pane under acid catalysis gave a mixture of per-O-isopropylidenated products whose reaction with 20% acetic acid in methanol yielded (73) as the only mono-isopropylidene derivative. Selective acetylation of the hydroxyl groups (Ac$_2$O, catalytic p-toluenesulfonic acid) gave (74) whose selective hydrolysis with 75% aqueous acetic acid yielded (75). Oxidation of the primary alcohol group by the Pfitzner-Moffatt procedure (DMSO, dicyclohexylcarbodiimide, TFA and pyridine) followed by deacetylation of the crude oxydation product (76) and removal of the benzyloxycarbonyl protective group by catalytic hydrogenolysis (Pd-black, aqueous AcOH) gave streptomycin (60).

Streptomycin B (77) was isolated from S. griseus extracts at an early stage of the streptomycin studies (171). It was difficult to separate it from streptomycin; however this could be achieved by counter-current distribution (172, 177) and by chromatography on alumina (173). That it was streptomycin-like was suggested early (174, 175) and subsequently confirmed (176) by chemical degradation. These results eventually led to the structure proposal as mannosidostreptomycin (77) (178).

Hydroxystreptomycin (78) is the 6-hydroxymethylstreptose derivative of streptomycin isolated from Streptomyces griseocarneus (179–181).

Bluensomycin (79), isolated from Streptomyces sp., is a streptomycin derivative in which one of the guanidine functions is replaced by a carbamate (170, 182–184).

The biosynthetic pathway leading to streptomycin has been fully elucidated (187–205). WALKER and WALKER (193) showed that the streptidine moiety of streptomycin arises from myoinositol (80) by the series of enzymatic transformations shown in Scheme 7, in which the

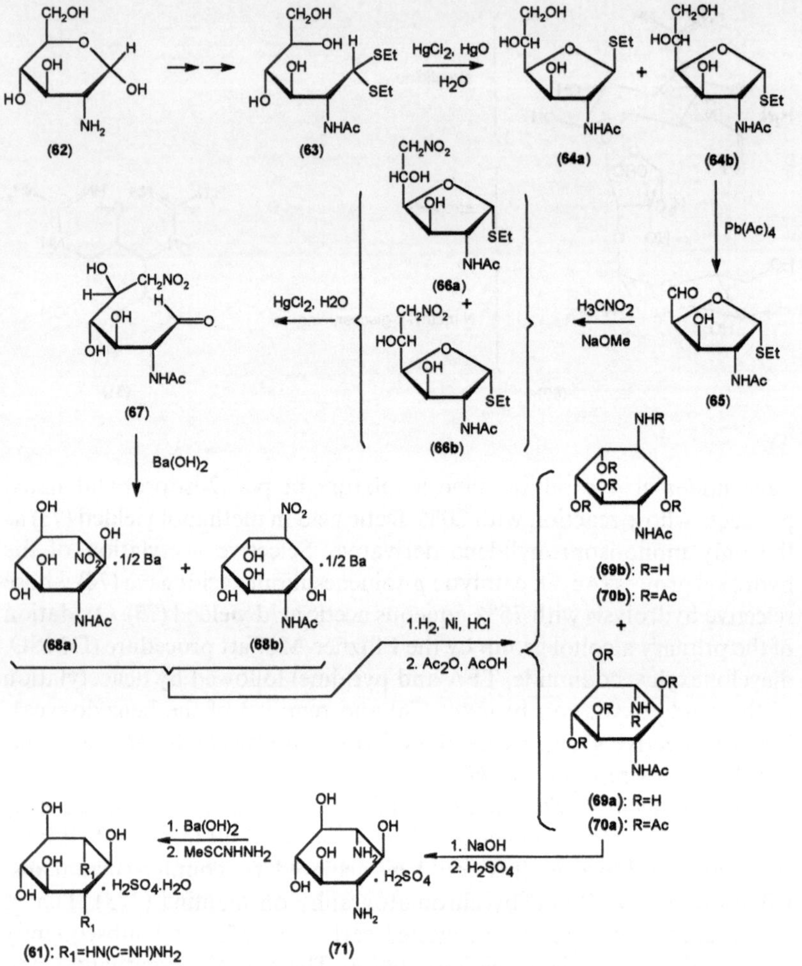

Scheme 6. Synthesis of streptidine

incorporation of guanidine functions is achieved by arginine amidine transfer mediated by transamidinase. Besides the direct incorporation of arginine or a "modified" arginine residue, this is probably the main pathway of guanidine incorporation into secondary metabolites. Streptomycin has been the subject of some excellent reviews (*209–217*) and a book (*218*).

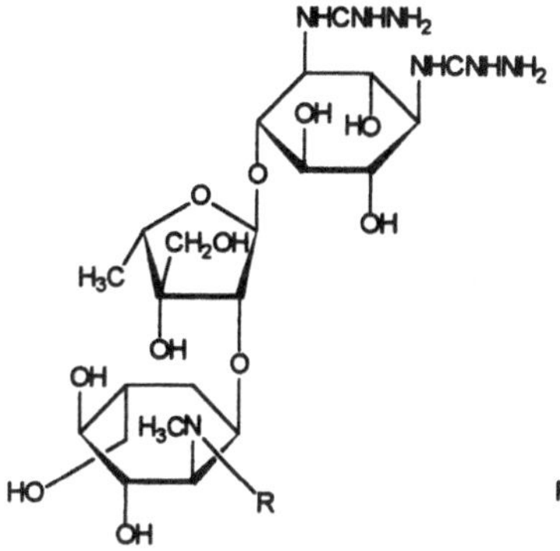

DSM R=H
(72) R=COOCH$_2$Ph

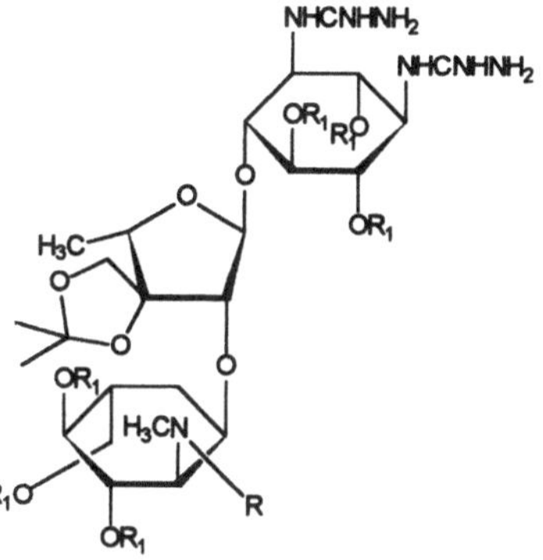

(73) R=COOCH$_2$Ph, R$_1$=H
(74) R=COOCH$_2$Ph, R$_1$=Ac

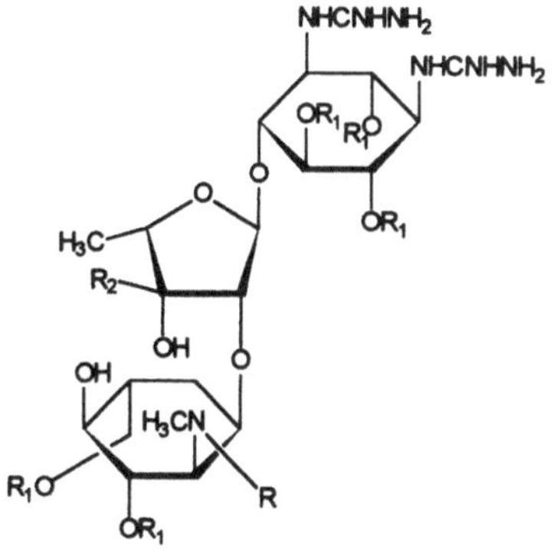

(75) R=COOCH$_2$Ph, R$_1$=Ac, R$_2$=CH$_2$OH
(76) R=COOCH$_2$Ph, R$_1$=Ac, R$_2$=CHO
(60) (streptomycin) R=R$_1$=H, R$_2$=CHO

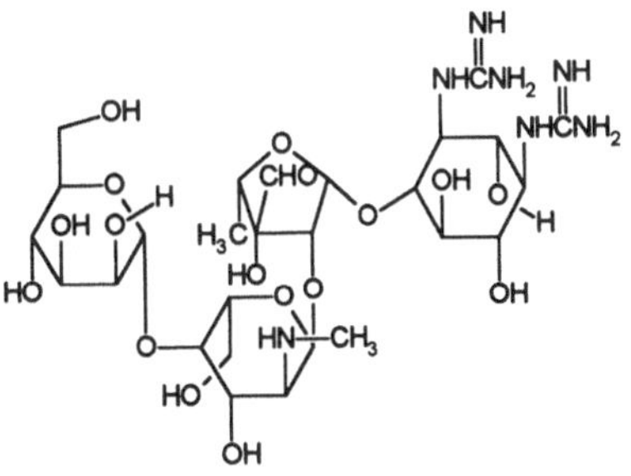

(77)

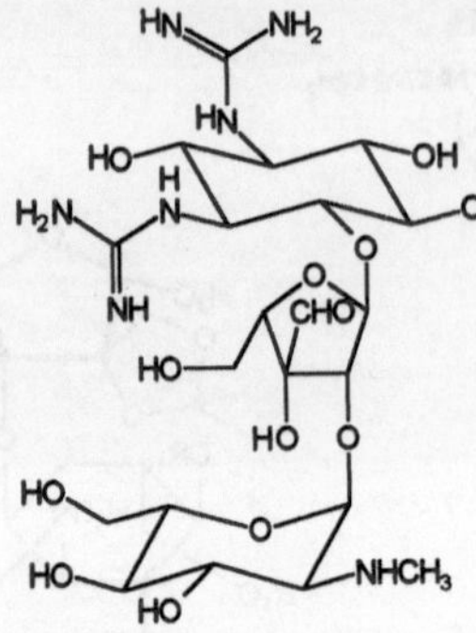

(78)

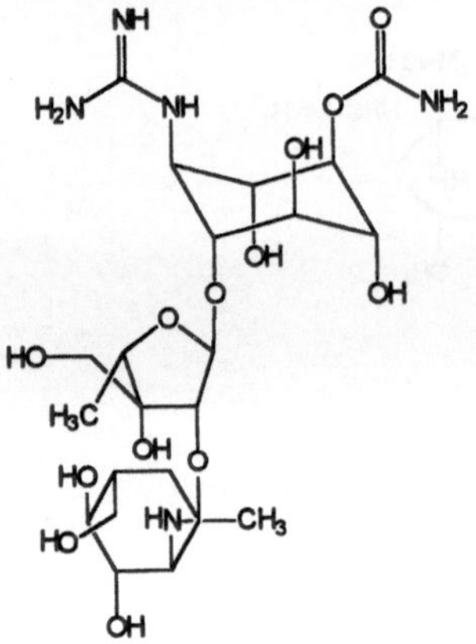

(79)

3.3. Viomycin and Related Tuberculostatic Compounds

Viomycin, isolated from *Streptomyces puniceus* and *Streptomyces floridae*, was a contemporary of streptothricin and streptomycin. Its basic polypeptide nature was soon verified (*219*) as well as its potent bacteriostatic and tuberculostatic properties (*220*). A series of attempts at structure elucidation of viomycin and its guanidine-containing amino acid viomycidine by chemical and spectral means only led to erroneous conclusions (*221–224*).

By complete and partial hydrolysis of viomycin, KITAGAWA *et al.* (*225*) proposed an incorrect amino acid sequence β-lysyl-seryl-L-α,β-

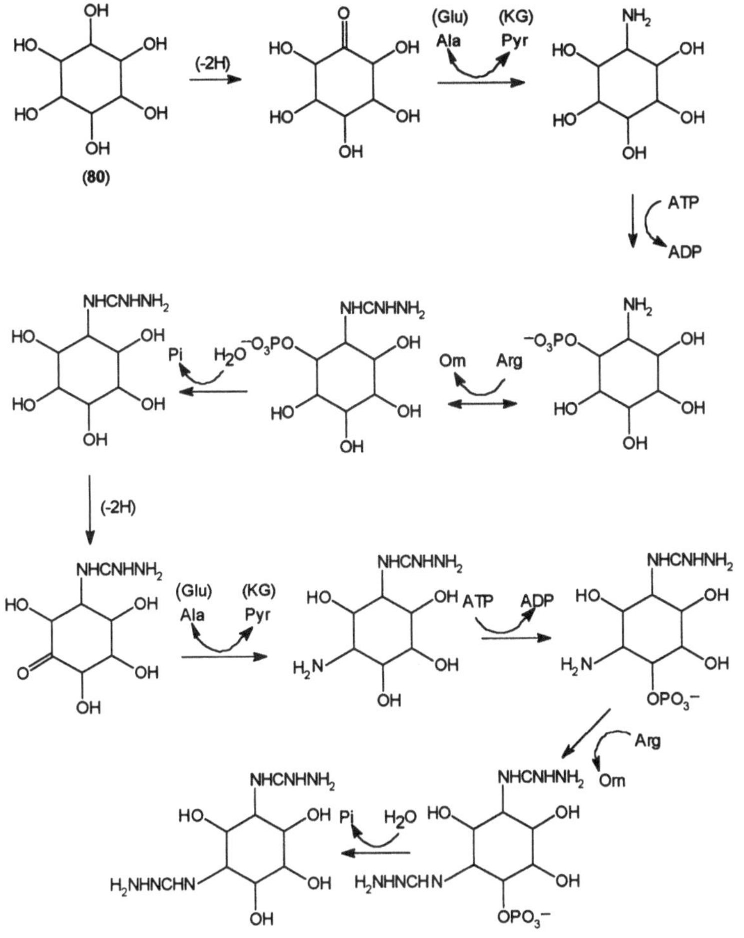

Scheme 7. Biosynthesis of streptidine from *myo*-inositol

diaminopropionyl-(-β-NH-)-viomycidyl-seryl-urea, where the β-amino group of the diaminopropionyl residue is involved in the peptide bond. BYCROFT et al. (*226*) demonstrated that a urea function was involved in the structure of the viomycin chromophore and that the guanidine function was part of a tetrahydropyrimidine ring. Hydrogenation of viomycin followed by total acid hydrolysis gave α-(2-iminohexahydro-4-pyrimidyl)glycine (**24**), one of capreomycidine stereoisomers leading to the correct formula for viomycidine (**81**). BYCROFT and coworkers proposed correctly that viomycidine is present in viomycin in its open form as in **(82)**, justifying their proposal on the basis that viomycin hydrochloride readily forms an *O*-methyl derivative on heating with methanol, and concluded that viomycidine was an artifact formed by reaction of the guanidine-carbinol system with the amino group of the glycine fragment. The isolation and X-ray structure of a substance named viocidic acid **(83)** from the hydrolysates of viomycin confirmed these proposals *(226, 233)*. The structures of viomycidine **(81)** and that of viocidic acid **(83)** were confirmed by others *(227–230, 234)*. BYCROFT *et al. (231)* subsequently proposed the correct structure **(84)** for the viomycin chromophore, but a wrong isomeric structure for viomycin *(235)*. The structure of the chromophore was later confirmed by synthesis *(232)*.

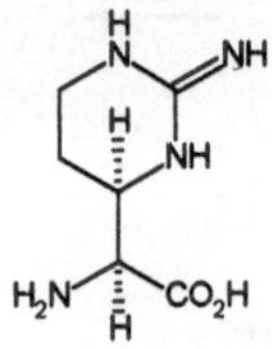

(24)

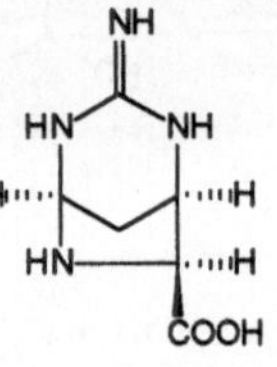

(81)

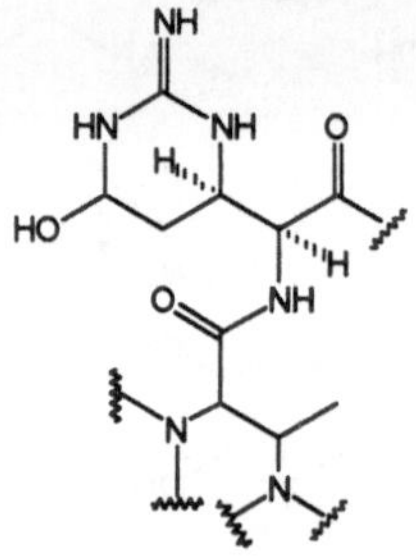

(82)

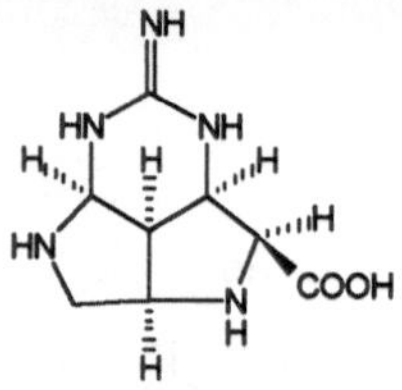

(83)

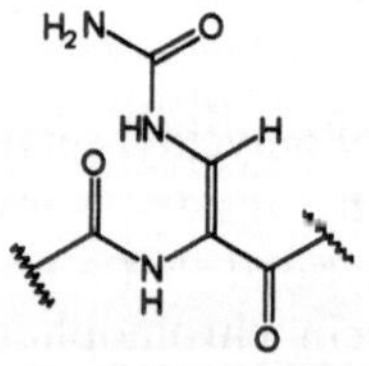

(84)

In a review of the chemistry of viomycidine, BYCROFT and collaborators *(237, 238)* attributed its positive response to the SAKAGUCHI and FEHLING reagents to the fact that its cyclic guanidino-carbinol moiety exists in equilibrium with the acyclic guanidine aldehyde tautomer (Scheme 8). On the basis of an X-ray analysis of tuberactinomycin O, YOSHIOKA *et al.* (*236*) came to the conclusion that viomycin was identical with tuberactinomycin B (**85**). This was confirmed by BYCROFT *et al.* (*239*) and by KITAGAWA and coworkers (*240*). Chemical modifications of viomycin have been carried out (*241–248*) in order to establish relationships between structure and biological activity. Although viomycin had potent tuberculostatic properties, its high toxicity has prevented its clinical use.

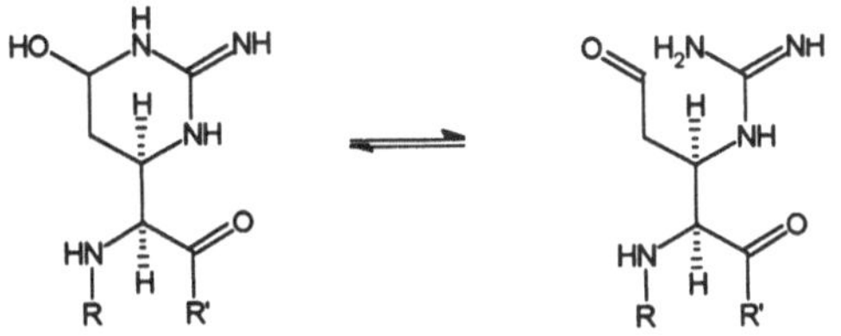

Scheme 8. Equilibrium between the open and the cyclized tautomeric forms of tuberactidine

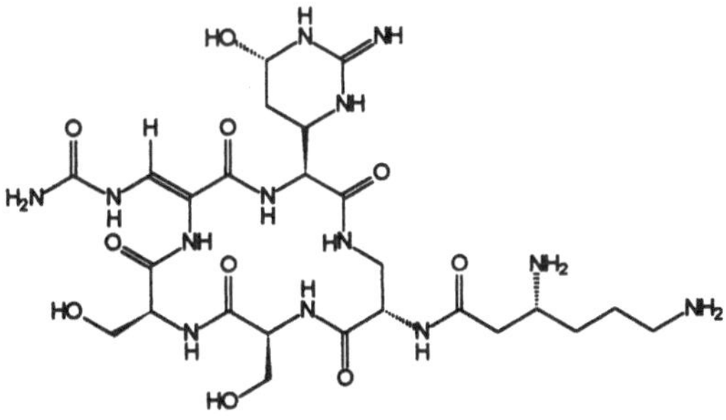

(**85**)

Biosynthetic experiments performed with viomycin (*100*) showed a high degree of [1-^{14}C]arginine, [1-^{14}C]ornithine and [*amidine*-^{14}C]arginine incorporation into viomycidine. Moreover, lysine appears to be the direct precursor of the β-lysine residue of viomycin, and it was suggested that the unusual amino acids (α,β-diaminopropionic acid, β-lysine and 6-hydroxy-capreomycidin) are formed prior to incorporation into the polypeptide.

Two other families of antibiotics already mentioned in connection with streptothricin and viomycin chemistry are the capreomycins from

Streptomyces capreolus and the tuberactinomycins from *Streptomyces griseoverticillatus* var. *tuberacticus*. Capreomycin is a mixture of different polypeptides named capreomycins IA, IB, IIA and IIB, which have two amino acids in common, α-(2-imino-hexahydro-4-pyrimidyl)glycine (capreomycidine) and α,β-diamino propionic acid, and different contents of serine, alanine and β-lysine (*249, 250*). Capreomycins IA and IIA contain serine, whereas capreomycins IB and IIB contain alanine. Capreomycins IIA and IIB do not contain β-lysine and all of them contain the characteristic chromophore found in viomycin (*251*). The structure of the basic amino acid, named capreomycidine, was tentatively proposed as (**86**) or (**24**), which differ from each other in the stereochemistry at the β-carbon (*252*).

(**86**)

Structure (**24**) was assigned to the stereoisomer of capreomycidine present in capreomycin after synthesis of the racemates of each diastereomer (Scheme 9) (*252, 253*). The absolute configuration of the capreomycidine (**24**) stereoisomer was established by chemical correlation with viomycidine (*252*). In a later publication (*254*) total synthesis of the capreomycidine stereoisomer (**24**) was reported confirming the absolute stereochemistry (Scheme 10) (*254*).

Structures (**106–109**) initially proposed (*255*) for capreomycins IB, IA, IIA and IIB were shown to be incorrect in the case of capreomycins IA and IB by Shiba *et al.* (*256*) who isolated N^α-2,4-dinitrophenyl-α,β-diaminopropionic acid from the hydrolysates of capreomycin rather than N^β-2,4-dinitrophenyl-α,β-diaminopropionic acid. They therefore inferred that β-lysine was linked to the β-amino group of diaminopropionic acid in both capreomycins IA and IB and proposed formulae (**110**) and (**111**), respectively. Further confirmation has come from detailed ^{1}H-NMR analyses of capreomycin and other polypeptides belonging to this class (*257*), as well as further degradations (*258*) and synthesis (*259*).

It has been verified recently that during the biosynthesis of capreomycin IA (**110**) the 2,3-diaminopropionate residue derives from serine *via* a dehydroalanyl intermediate. Moreover, it was observed that labeled

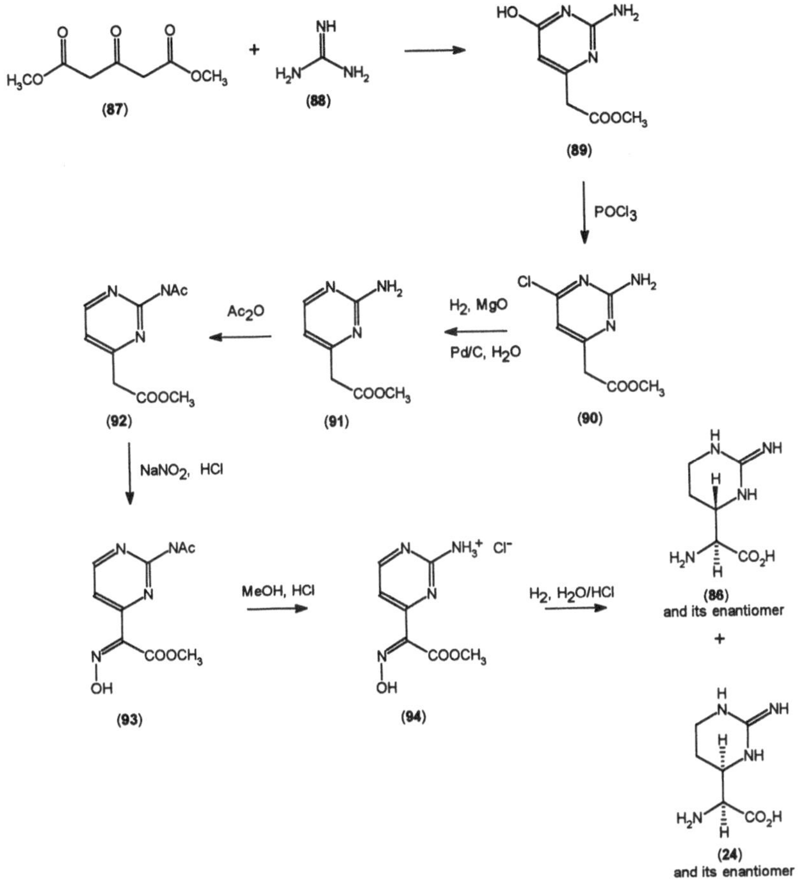

Scheme 9. Synthesis of racemic capreomycidines

serine is efficiently incorporated into the serine residue of capreomycin IA (110) and labeled alanine into the alanine residue of capreomycin IB (111), discarding the possibility of interconversion between the two poly-peptides (260).

The second related family of polypeptides comprises the tuberacti-nomycins (261–263). After hydrolysis and separation by ion-exchange chromatography, L-serine, L-α,β-diaminopropionic acid, a guanidine amino acid and α-hydroxy-L-β-lysine were obtained (264). The guanidine derivative was degraded with 15% HBr at 50 °C to give viomycidine (81) (identified by X-ray analysis) and tuberactidine. Tuberactidine exhibited

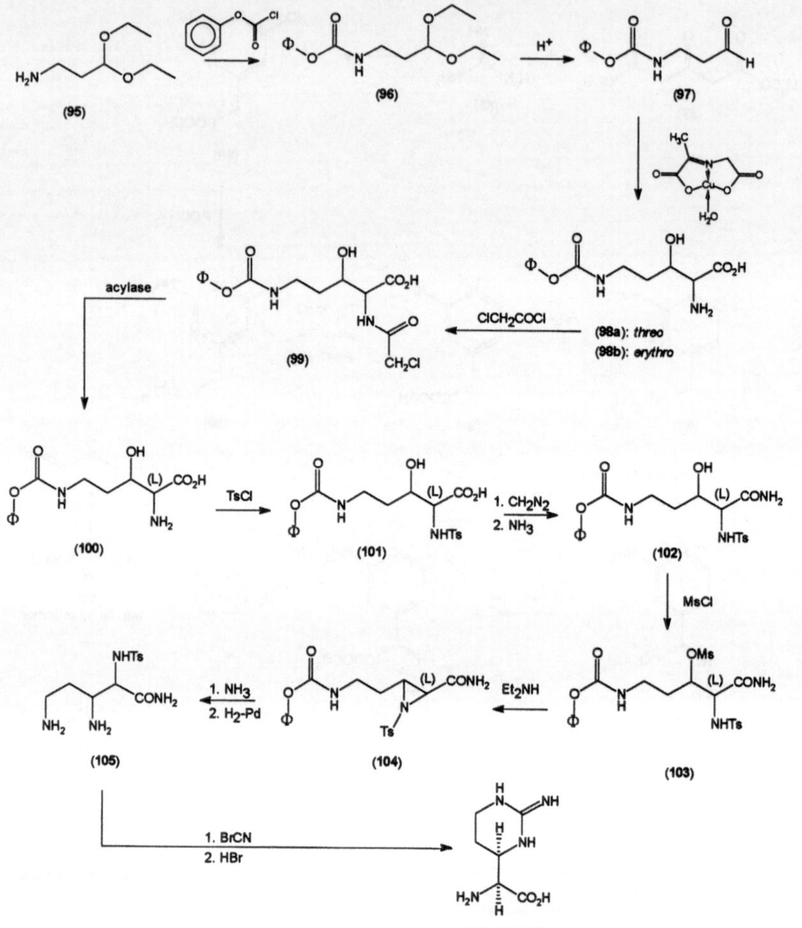

Scheme 10. Synthesis of (2*S*),(3*R*)-capreomycidine

a positive Cotton effect at 221 nm ([Φ]$_{+807}$ pK, 0.1 N HCl), characteristic of L-amino acids; its structure was determined as (4*S*,6*R*,7*S*) α-(2-
imino-4-hydroxyhexahydro-6-pyrimidyl)glycine (**112**) by ¹H-NMR spectroscopy. Tuberactidine is, in fact, the open form of viomycidine (**81**), itself
in equilibrium between the cyclol form and the guanidino-aldehyde forms
shown in Scheme 8. This could be verified by isolation of (**113**) after
sodium borohydride reduction of tuberactinomycin, followed by acid
hydrolysis (*264*).

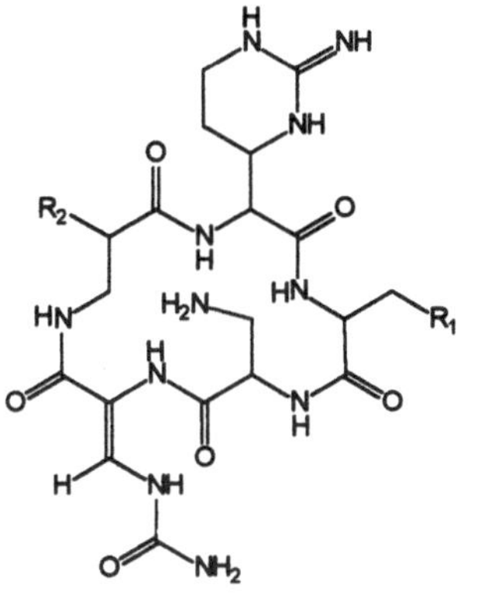

(106) R₁=H, R₂=β-lysine
(107) R₁=OH, R₂=β-lysine
(108) R₁=R₂=H
(109) R₁=OH, R₂=H

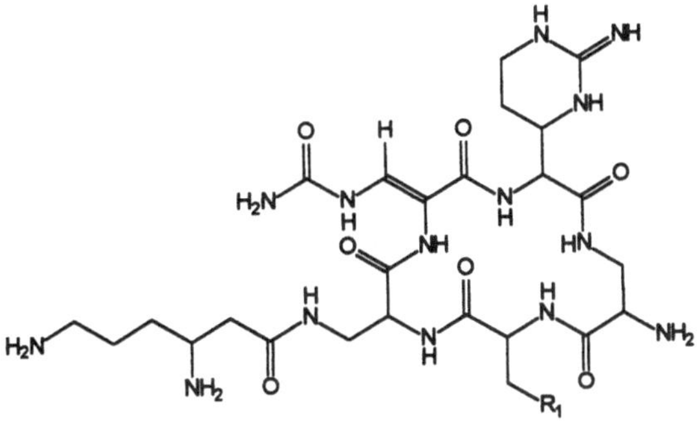

(110) R₁=OH
(111) R₁=H

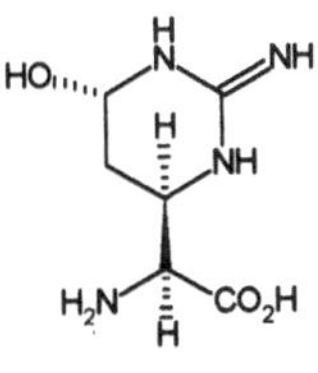

(112)   (113)

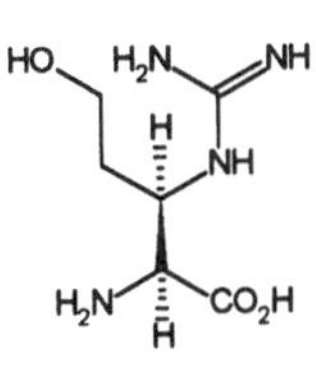

Detailed study of the guanidine amino acids and the other constituents of the tuberactinomycins confirmed these conclusions (265).
Interestingly, a dimer of tuberactidine was isolated after hydrolysis of
tuberactinomycin A. On the basis of pKa measurements this substance
(114) did not contain a free amino group, while ^{1}H-NMR and IR spectra
were indicative of the orthoformyl and ester (or lactone) functions,
respectively. Curiously, (114) is interconvertible with the dihydrobromide
of (115) which was also isolated. As stated, this equilibrium is closely
related to the lactone-hemilactal equilibrium in tetrodotoxin (265).

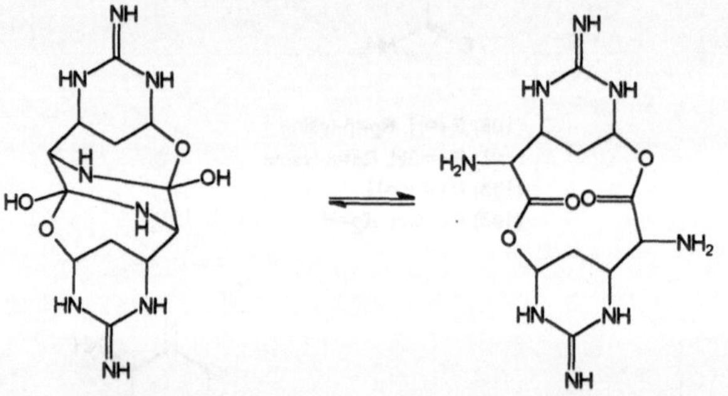

(114)   (115)

Structures of the tuberactinomycins A, B (viomycin), N and O were
unambiguously determined by X-ray analysis (236) and by chemical
means (266) as (116), (85), (117) and (118), respectively. Studies dealing
with their synthesis (267) and chemical modifications (268–270) have also
appeared.

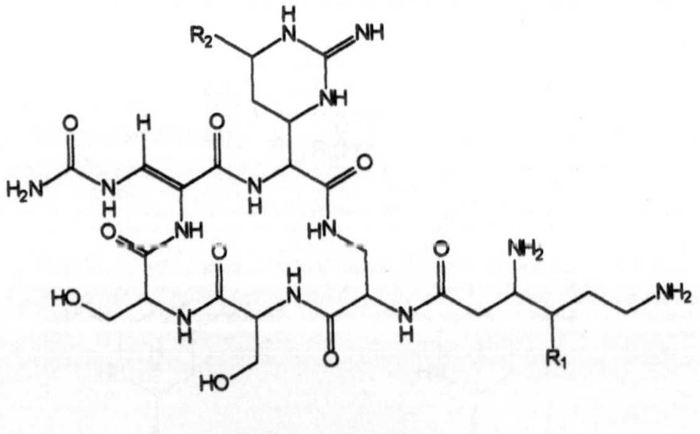

(116) R$_1$=OH, R$_2$=OH
(117) R$_1$=OH, R$_2$=H
(118) R$_1$=H, R$_2$=H

Other such tuberculostatic compounds, LL-BM547α (**119**) and LL-BM547β (**120**), were isolated from a *Nocardia* strain (*271*). They were identified by chemical degradation (acid hydrolysis) and ^{13}C-NMR analysis.

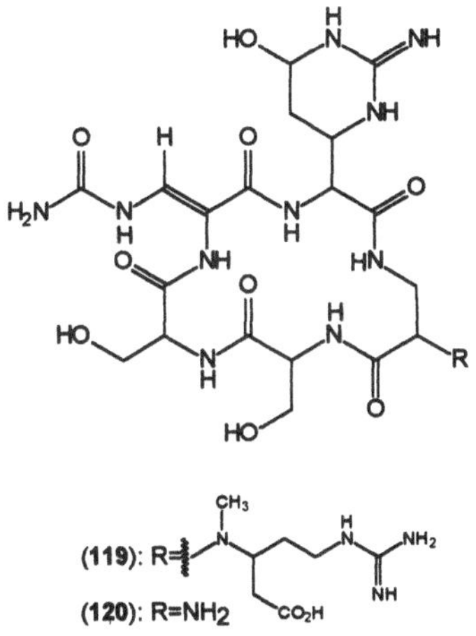

(**119**): R=

(**120**): R=NH$_2$

PAUNCZ (*272*) has developed a TLC technique using cation-exchange resin coated plates for the separation of basic water-soluble antibiotics. Viomycin, streptamine, capreomycidine, streptidin, streptomycin, mannosidohydroxystreptomycin and deoxystreptamine were efficiently separated by this method.

3.4. Blasticidin S and Related Compounds

Blasticidin S is an antifungal agent effective against rice blast disease (promoted by *Piricularia oryzae*), which has been isolated from *Streptomyces griseochromogenes* (*273*). Chemical studies by OTAKE, YONEHARA and coworkers (*274–278*) led to isolation of the nucleoside moiety cytosinine (**121**, as subsequently formulated) and a guanidine-containing residue, blastidic acid (**124**). Ozonolysis of *N,N'*-diacetylcytosinine methyl ester (**122**) followed by oxidation yielded *erythro*-β-hydroxyaspartic acid (**123**) whose stereochemistry followed from its diazotization to give *meso*-tartaric acid. Hence the gross structure of blasticidin S was (**125**).

Early confusion about the inference to be drawn from the isolation of *erythro*-β-hydroxy-L-aspartic acid (**123**) concerning cytosinine's stereochemistry at C-4' and C-5' which, as pointed out by FOX and WATANABE

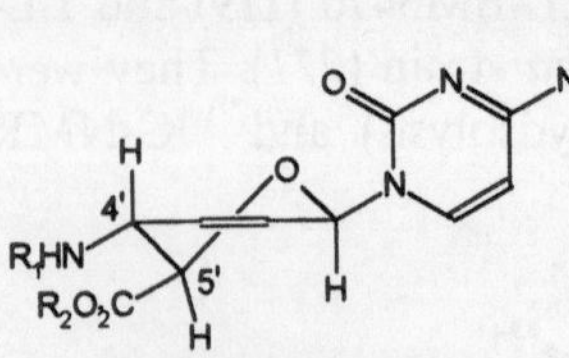

(121): $R_1=R_2=R_3=H$
(122): $R_1=R_3=COCH_3$, $R_2=CH_3$

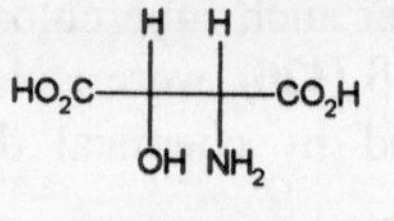

(123)

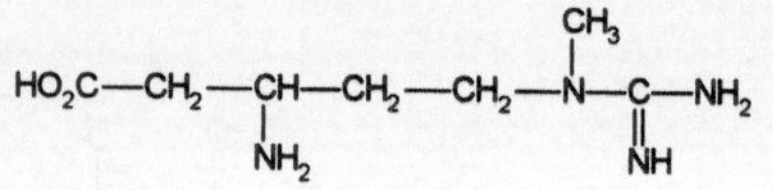

(124)

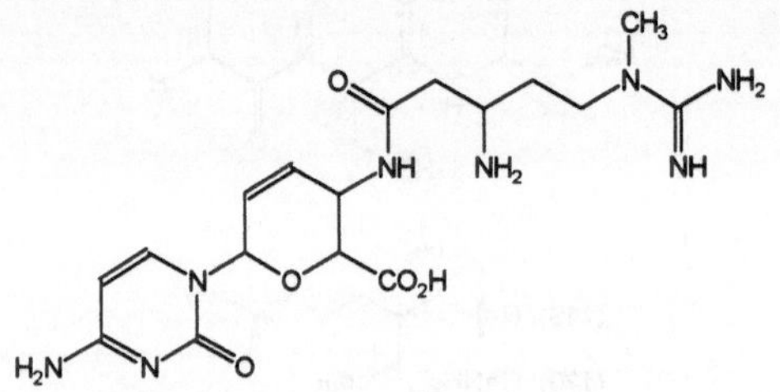

(125)

(280), should be (C-4′ S, C-5′ S) rather than (C-4′ R, C-5′ R) as stated originally (276) was cleared up (281) leading to formula (121) for cytosinine and (126) for blasticidin S, the stereochemistry at the glycosidic linkage being based on ORD measurements. Structure (126) has been confirmed by an X-ray analysis (279).

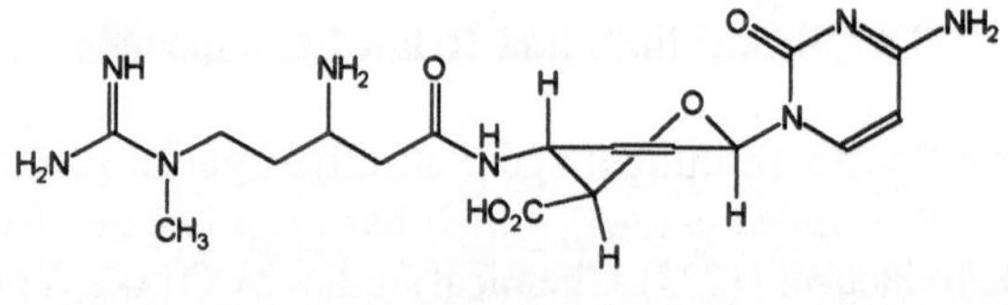

(126)

Other antibiotics related to blasticidin S have been isolated and identified: mildiomycin (127, R = OH) and mildiomycin D (127, R = H) isolated from *Streptoverticillium rimofaciens* ($282–286$), miharamycin A (128, R = OH) and miharamycin B (128, R = H) isolated from *Streptomyces miharaensis*, which were also active against rice blast disease (287), bagougeramines A (129) and B (130) from *Bacillus circulans* bacterial strains which contain guanidino-D-alanine as the amino acid chain

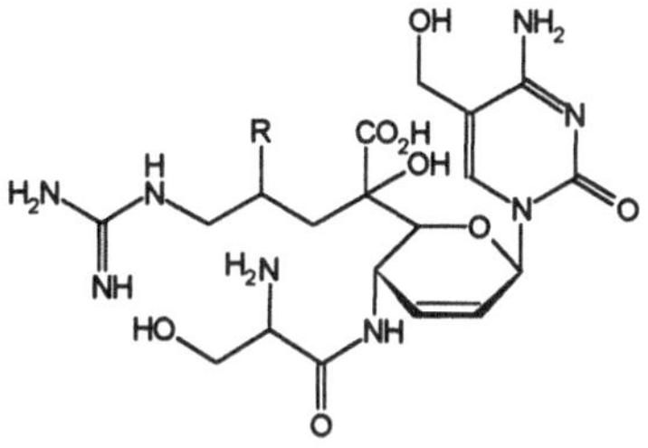

(127)

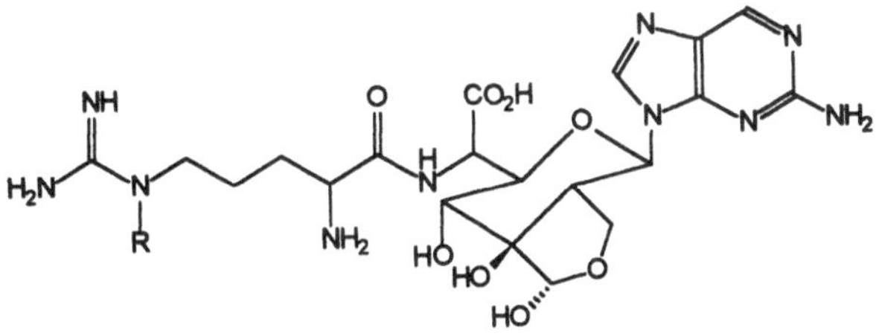

(128)

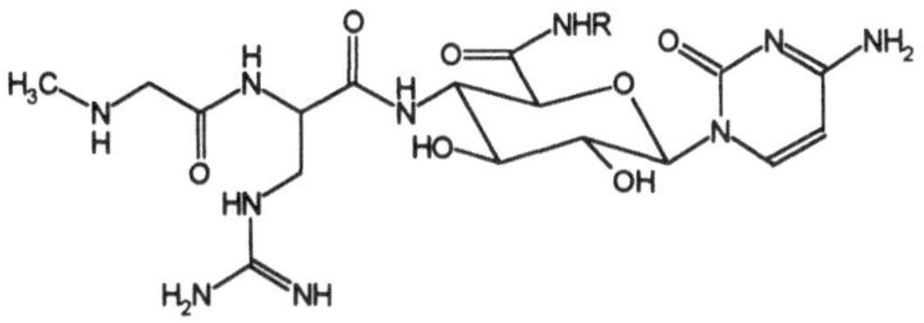

(129) R=H
(130) R=CH₂CH₂CH₂NHCH₂CH₂CH₂CH₂NH₂

(*288, 289*), arginomycin (**131**), produced by *Streptomyces arginensis*, in which a 3,6-dimethylarginine residue is linked to the nucleoside moiety (*290*), and Sch 36605 (**132**), an anti-inflammatory nucleoside derivative isolated from an unidentified strain of Streptomycete (*291, 292*). Finally, blasticidin H (**133**) and demethylblasticidin S (**134**) represent intermediates in blasticidin S biosynthesis (*293, 294*), which were also isolated from *Streptomyces griseochromogenes*.

Blasticidin S showed a high degree of incorporation of [2-^{14}C]cytosine, [U-^{14}C]cytidine, [*methyl*-^{14}C]-L-methionine, [*amidine*-^{14}C]-L-arginine and [U-^{14}C]-L-arginine, and good incorporation of [U-^{14}C]-D-glucose, [1-^{14}C]-D-glucose and [6-^{14}C]-D-glucose during its biosynthesis (*295, 296*). Arginine is the direct precursor of the L-β-arginine

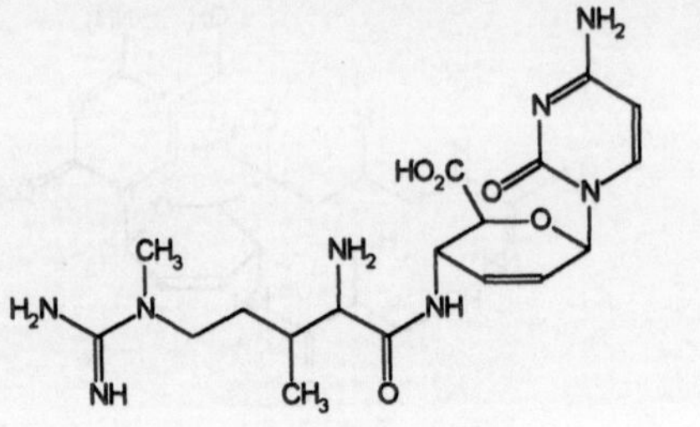

(131)

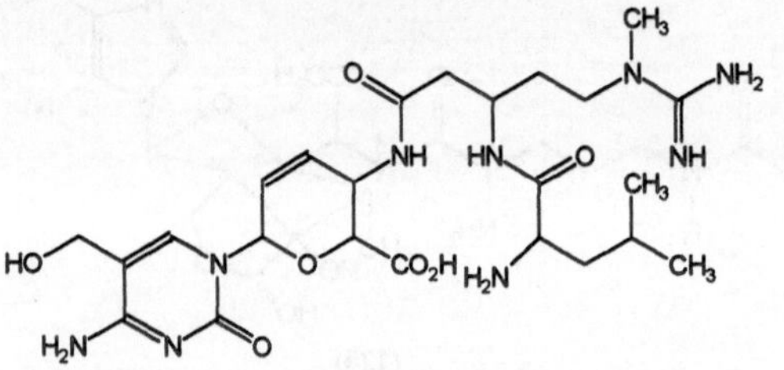

(132)

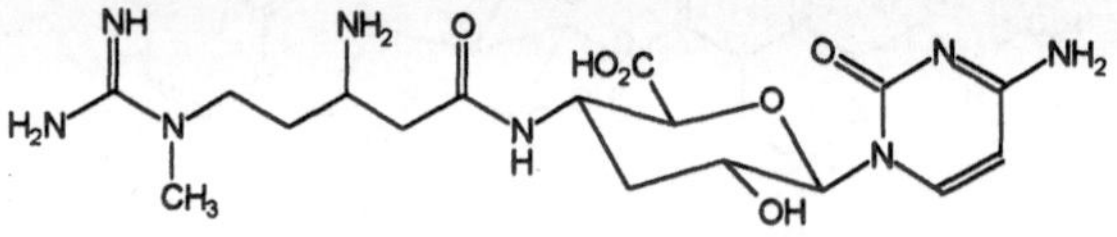

(133)

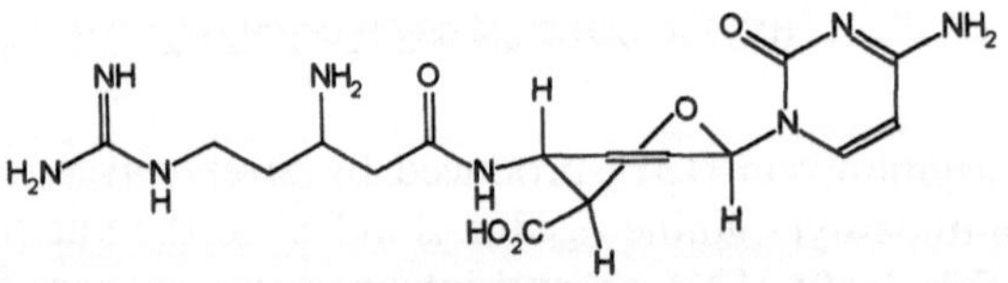

(134)

residue (297). The nucleoside moiety in blasticidin S derives from cytosylglucuronic acid (CGA) (298), and it was recently established (299) that N-guanyl methylation of blastidic acid is the last step in the biosynthesis of blasticidin S. HERBERT (111) has reviewed the biogenetic pathway of blasticidin S.

Several syntheses of the nucleoside moiety of blasticidin S have appeared (300–303).

3.5. Netropsin (Congocidine)

Netropsin (also called congocidine or T-1384) was isolated from *Streptomyces netropsis*. FINLAY *et al.* (*304*) proposed an incorrect molecular formula which was eventually revised to $C_{18}H_{26}N_{10}O_3$ by COSAR *et al.* (*305*) and by JULIA and PRÉAU-JOSEPH (*306*). In addition to the presence of two UV bands at 236 and 296 nm, netropsin gave positive responses to the SAKAGUCHI and EHRLICH tests. Alkaline hydrolysis of netropsin gave two monobasic salts, $C_{15}H_{20}N_6O_3$, which was responsible for the UV absorptions, and $C_3H_5N_3O$. COSAR *et al.* (*305*) suggested that the latter was glycocyamidine, but VAN TAMELEN and co-workers (*307*) have isolated a related compound $C_3H_7N_3O_2$ which was shown to be guanidinoacetic acid. Hydrolysis of the C_{15} base gave two moles of a pyrrole derivative (*308*). On the basis of these results structure (136) was proposed for the C_{15} base and the incorrect structures (137) and (138) were proposed for netropsin and its deaminated product, respectively. In order to confirm these proposals, the netropsin degradation products (135), (136), and (138) were synthesized (*309*).

VAN TAMELEN and POWELL (*310*) agreed with the proposed structure of netropsin (137) but JULIA and PRÉAU-JOSEPH (*311, 312*) revised the

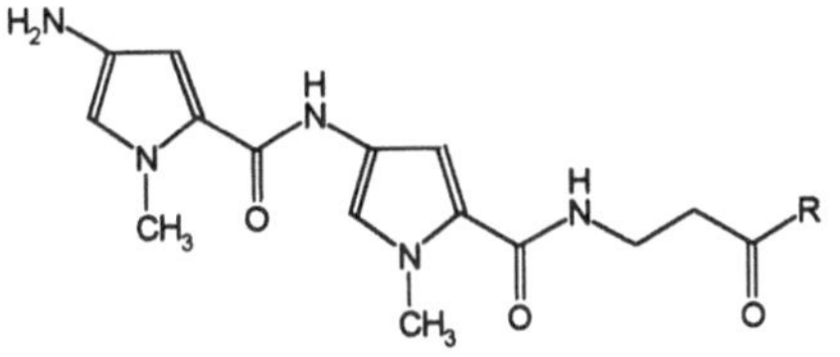

(135) R=OH
(136) R=NH₂

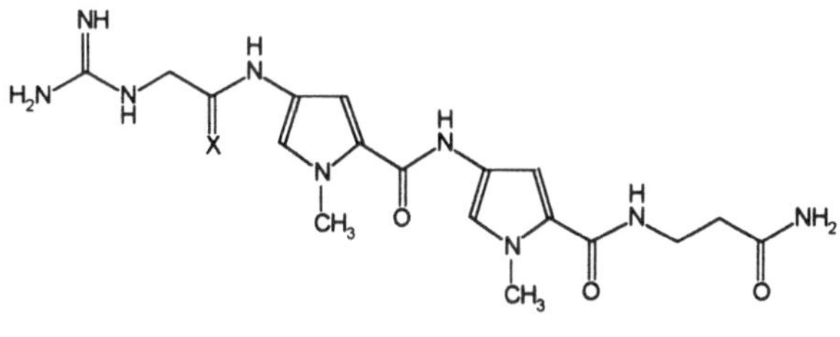

(137): X=NH
(138): X=O

structure to (139), confirming and explaining the formation of glycocyamidine during alkaline hydrolysis of netropsin by the mechanism
outlined in Scheme 11. The proposed formula (139) was verified by total

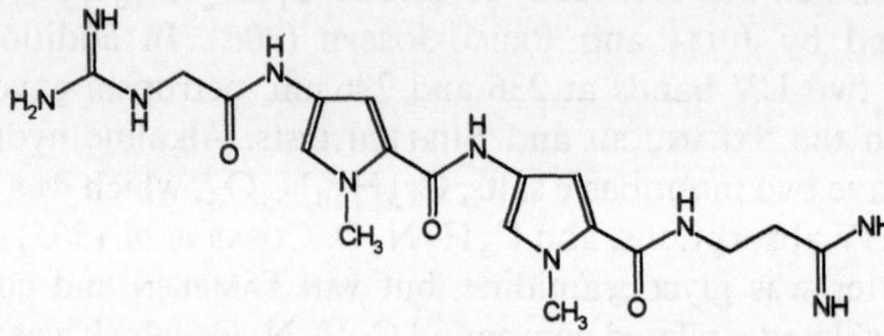

(139)

synthesis (Scheme 12). NAKAMURA *et al.* (*313*) independently arrived at the
same conclusions. Netropsin has antibacterial and trypanocidal activities
(*304*) and has also been utilised as a probe in studies of DNA drug
interactions (*314–316*).

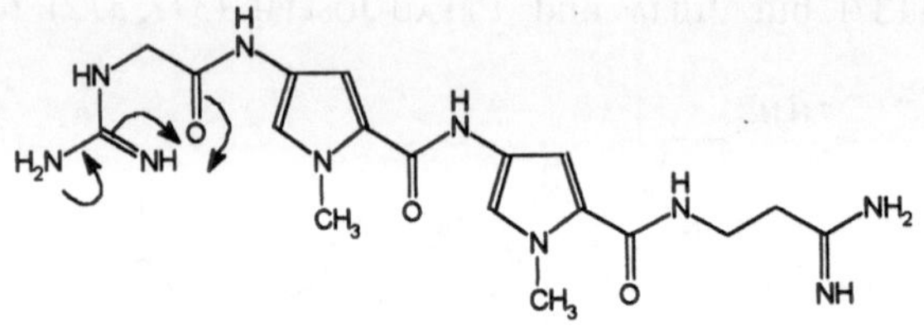

Scheme 11. Mechanism of glycocyamidine elimination from netropsin in alkaline medium

3.6. Macrocyclic Lactone Antibiotics

A variety of macrocyclic lactone antibiotics possessing an alkyl
guanidine side chain has been isolated from various actinomycete strains.
The first member of this group of guanidine derivatives was primycin
(144), isolated from cultures *Streptomyces primycini* originating from the
intestinal tract and the faeces of wax moth larvae (*Galeria melonella*)
(*317*). The structure of primycin was established by chemical degradation
to the secoprimycins (*318, 319*), mass spectrometrical analysis of these
(*320*) and the degradation of permethylated primycin (*321*). Primycin
showed strong activity against various *Candida albicans* and *Staphylococcus aureus* strains (*322*).

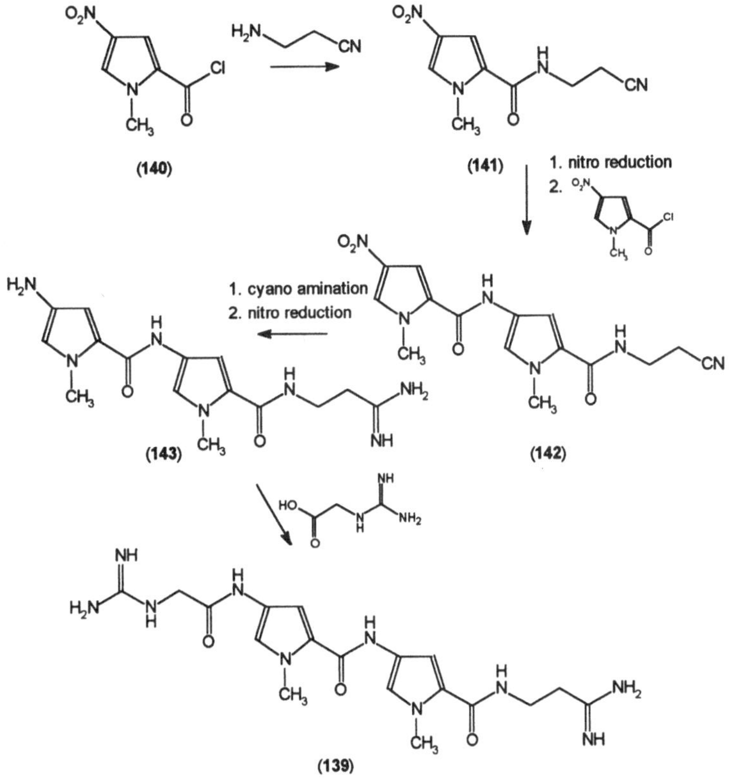

Scheme 12. Synthesis of netropsin

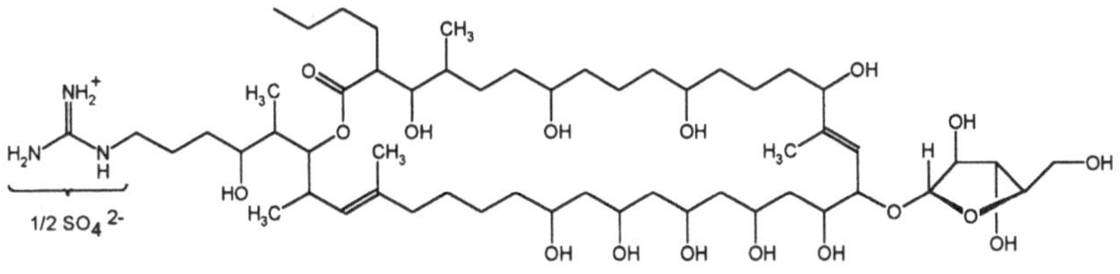

(144)

The azalomycins, active against various fungi, bacteria and yeasts, were isolated from *Streptomyces hygroscopicus* var. *azalomyceticus* culture filtrates (*323, 324*) and characterized by UV and IR spectroscopy and by chemical reactions (*325*). The structure of the first of these, azalomycin F_{4a} (**145**), was established much more recently (*326–329*) by a complete spectroscopic study which included FAB mass spectra, field desorption mass spectra, UV, IR, ^{13}C-NMR and ^{1}H-NMR spectroscopy (*326*) and

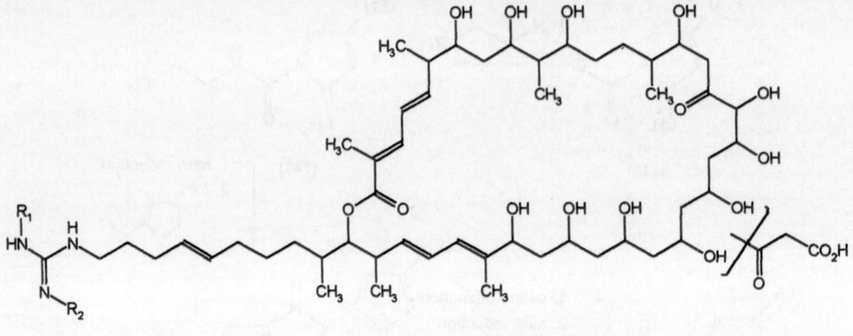

(145): R₁=CH₃, R₂=H
(146): R₁=R₂=H
(147): R₁=R₂=CH₃

also by chemical degradation (*327, 328*). Azalomycins F_{3a} (**146**) and F_{5a} (**147**) are the unmethylated and *N,N'*-dimethyl guanidine analogues of azalomycin F_{4a} (*329*).

Similarly the related scopafungin, isolated from *Streptomyces hygroscopicus* var. *enhygrus* var. nova and under the name niphimycin from *S. hygroscopicus*, var. B-255, was identified as (**148**) (*50, 330, 331, 336–338*).

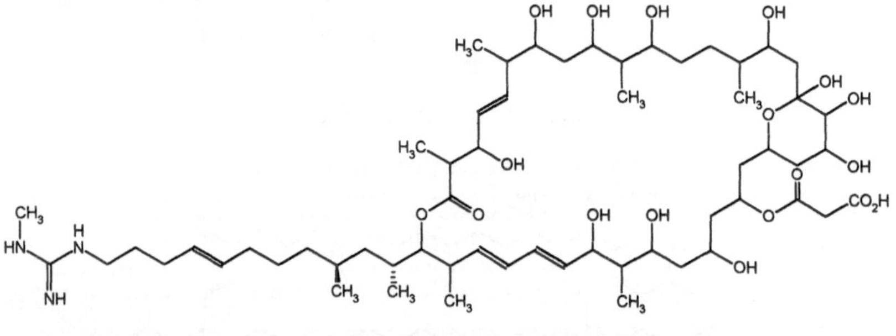

(148)

Copiamycin (**149**), neocopiamycin (**150**) and the demalonyl analogue of copiamycin (**151**) were isolated from *Streptomyces hygroscopicus* var. *cristallogenes* (*332–335*). Neocopiamycin (**150**) has stronger antibacterial and antifungal activity but lower toxicity than copiamycin (**149**). The demalonyl derivative of copiamycin (**151**) is more active against bacteria than neocopiamycin, but has a comparable antifungal spectrum. The position of the malonyl group in niphimycin (**148**) and in copiamycin (**149**) was unambiguously established by ^{1}H-NMR experiments (*338*).

(149): R₁=H, R₂=CO-CH₂-CO₂H, R₃=CH₃
(150): R₁=H, R₂=CO-CH₂-CO₂H or R₁=CO-CH₂-CO₂H, R₂=H , R₃=H
(151): R₁=R₂=H, R₃=CH₃

Further members of this class of compounds are the guanidylfungins A (**152**) and B (**153**) from *Streptomyces hygroscopicus* No. 662 (*339, 340*), guanidolide A (**154**), from *S. hygroscopicus* var. *crystallogenes* (*341*), RP 63834 (**155**), the largest member of this class of compounds, from an unidentified species of streptomycete (*342*), MBA 028-24A (**156**) and MBA 028-24B (**157**), also from an unidentified species of streptomycete (*343*) and malolactomycin A (**158**) and malolactomycin B (**159**) from *Streptomyces* sp. 83–634 (*344*). Structures were determined by extensive analysis of spectral data, mainly various 2D-NMR and mass spectral techniques.

Studies on the biosynthesis of azalomycin F₄ₐ (*345*) and guanidylfungins (*340*) have shown that all methyl substituents on the carbon skeleton arise from the incorporation of propionate units. The origin of the guanidine function, as well as its methyl substituents, has not been elucidated. Some biologically active derivatives of these macrocyclic lactones have been synthezised (*346–349*).

(152): R=CH₃
(153): R=H

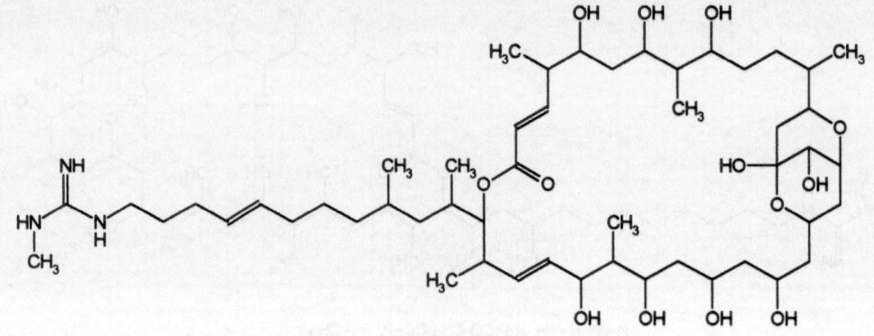

(154)

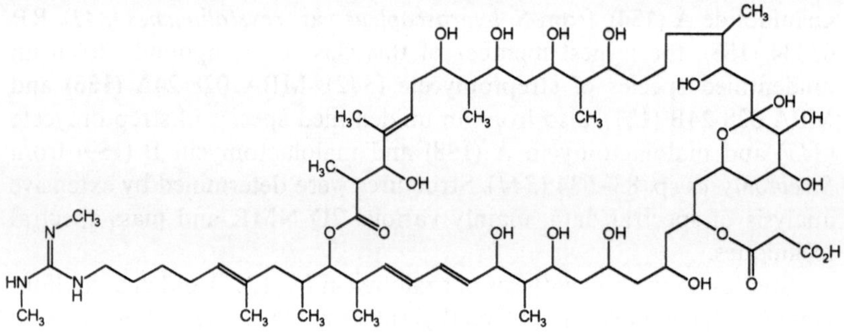

(155)

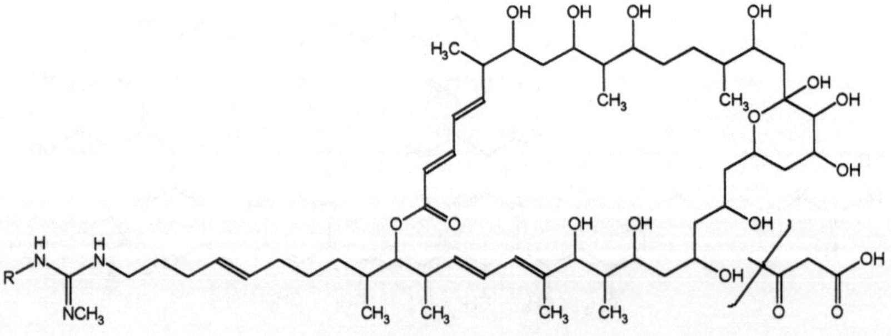

(156): R=H
(157): R=CH₃

(158): R$_1$=COCH$_2$CO$_2$H, R$_2$=H
(159): R$_1$=H, R$_2$=COCH$_2$CO$_2$H

3.7. Peptide Guanidines

Stendomycin was isolated from *Streptomyces antimycoticus* (*350*). This unusual polypeptide (160) consists of a mixture of long chain fatty acid homologues linked to the terminal amine function of a peptide moiety (*351*). The fatty acids are a mixture of isomyristic (or 13-methyltridecanoic) and isotridecanoic (or 11-methyllauric) acids and their lower homologues (*351*). The amino acid constitution of stendomycin was determined by chemical degradation (*352–354*) and mass spectrometry (*355*). Many unusual amino acids were present: dehydrobutyrine (ΔAbu), D-threonine, D-isoleucine, *N*-methyl-L-threonine and a basic component (see 161). The D-isoleucine residues were replaced by valine in part of the antibiotic mixture. The basic component, named stendomycidine [B in (160)], was identified as the N_1,N_3-dimethyl derivative of 2-imino-hexahydro-4-pyrimidyl-glycine (capreomycidine). Its carboxyl

(160)

group esterifies the hydroxyl group of D-threonine, while its α-amino substituent is linked to the carboxyl of D-isoleucine as in (161) (353, 357).

The structure of the major component (161) of the stendomycin mixture was established by careful analysis of its partial and total hydrolysis products (356). Circular dichroism spectroscopy indicated that the lactone ring prefers a compact globular conformation which hides the amide bonds, while the numerous hydrophobic amino acid side chains and the fatty acid moiety are on the outside of the molecule (356). ¹H-NMR conformational studies (358) showed β and β-like turns for the lactone ring and a left-handed α-helical segment for the series of D-amino acids in the linear fragment.

(161)

The bleomycins are a group of polypeptide antibiotics which exhibit strong antitumor activity. Some members of this class are utilised in clinical anti-cancer therapy. These complex molecules have been the subject of much research and a complete review of bleomycin chemistry is beyond the scope of this article. Furthermore, the guanidine derivatives of bleomycins are not the more important members of this family of compounds. All bleomycins possess the same polypeptide framework, differing in the composition of the terminal positively charged ion. The guanidine derivatives are bleomycin B₂ (162a), bleomycin B₄ (163a), phleomycin D1 (**162b**) and phleomycin D2 (**163b**). Various reviews of the bleomycins have appeared. UMEZAWA has reviewed their chemistry, biological activities (*359, 360*), as well as their isolation, structure determination and mode of action (*361, 362*). Others have reviewed their mode of action resulting from interaction with DNA (*363*), their chemistry, biochemistry and biological aspects (*364, 365*), the chemistry involved in their mode of action (*366*) and the mechanisms by which bleomycin induces DNA degradation (*367*). The most important member of this group, bleomycin A_2 (**164**), not itself a guanidine derivative, has been synthesized (*368, 369*), and syntheses of bleomycin analogues have appeared (*370*). This important family of compounds has provided some insight into DNA-drug interactions (*371* and references therein). The absolute stereochemistry of the thiazolinylthiazole moiety of phleomycin D1 (**162b**) has been determined by chemical degradations and analysis of circular dichroism spectra (*372*).

The edeines are polypeptides originating from *Bacillus brevis* Vm4, which exhibit a specific inhibitory effect on bacterial DNA synthesis (*373, 374*). They were isolated as a mixture of four main compounds, edeines A, B, C and D, which could be separated by CM-cellulose 70 and Sephadex G-25 chromatography. Subsequent purification of the edeine complex was achieved by countercurrent distribution, which considerably improved the antibiotic activity (*375, 377*). Hydrolysis of edeine A gave glycine, isoserine, β-tyrosine, 2,3-diaminopropionic acid and spermidine (*376*). Two unknown compounds were also isolated: the first, a diaminodicarboxylic acid is 2,6-diamino-7-hydroxyazelaic acid (**165**) based on chemical degradation (periodate oxidation) and ^{1}H-NMR analysis. The second, also a diaminodicarboxylic acid, contains the elements of one H_2O unit less than (**165**), exhibits UV absorption at 200 nm (ε = 12000) and was supposed to be (**166**). ^{1}H-NMR analysis of (**166**) indicated *E* stereochemistry for the double bond. Careful hydrolysis of edeine A under acidic conditions showed that (**166**) was an artifact formed during the hydrolysis. Furthermore, (**165**) appears to be a mixture of stereo-isomers at the different centers of asymmetry (*376*).

(**165**)   (**166**)

Structural analysis of edeine B (*377*) has shown that the only difference between the two polypeptides is the guanyl substitution in the terminal spermidine residue of edeine B. This new base, *N*-formamidino-*N'*-(3-aminopropyl)-1,4-diaminobutane, or guanylspermidine (**170**), has been previously isolated from leeches and marine worms together with its related putrescine and cadaverine analogues (*379–389*). The chemistry of these compounds has been reviewed (*10, 11*). The identity of guanylspermidine (**170**) was firmly established by total synthesis from 1,4-diaminobutane (**167**) (Scheme 13) (*377*).

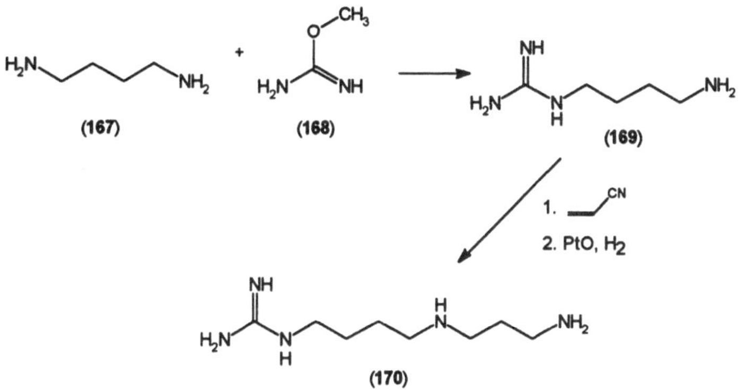

Scheme 13. Synthesis of guanylspermidine

The structures of edeines A and B were determined by a series of enzymatic hydrolyses with carboxypeptidase B (*378*). It was verified that each edeine occurs as a mixture of isoseryl-2,3-diaminopropionyl $\alpha \to \beta$, $\alpha \to \alpha$ interconvertible biologically inactive isomers, the edeines A_2 and B_2. Since edeines A (**171**) and B (**172**) have approximately equal antibiotic activity, the substitution of the terminal amino group by guanidine appears to produce only a slight change in the structural and functional characteristics of the antibiotic (*378*).

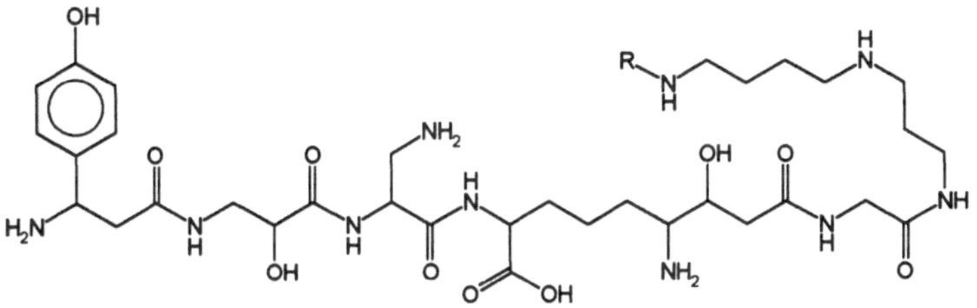

(**171**): R=H
(**172**): R=(C=NH)NH$_2$

Enduracididine (**173**) and its respective D-isomer (**174**) have been isolated from the enduracidin complex found in *Streptomyces fungicidicus*

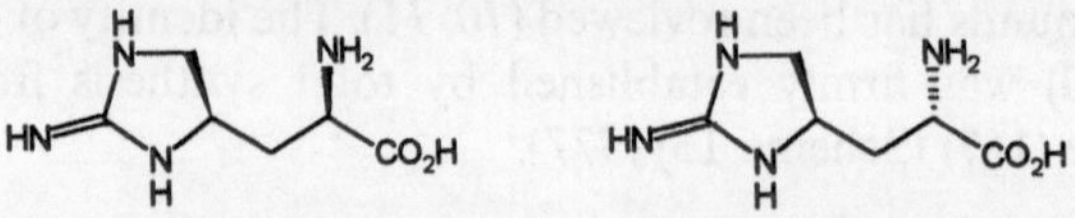

(173)

(174)

(390–392). The presence of a guanidine nucleus was suggested by pKa titrations, infrared spectra and by permanganate oxidation of both amino acids. Their complete structure was assigned after ^{1}H-NMR and mass spectral analyses as well as X-ray crystallographic and optical rotatory dispersion measurements (392).

Both enduracididine isomers are present in enduracidin A (175) and in enduracidin B (176) (393–395). The structures of the polypeptides were determined by hydrolysis, separation and amino acid identification, as well as by EDMAN degradation. Each enduracidin possesses a fatty acid linked to the N-terminus: in enduracidin A (175) this is 10-methylundeca-2-cis-4-trans-dienoic acid. This is the only difference between these two peptides. Moreover, the enduracidins contain an unusual amino acid, α-amino-3,5-dichloro-4-hydroxyphenylacetic acid (396).

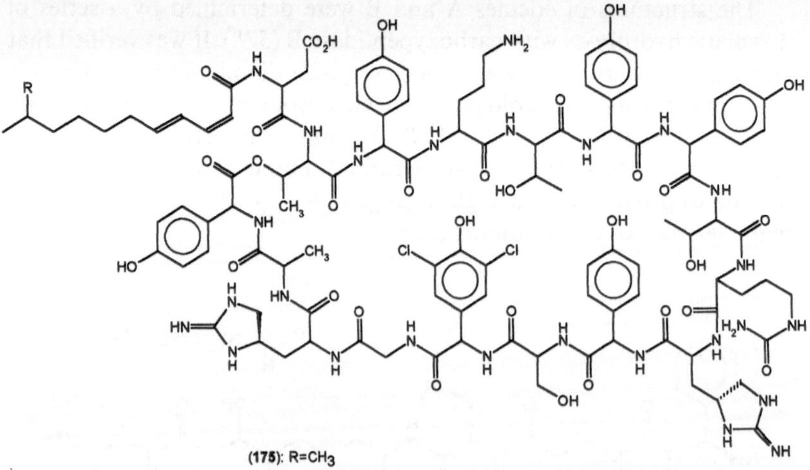

(175): R=CH$_3$
(176): R=CH$_2$CH$_3$

The arphamenines A (177) and B (178) were isolated from Chromobacterium violaceum by OHUCHI and co-workers (397). Structures were determined by spectroscopic analysis of the natural products and by mass spectral analysis of the derivatives (179) and (180) (46). Biosynthetic studies on the arphamenines showed that arginine, acetic acid and

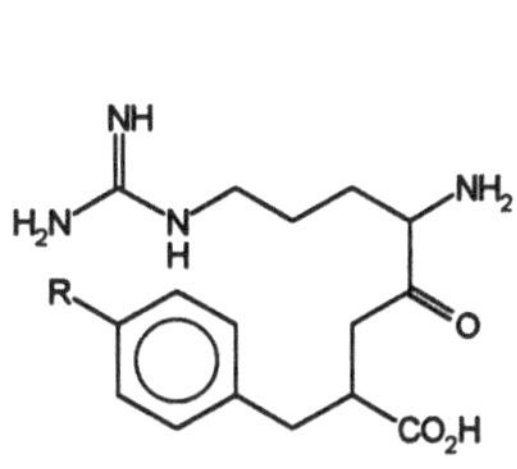

(177): R=H
(178): R=OH

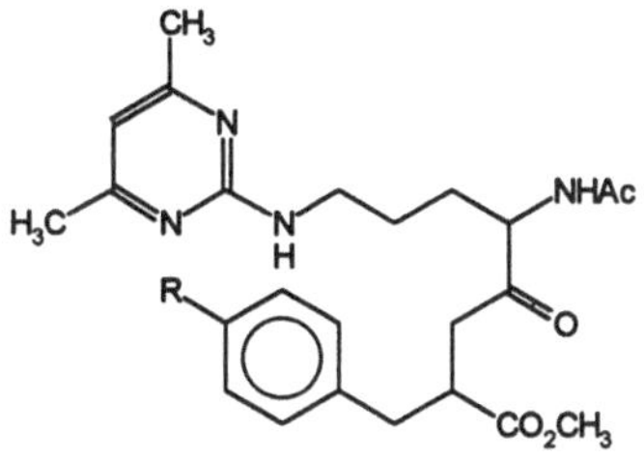

(179): R=H
(180): R=OH

phenylalanine are well incorporated into arphamenine A, and arginine, acetic acid and tyrosine into arphamenine B (398). Further biosynthetic experiments (399, 400) have shown that β-phenylpyruvic (181), benzylmalic (182), benzylfumaric (183) and benzylsuccinic (184) acids are intermediates in the biosynthesis of arphamenine A, thus establishing its complete biogenetic pathway (Scheme 14).

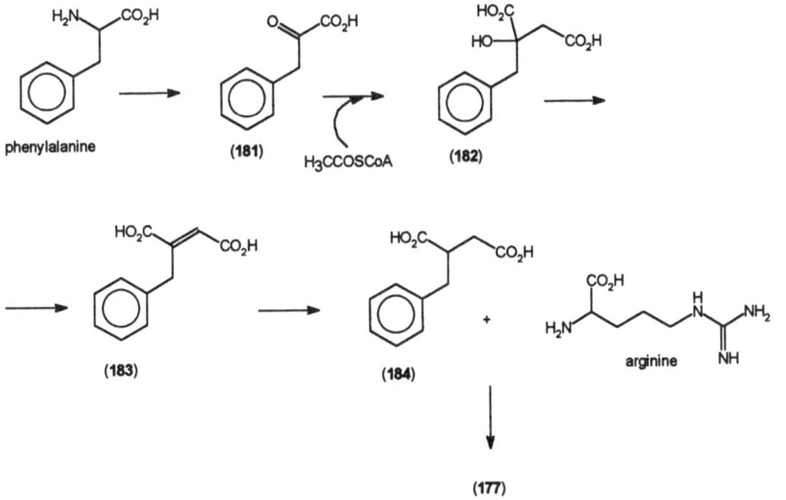

phenylalanine   (181)   H3CCOSCoA   (182)

(183)   (184)   arginine

(177)

Scheme 14. Biosynthesis of arphamenine A from phenylalanine

Two unusual arginine derivatives, fosfazinomycins A (185) and B (186), were isolated from *Streptomyces lavendofoliae* (401). The structures were determined by spectral analysis, including measurements of the ^{1}H-^{31}P and ^{13}C-^{31}P couplings. Acid hydrolysis of fosfazinomycin A gave

(187) and (188). The basic moiety of fosfazinomycin B (189) was obtained in analogous fashion. The presence of a phosphonic acid linked to an arginine N-methylhydrazined derivative in fosfazinomycins is unique.

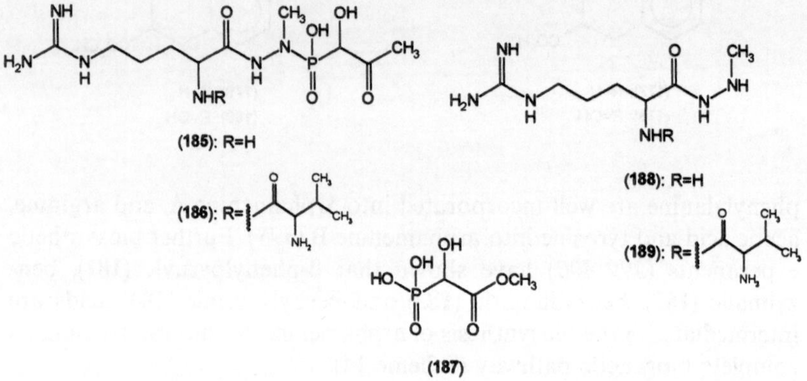

(185): R=H

(186): R=

(187)

(188): R=H

(189): R=

Histargin (190) was isolated from *Streptomyces roseoviridis* (402). The structure was elucidated by spectral analysis of the natural product and its 2-amino-3,5-dimethylpyrimidyl derivative (47); the absolute stereochemistry was established by its total synthesis from L-histidine, L-arginine and dibromoethane (47). MORIGUCHI et al. (403) have developed an alternative route for the synthesis of histargin and some derivatives (Scheme 15).

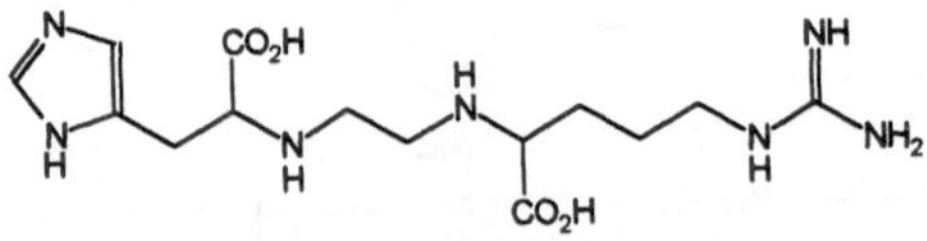

(190)

The rhizocticins A (195), B (196), C (197) and D (198) are unusual di- and tripeptides that were isolated from a *Bacillus subtilis* strain (404). The structures were determined by extensive analysis of spectral data, including ^{31}P-NMR spectroscopy, GC-MS and also by enzymatic degradation with carboxypeptidase A, thermolysin or trypsin.

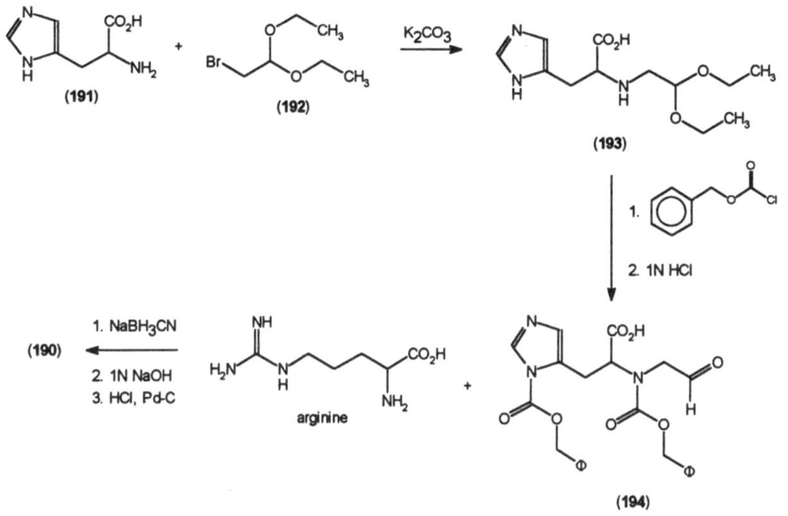

Scheme 15. Synthesis of histargin

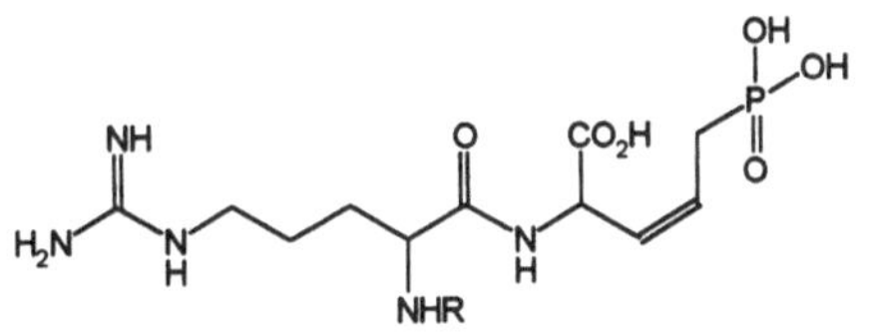

(195): R=H   (197): R=L-isoleucyl

(196): R=L-valyl   (198): R=L-leucyl

The resorcinomycins A (199) and B (200) were isolated from *Strepto-verticillium roseoverticillatum* (*405*). Structures (199) and (200) were established by spectral data analysis and by acid hydrolysis of (199), which yielded the α-guanidino acid (201). The (S) absolute configuration of resorcinomycin A was established by comparing the circular dichroism spectrum of the natural product with that of a synthetic sample of *N*-[(S)-α-guanidinophenylacetyl]-glycine (207) (Scheme 16). Resorcinomycin A is especially active against mycobacteria (*406*).

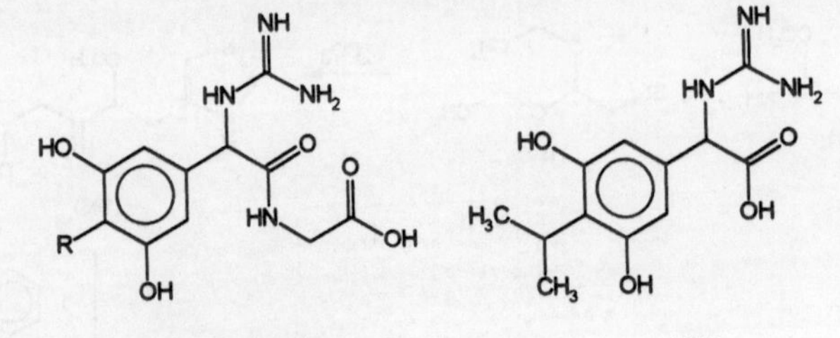

(199): R= isopropyl
(200): R= ethyl

(201)

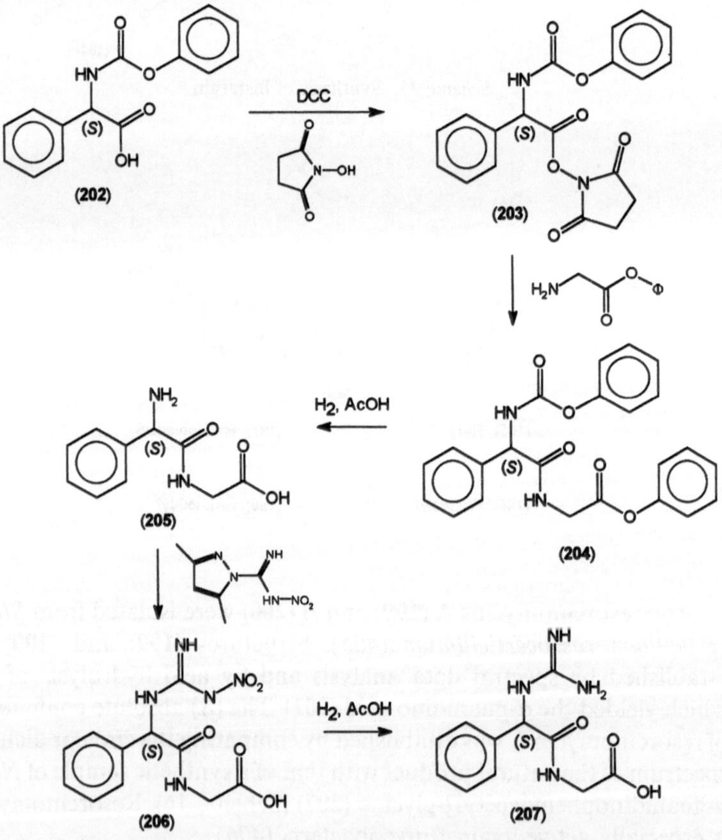

Scheme 16. Synthesis of N-[(S)-α-guanidinophenylacetyl]-glycine

FUNABASHI *et al.* have isolated TAN-1057 A (**208**) and the corresponding unnatural diastereomer TAN-1057 B (**209**) from a *Flexibacter* strain (*407, 408*). Their structures were determined by extensive spectroscopic analysis, including CD, SIMS, ^{1}H-, ^{13}C-, ^{1}H-^{1}H COSY, ^{1}H-^{13}C COSY and COLOC NMR spectra, and also by chemical degradation. Acid hydrolysis of TAN-1057 A (**208**) gave four amino acid derivatives, i.e. 3(*S*)-β-homoarginine and three 2,3-diaminopropionic acid derivatives (**210**), (**211**) and (**212**). Compounds (**210**) and (**211**) exist as 4:1 equilibrium mixture in aqueous solution, while (**212**) has been shown to be racemic 3-amino-2-methylaminopropionic acid. The identity of (**210**) was confirmed by its total synthesis from 2(*S*)-N^2-carbobenzoxy-2,3-diaminopropionic acid (*408*). The free guanidine and amino groups were attributed to the L-β-homoarginine moiety by derivatization methods. Compounds (**208**) and (**209**) are unstable in aqueous alkaline media, being rapidly hydrolysed to the corresponding open form (**213**). It was also demonstrated that (**208**) is the natural product, (**209**) resulting from epimerization in alkaline or acidic media.

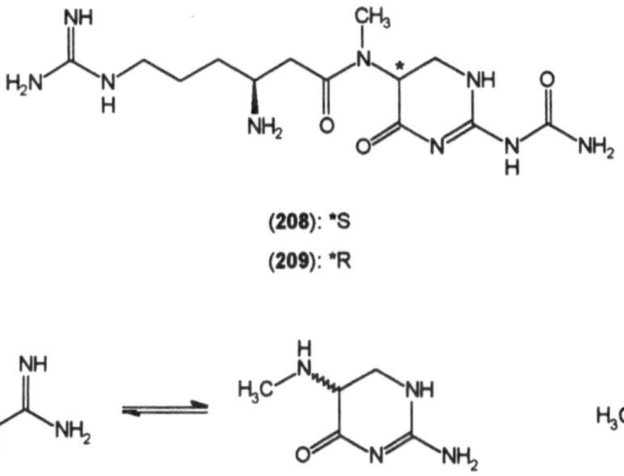

(**208**): *S
(**209**): *R

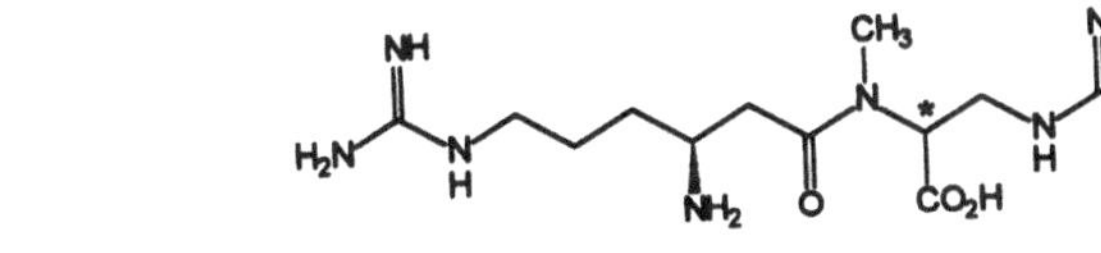

(210)   (211)   (212)

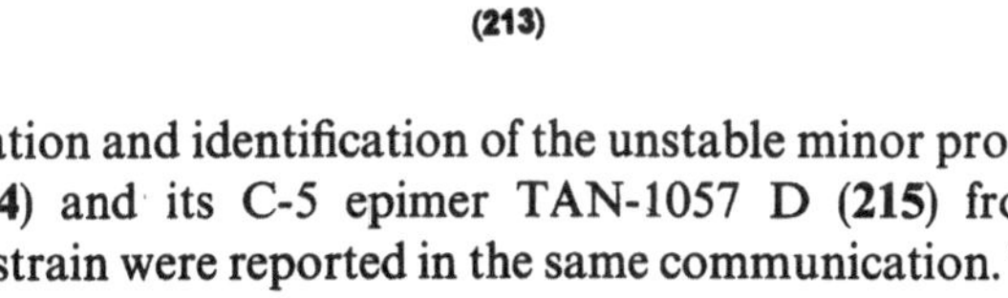

(213)

The isolation and identification of the unstable minor products TAN-1057 C (**214**) and its C-5 epimer TAN-1057 D (**215**) from another *Flexibacter* strain were reported in the same communication. These latter

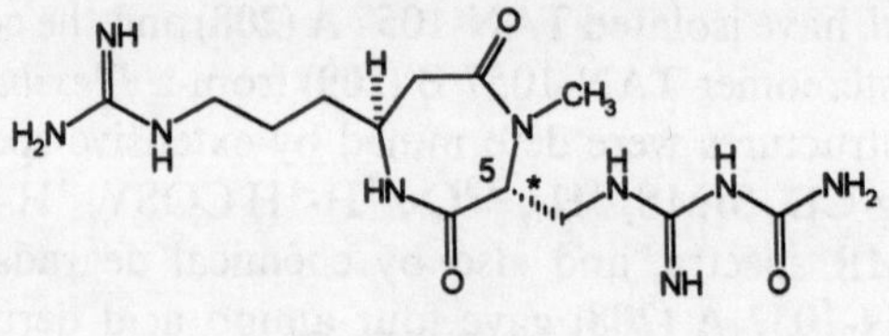

(214): *S
(215): *R

compounds were easily converted into (208) and (209) in alkaline solution.

AOYAGI et al. (409, 410) have reported the isolation of benarthin (216) from *Streptomyces xanthophaeus*. Benarthin is a pyroglutamyl peptidase inhibitor whose structure determination was based on the analysis of spectral data. Hydrolysis of benarthin (216) gave 2,3-dihydroxybenzoic acid, L-threonine and L-arginine. The total synthesis of benarthin has been achieved (Scheme 17) (411).

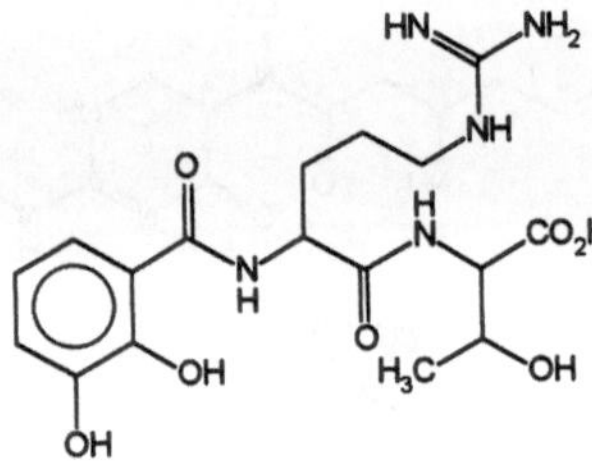

(216)

KAMIYAMA et al. (412) isolated an argininal derivative Ro 09-1679 (223) from *Mortierella alpina*. Its structure was fully established by analysis of spectroscopic data, namely FABMS and HMBC.

The structure of lavendomycin (224), isolated from *Streptomyces lavendulae* (413), was established by analysis of spectral data, EDMAN degradation (414) and synthesis (415). Lavendomycin possess a variety of rare aminoacids, such as (S)-pipecolinic acid, didehydroaminobutyric acid, (S,S)-α,β-diaminobutyric acid and (2S)-3-methylarginine.

Biphenomycin C (225) seems to be the biogenetic precursor of biphenomycin A (226) (416). Both compounds were isolated from *Streptomyces griseorubiginosus* together with biphenomycin B (227) (416, 417). The structures of these compounds were determined by spectral analysis, chemical degradation and synthesis (416, 418–420). Biphenomycin A has

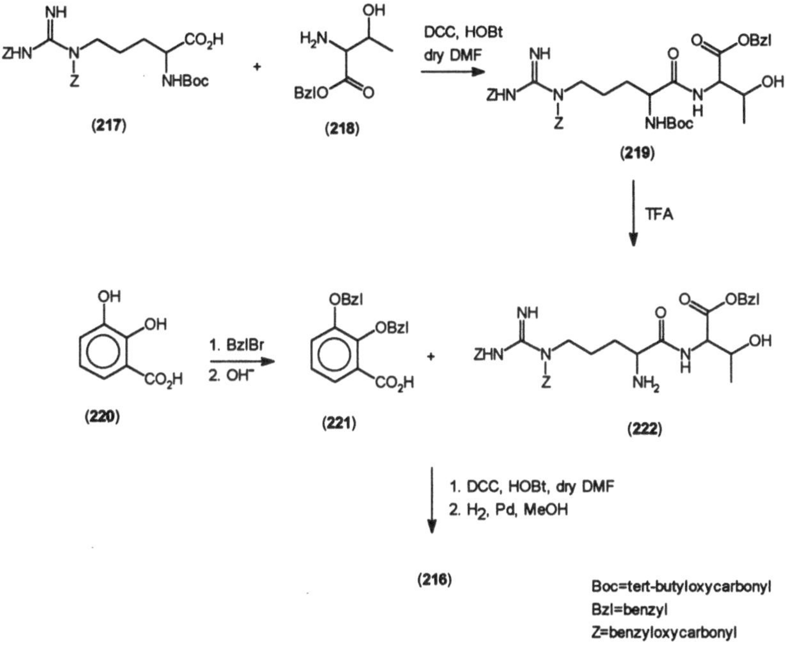

Scheme 17. Synthesis of benarthin

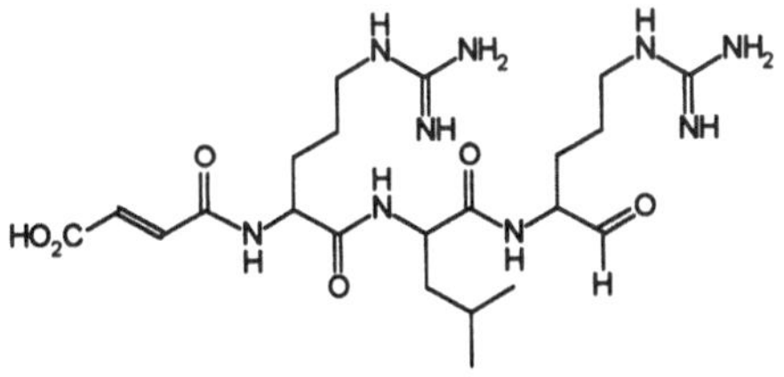

(223)

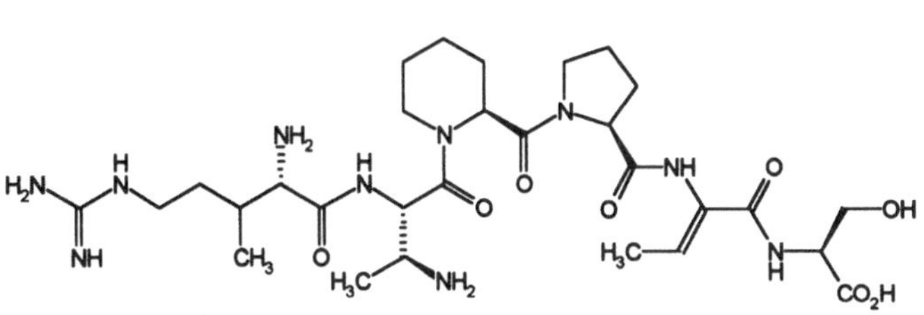

(224)

(**225**): R₁=OH, R₂= —NH ...

(**226**): R₁=R₂=OH
(**227**): R₁=H, R₂=OH

been the subject of detailed CD, NMR and computational analysis (*421–423*).

The pheganomycins (**228a–e**), isolated from *Streptomyces cirratus*, showed activity against streptomycin-, kanamycin- and viomycin-resistant strains of *Mycobacterium smegmatis*. They were identified by chemical degradation, spectroscopic analysis and peptidic synthesis (*424*).

(**228a**): R₁=OH, R₂=OH
(**228b**): R₁=OH, R₂=L-Asp
(**228c**): R₁=OH, R₂=L-Asp-L-Arg
(**228d**): R₁=OH, R₂=L-Asp-L-Pro-Gly-L-Thr
(**228e**): R₁=H, R₂=L-Asp

Several oligopeptide protease inhibitors containing guanidine amino acids isolated from a variety of microorganisms are notable for their unusual biological activities. These are now described.

The chymostatins and elastatinal contain the capreomycidines (see Sect. 3.3) as part of their structure. The chymostatin complex is a mixture

of different polypeptides (*425, 426*). Partial hydrolysis of the chymostatins gave three unknown compounds (**229**), (**230**) and (**231**), plus leucine,
valine and isoleucine. Hydrolysis of (**229**) gave DL-phenylalanine and a
basic amino acid. The structure of the latter was established as a mixture
of the 2(*S*)- and 2(*R*)-capreomycidines by derivatization, ^{1}H-NMR
analysis, optical rotatory dispersion measurements and epimerization of
each optically pure compound into its respective epimer by treatment
with DCl. The component (**230**) was shown to be a mixture of three
components, containing either L-leucine (**230a**), L-valine (**230b**) or L-
isoleucine (**230c**) at the caproeomycidine C-terminal amino acid. The
third component of chymostatin was identified as phenylalaninal (**231**) by
chemical and spectroscopic means. By further chemical degradation
(oxydation, hydrolysis, EDMAN degradation and hydrazinolysis), the
structures of chymostatin A (**232**), chymostatin B (**233**) and chymostatin
C (**234**) were established (*426*).

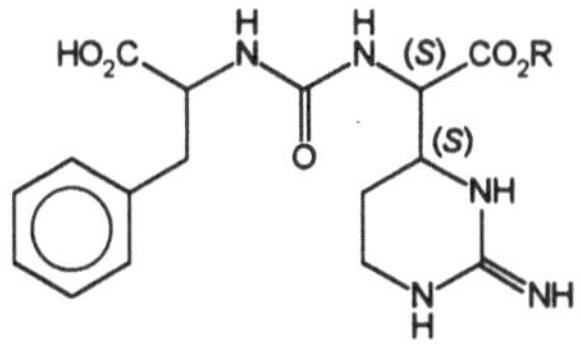

(**229**): R=H
(**230a**): R=L-leucine
(**230b**): R=L-valine
(**230c**): R=L-isoleucine

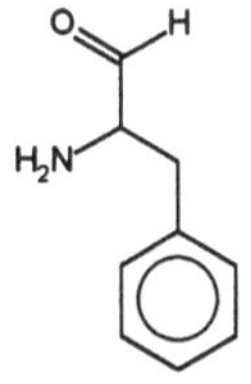

(**231**)

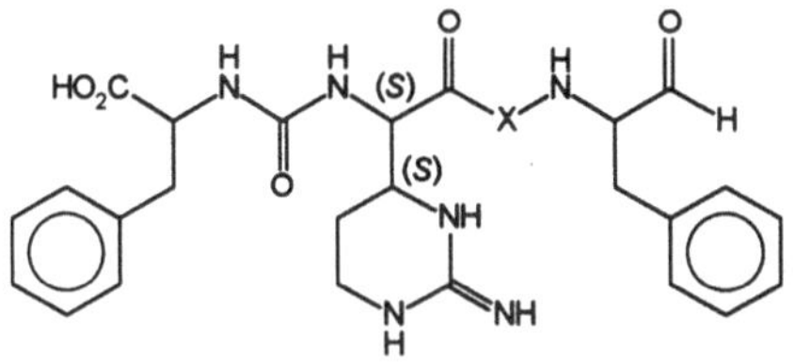

(**232**): X= L-leucine
(**233**): X= L-valine
(**234**): X= L-isoleucine

Elastatinal (**235**) was isolated from *Streptomyces griseoruber* by a
series of ion-exchange chromatographies (*427*). Its structure was established after oxidation (on the terminal alaninal aldehyde), hydrolysis of
the corresponding acid (which yielded glutamic acid, alanine, leucine and

a mixture of diastereomeric capreomycidines) and mild hydrolysis (giving a mixture of hydantoins composed of leucine and the capreomycidines). The major component of this mixture was synthesized from L-leucine and 2(S),3(S)-capreomycidine (428).

(235)

The leupeptins have been isolated from various actinomycete strains (429). Leupeptin Pr-LL (236) and leupeptin Ac-LL (237) were identified by amino acid sequence analysis (429), by analysis of spectral data (430) and synthesis (431). The two compounds possess anti-plasmin, anti-inflammatory and anti-blood coagulation activities, besides inhibiting trypsin, papain, plasmin, thrombokinase and dipeptidyl aminopeptidase III (DAPIII). Leupeptins also showed inhibitory effects on bovine pancreatic trypsin, on calcium-active neutral protease and other proteins (432, 433 and references therein).

(236): R=CH$_2$CH$_3$
(237): R=CH$_3$

Antipain (238) was also isolated from various actinomycete strains (434, 435). Its structure was established by chemical degradation. It exhibits inhibitory effects against an array of proteases (434).

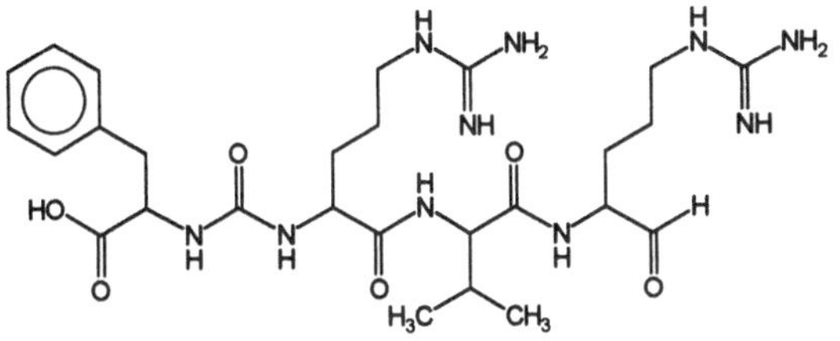

(238)

E-64 (239), a thiol protease inhibitor, was isolated from *Aspergillus japonicus* (*436*). Its structure was elucidated by analysis of spectral data and synthesis (Scheme 18) (*437*). It exhibits a papain inhibitory effect (*438*).

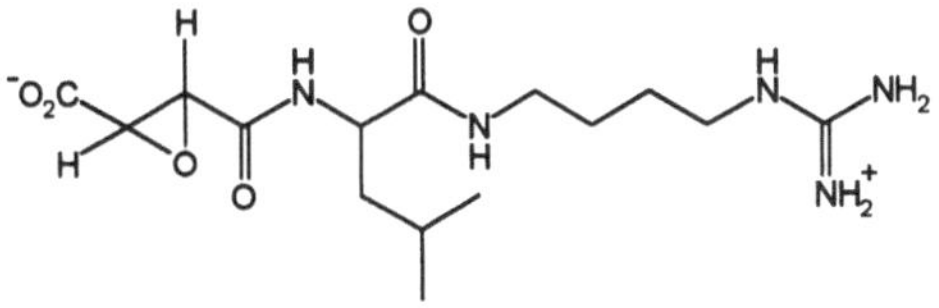

(239)

α-MAPI (247) and β-MAPI (248) have been isolated from *Streptomyces nigrescens* by MURAO and WATANABE (*439, 440*). Their structures were established by chemical degradation and spectroscopic analysis (*441*). Both MAPIs show inhibitory effect against α-chymotrypsin and thiol proteinases (*439, 440*).

An inhibitor of angiotensin converting enzyme, L-681,176 (249), was isolated from a *Streptomyces* strain (*442*). Its structure was fully elucidated by analysis of the spectral data (*443*).

Strepin P-1 (250), isolated from *Streptomyces tanabeensis*, is a cysteine proteinase inhibitor (*444*), which also inhibits calpain I, calpain II and papain. The structure of strepin P-1 (250) was determined by spectrometric analysis and enzymatic degradation.

Finally, MURAO et al. (*445*) have isolated thiolstatin D (251) from *Bacillus cereus*. The structure was deduced by acid and alkaline hydrolysis, as well as by ^{1}H- and ^{13}C-NMR spectral analysis.

Interestingly, most of the proteinase inhibitors reported so far contain a guanidine amino acid and an aldehyde terminus, this latter being in part responsible for the inhibitory activities displayed by these compounds. The chemistry and biological activities of proteinase inhibitors have been extensively reviewed by UMEZAWA (*446, 447* and references therein).

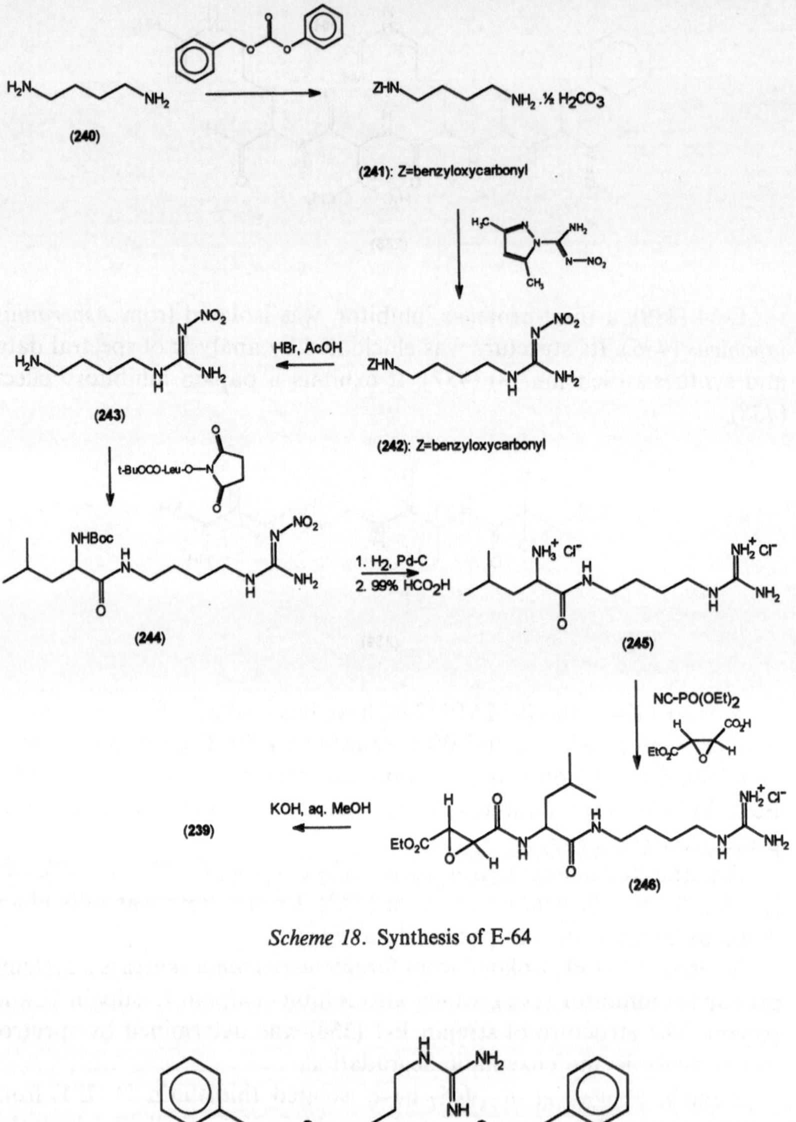

Scheme 18. Synthesis of E-64

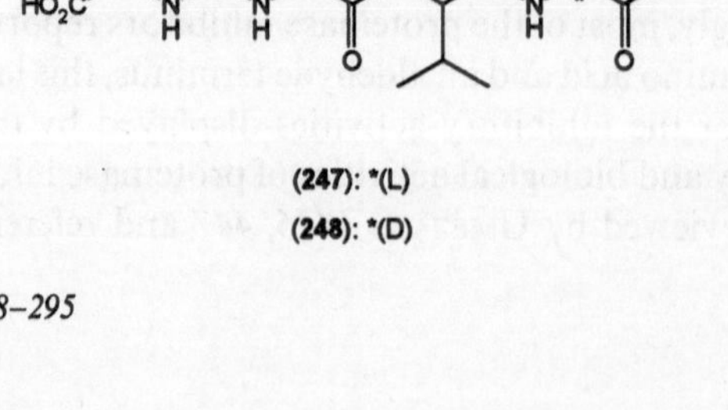

(247): *(L)
(248): *(D)

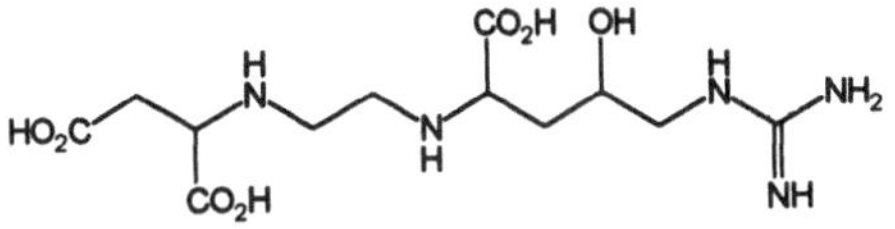

(249)

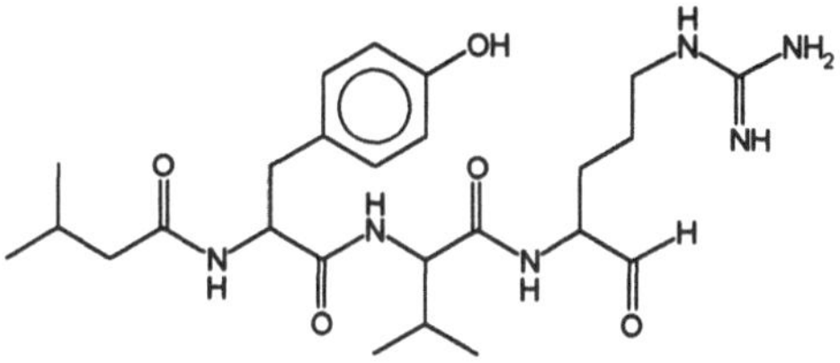

(250)

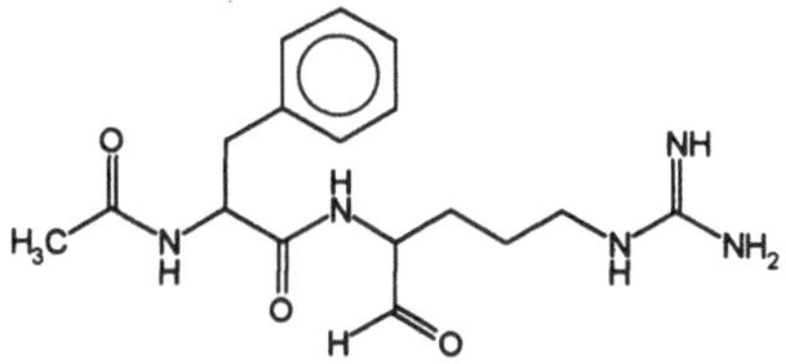

(251)

3.8. Other Guanidine Derivatives Originating from Microorganisms

Arglecin was isolated by UMEZAWA *et al.* (*448, 449*) from *Streptomyces toxytcitrini* and *Streptomyces lavendulae*. Structure (252) was eventually established on the basis of ^{1}H-NMR analysis and chemical degradation (*450*). For example, hydrogenation of the *N*-acetyl derivative yielded a tetrahydro derivative. The UV spectrum of arglecin was very similar to that of flavacol (253). Bromination of arglecin followed by Zn/AcOH treatment gave the diketopiperazine (254) which hydrolysed with 48% HBr gave leucine and arginine from which (254) could be resynthesized (*450*). These results exclude alternative structures (255) and (256). The optical inactivity of arglecin (252) is explained by tautomerism with the fully aromatic form (257).

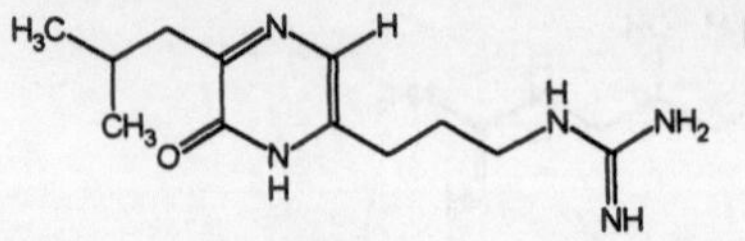

(252)

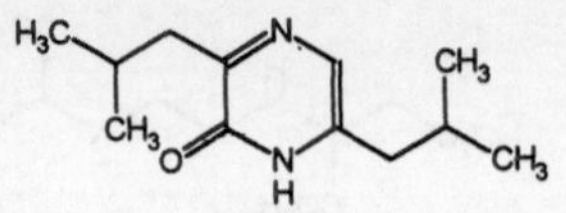

(253)

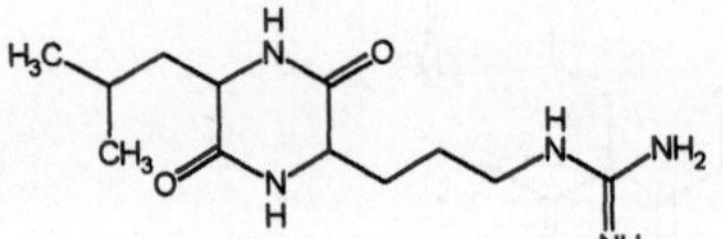

(254)

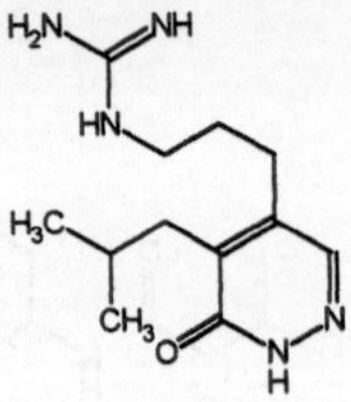

(255)

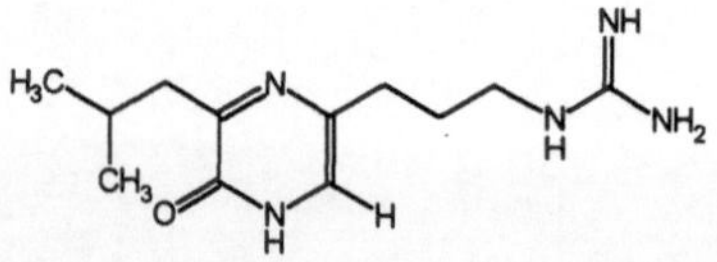

(256)

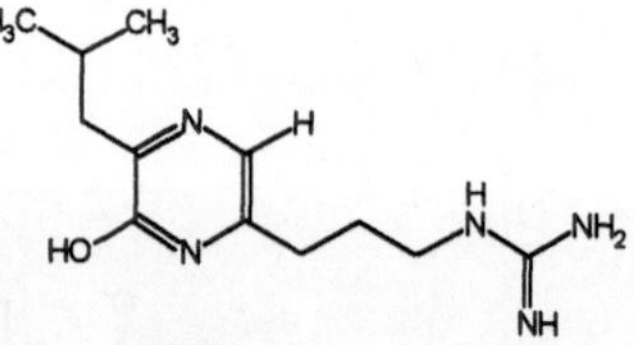

(257)

The structure of the lower homologue argvalin (**258**), isolated from *Streptomyces filipensis*, was established in a completely analogous manner (*451*) by ^{1}H-NMR spectral analysis and chemical degradation. In this case the diketopiperazine (**259**) with 48% HBr afforded valine and arginine from which (**259**) could be resynthesized.

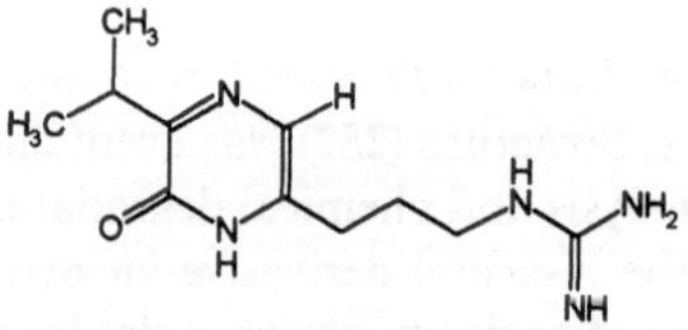

(258)

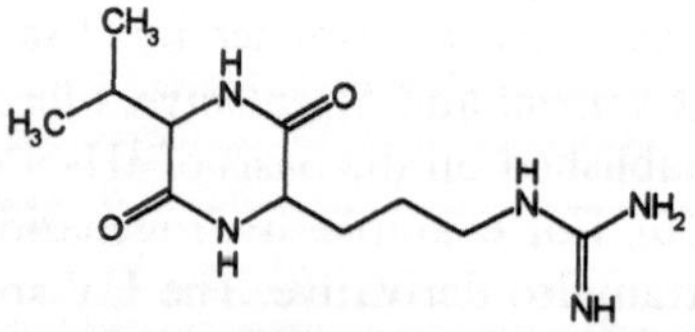

(259)

Streptomyces toxycitrini has shown high incorporation of leucine and arginine during the biosynthesis of arglecin (*452*). When L-arginine or L-leucine were replaced by their D-isomers, poor or no production of arglecin was observed. Interestingly, the addition of arginine and/or leucine analogues (as, for example, L-homoarginine and L-norleucine) led to the production of corresponding compounds. Long term incubation of *S. toxycitrini* increases the production of 3-isobutyl-6-(3-aminopropyl)-2(1H)-pyrazinone (**260**), which appears to be product of catabolism of arglecin. No argvaline was produced by *S. toxycitrini* on a medium containing arginine and valine. Hence, the specificity of the *S. toxycitrini* organism apparently differs from that of *S. filipensis*.

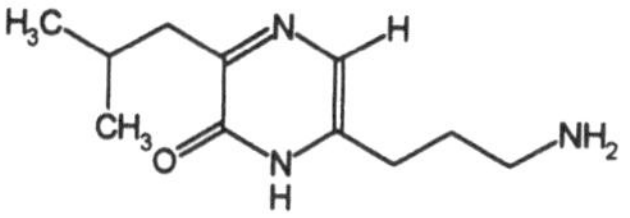

(260)

Both arglecin (**252**) and argvalin (**258**) have been synthesized by almost identical routes (the synthesis of arglecin is outlined in Scheme 19), starting from an appropriate diketopiperazine (**261**) for the synthesis of argvalin (*453, 454*).

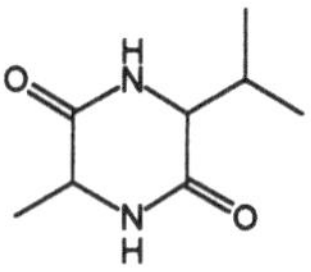

(261)

A *Nocardia* strain has provided unexpectedly complex guanidine derivatives. LL-BM123α (**272**) was the first member of this series and was identified by chemical and spectroscopic means (*455*). The nature of the glucoside linkages between the sugar residues was established by ^{13}C-NMR analysis.

Other members of this series possess diverse structural features (*456, 457*) such as a 4-hydroxycinnamoylspermidine moiety linked to a diaminated sugar and varied urea and guanidine substituents on the sugar residues. The structures of LL-BM123β (**273**), LL-BM123γ$_1$ (**274**) and LL-BM123γ$_2$ (**275**) were established by hydrolytic degradation and spectroscopic analyses (mainly ^{13}C-NMR and mass spectrometry) (*458*) although assignment of the ^{13}C signals of the guanidine and ureido carbonyl carbons was suggestive only. Some synthetic analogues were prepared in the search for stronger antibiotic properties (*459, 460*).

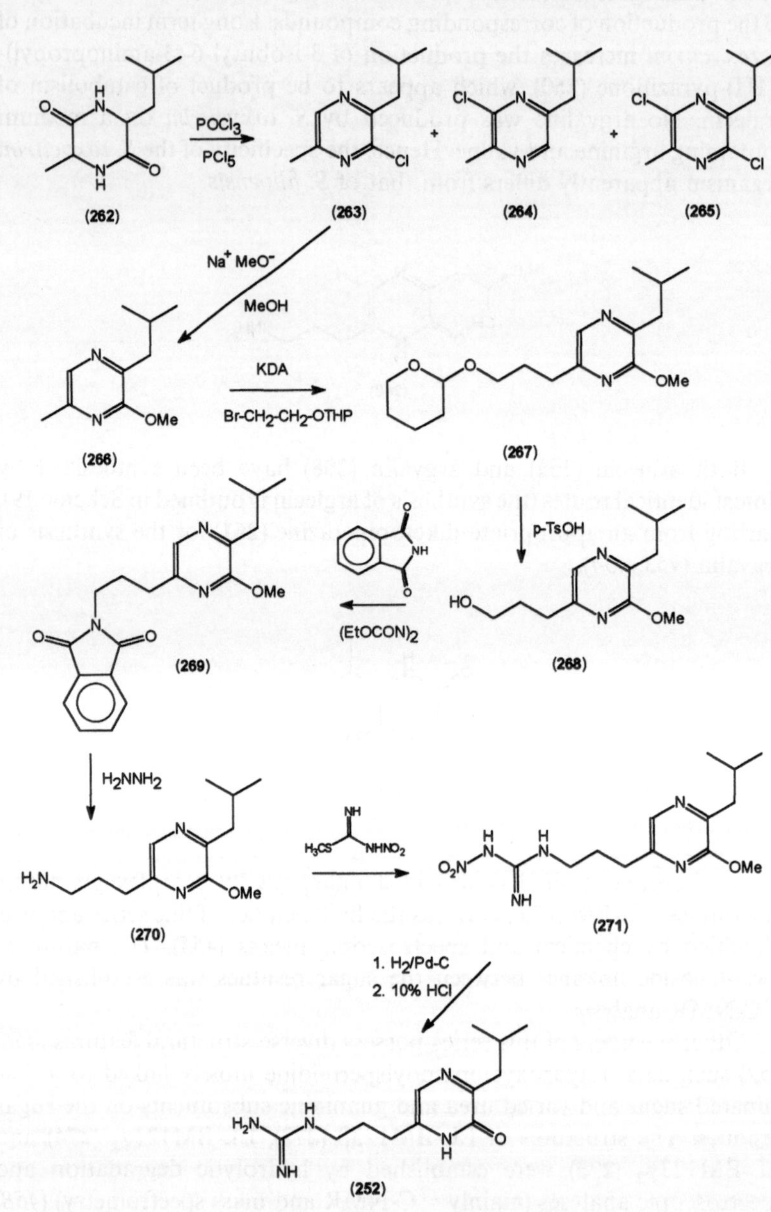

Scheme 19. Synthesis of arglecin

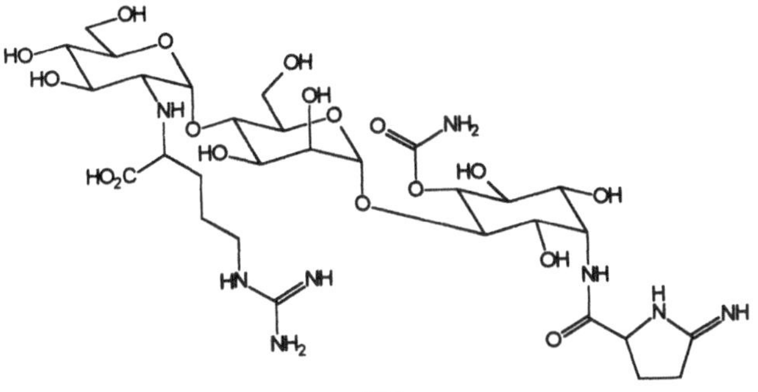

(272)

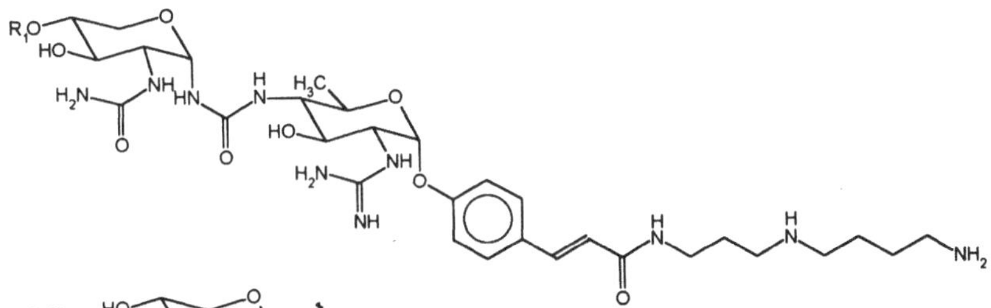

(273): R₁=

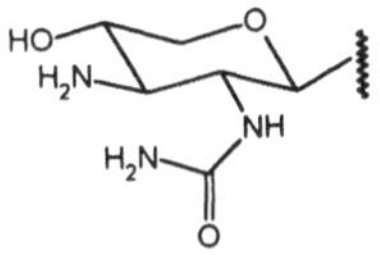

(274): R₁=

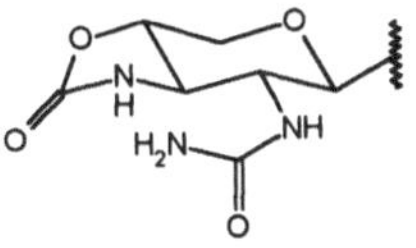

(275): R₁=

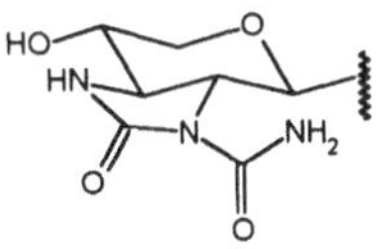

A complex of LL-BM123β (273), LL-BM123γ₁ (274) and LL-BM123γ₂ (275), named cinodine, has shown good incorporation of L-tyrosine, L-phenylalanine and p-coumaric acid into the cinnamoyl moiety during biosynthesis. Moreover, all sugar residues efficiently incorporated 2-glucosamine. The spermidine moiety arises from arginine, ornithine or putrescine. Particularly efficient was the incorporation of [amidine-¹⁴C] arginine, probably into the guanidine and ureido side chains of the sugar residues (461, 462).

The octacosamicins A (276) and B (277) are long chain fatty acid guanidine derivatives isolated from an Amycolatopsis azurea strain (463). Their structures were determined by various NMR techniques, including ¹H-¹H COSY, ¹H-¹³C COSY, and HMBC (464).

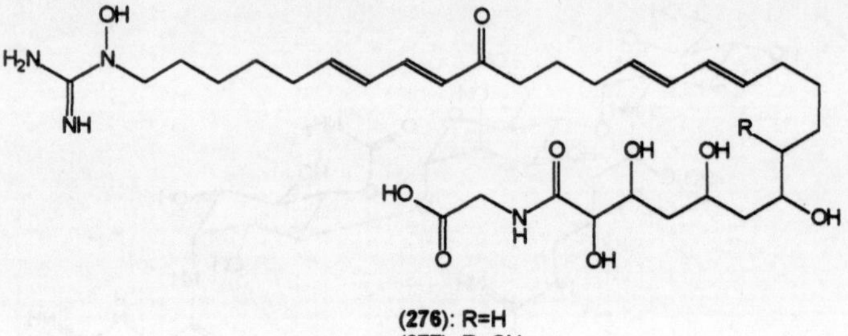

(276): R=H
(277): R=CH₃

An unusual antifungal component Sch 40873 (**278**) has been isolated from an *Actinomadura* strain (*44*) and identified by both spectroscopic analysis, mainly mass spectrometry, and by chemical degradation. Hydrolysis with 6N HCl gave (**279**), Jones oxidation gave (**280**) and HNO₃ oxidation resulted in decomposition to pimelic acid (**281**), suberic acid (**282**) and azelaic acid (**283**).

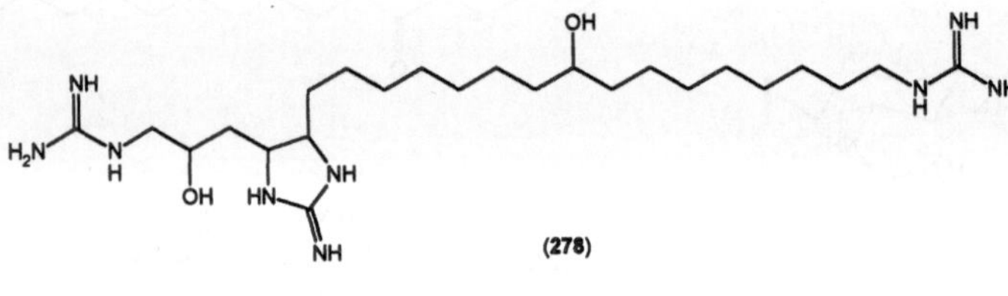

(278)

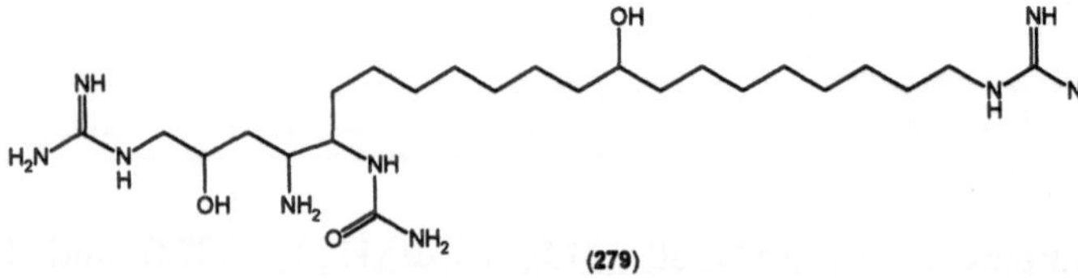

(279)

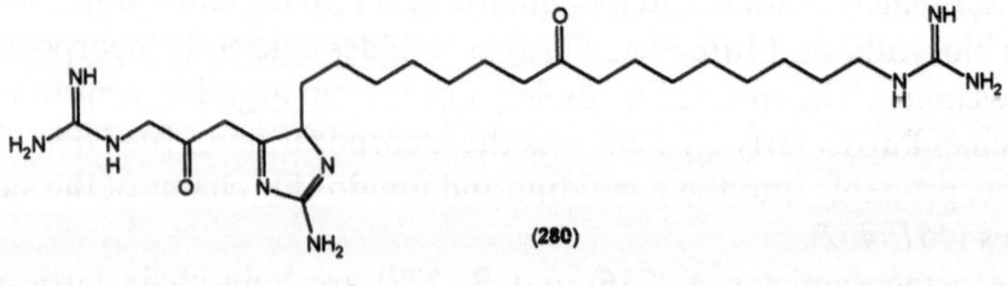

(280)

(281) 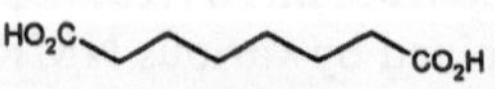   **(282)**    **(283)**

Ficellomycin (**284**), isolated from *Streptomyces ficellus* by ARGOUDELIS *et al.* (*465*), possesses a 1-azabicyclo-[3.1.0] hexane ring system (*466*). The structure was established by analysis of the spectra of the natural product and the corresponding methyl ester. The aziridine ring seems to be stabilized by the neighboring guanidine group and it may be responsible for the biological activities (as an alkylating agent) reported for ficellomycin.

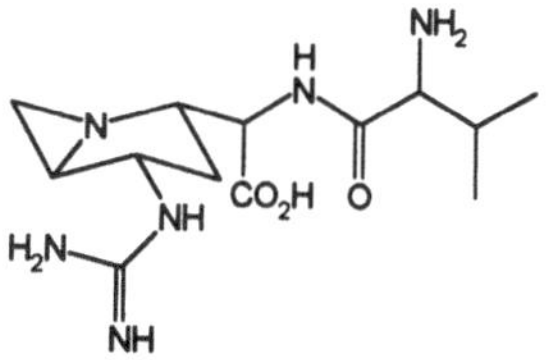

(**284**)

Lydicamycin (**285**) isolated from *Streptomyces lydicus* has a poly-functionalized skeleton with many interesting structural features: i) a *N*, *N*-disubstituted guanidine included in a pyrrolidine ring; ii) a *cis*-junction in the bicyclic moiety, and; iii) a tetramic acid. The structure of lydicamycin (**285**) was established by ^{13}C, ^{1}H-, ^{1}H-^{1}H COSY, ^{1}H-^{13}C HETCOR, HOHAHA, HMBC and NOESY NMR analysis (*467*).

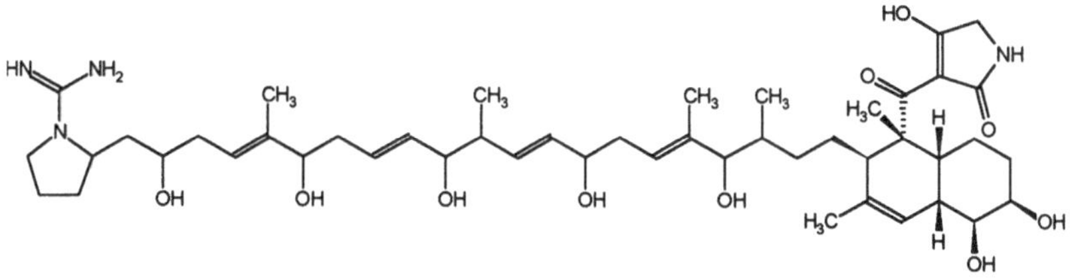

(**285**)

Xenocoumacin-1 (**286**) has been isolated from a *Xenorhabdus* strain (*468*) together with the pyrrolidine derivative (**287**). Both were identified by analysis of spectral data using various 1D- and 2D-NMR techniques. The authors also proposed a biogenetic pathway for these compounds based on the condensation of leucine with four acetate units to give a substituted salicylic system which subsequently undergoes a series of stepwise transformations (lactone formation, new acylation with acetate and further condensation with arginine or proline). Xenocoumacin-1 (**286**) has potent antibacterial and antiulcer activities (*468*).

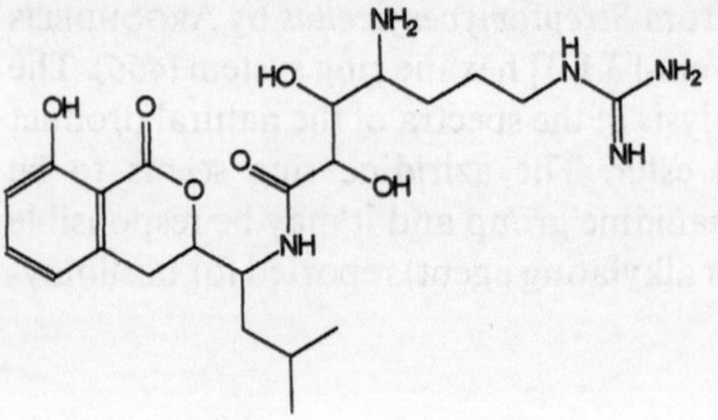

(286)

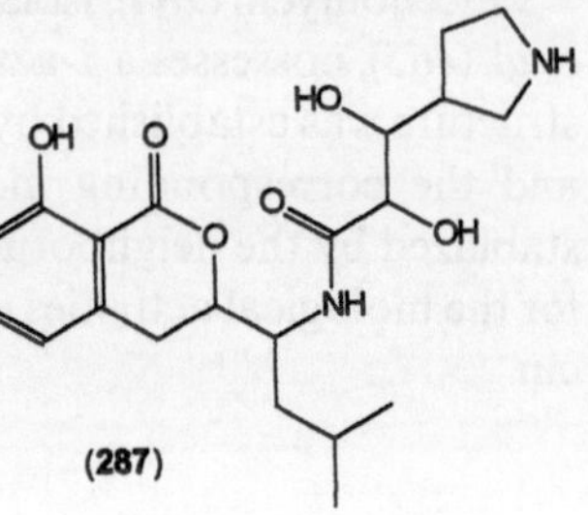

(287)

4. Guanidine Secondary Metabolities from Marine and Freshwater Organisms

Marine guanidine natural products have been reviewed by CHEVOLOT (*10*) and by CHRISTOPHERSEN (*469*). The synthesis of some marine guanidines has also been discussed (*470*). Since the appearance of these reviews, a great variety of marine guanidines has been isolated, identified and, in some cases, synthesized. Moreover, structures of compounds reviewed earlier have been revised. Although some compounds are of microbial origin, they appear in this section because of their historic background. This is the case for tetrodotoxin and saxitoxin, among others.

Many compounds considered by CHEVOLOT and by CHRISTOPHERSEN as guanidine derivatives will not be included in the present review because they possess a 2-aminoimidazole nucleus rather than a true guanidine unit. This includes some derivatives of the compounds biogenetically related to oroidin as well as oroidin itself, the zoanthoxanthins and so on. The reader may consult the above mentioned reviews, as well as FAULKNER'S reviews in Natural Product Reports (*471–478*) for a complete list of 2-aminoimidazole derivatives from marine sources.

4.1. Marine and Freshwater Microorganisms

Tetrodotoxin (**3**) and saxitoxin (**6**) are certainly among the most interesting discoveries in natural products chemistry. Today recognised as extremely valuable tools in neurochemistry, neurophysiology and biochemistry, tetrodotoxin and saxitoxin have been the subject of various reviews, dealing with their chemistry, biological activities and biochemistry among other topics. Thus an extensive review of tetrodotoxin and saxitoxin is beyond the scope of this work and only selected reports which have appeared prior to 1993 will be discussed. SCHEUER (479) early reviewed the chemistry of these compounds. CHEVOLOT, SHIMIZU and MOORE have also discussed many aspects of them in an Academic Press Series edited by SCHEUER (10, 480, 481). Methods for isolation of saxitoxin and tetrodotoxin have been reviewed (482, 483). The structure elucidation and synthesis of these compounds have been reviewed by different authors (484–487). More recently, an issue of the Annals of the New York Academy of Sciences (488) reported much of the early work dealing with these compounds. The topics reviewed include historical aspects (489), chemistry and biochemistry (490–492), distribution in nature (493) and mechanism of action studies (494–497). NARAHASHI (498) has also reviewed the last topic.

After structure elucidation of tetrodotoxin (3) in 1964 (24, 499–505), hundreds of papers dealing with this compound have appeared. For example, over 650 entries dealing with tetrodotoxin appeared in Chemical Abstracts in the decade 1982–1992. It is now well established that tetrodotoxin and its derivatives were widespread in nature (493, 506–523) and that they are of bacterial origin (508, 512–517). Some of the new tetrodotoxin derivatives isolated are 4-epitetrodotoxin (290a), isolated from Takifugu pardalis and T. poecilonotus (506), together with tetrodonic acid (288), 4,9-anhydro-4-epi-tetrodotoxin (289a) and tetrodotoxin (3); 6-epi-tetrodotoxin (290b), 11-deoxytetrodotoxin (290c) and 11-deoxy-4-epi-tetrodotoxin (290h) which accompany (3), (289a), 4,9-anhydro-4-epi-11-deoxytetrodotoxin (289b), 4,9-anhydro-4-epi-6-epi-tetrodotoxin (289c) and (290a) in the newt Cynops ensicauda (507, 508), 11-nortetrodotoxin-6(R)-ol (290d), isolated from the puffer fish Fugu niphobles (509) and 11-oxotetrodotoxin (290e) isolated from the puffer fish Arothron nigropunctatus (511).

The structure of chiriquitoxin (290f), early isolated from the Costa Rican frog Atelopus chiriquensis (523), has been elucidated recently (524). Unlike tetrodotoxin, chiriquitoxin (290f) exists in solution mainly in the hemilactal form. The 13,6-lactone chiriquitoxin derivative (290g) was obtained on keeping chiriquitoxin in 1% TFA, 4% CD_3CO_2D/D_2O for one month at 5°C. The configurations of (290g) at C-6 (S), C-11 (R) and C-12 (S) were suggested by NOE measurements. Biosynthesis of chiriquitoxin (290f) may involve tetrodotoxin (3) and glycine. Furthermore, chiriquitoxin is as toxic as tetrodotoxin; this suggests that the sodium channel protein has specific binding sites for the C-12 NH_2 and/or the C-13 CO_2H groups (524). More recently, KOTAKI and SHIMIZU (525) have reported the structure of 1-hydroxy-5,11-dideoxytetrodotoxin (291) isolated from the newt Taricha granulosa. The N-hydroxy substitution could be established by careful ^{13}C-NMR and FABMS analysis.

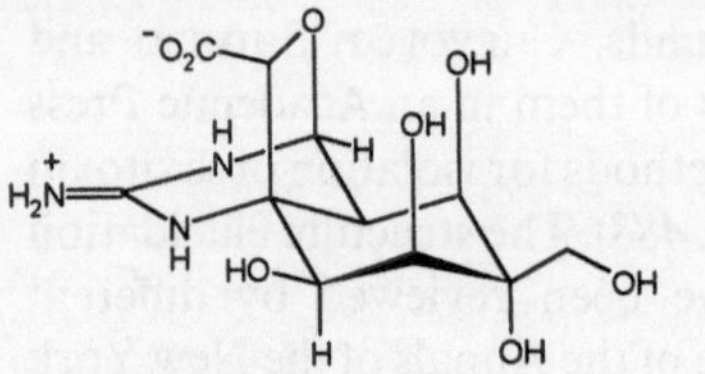

(288)

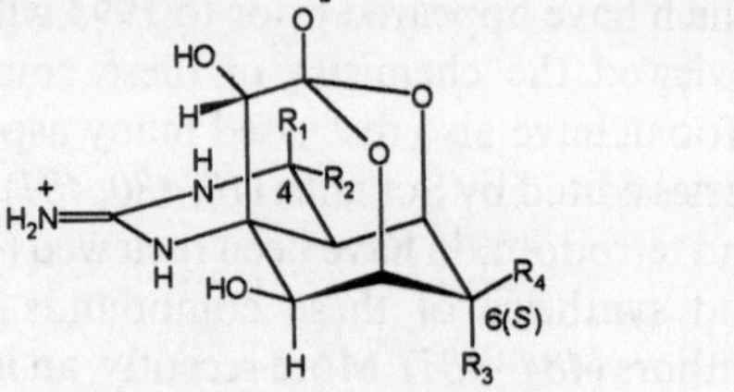

(3): R$_1$=H, R$_2$=R$_3$=OH, R$_4$=CH$_2$OH
(290a): R$_1$=R$_3$=OH, R$_2$=H, R$_4$=CH$_2$OH
(290b): R$_1$=H, R$_2$=R$_4$=OH, R$_3$=CH$_2$OH
(290c): R$_1$=H, R$_2$=R$_3$=OH, R$_4$=CH$_3$
(290d): R$_1$=R$_3$=H, R$_2$=R$_4$=OH
(290e): R$_1$=H, R$_2$=R$_3$=OH, R$_4$=CH(OH)$_2$
(290f): R$_1$=H, R$_2$=R$_3$=OH, R$_4$= CH(OH)CH(NH$_2$)CO$_2$H

(290g): R$_1$=H, R$_2$=OH, R$_3$,R$_4$= 11(R) 12(S) 13
—CH(OH)CH(NH$_2$)CO—O—

(290h): R$_1$=R$_3$=OH, R$_2$=H, R$_4$=CH$_3$

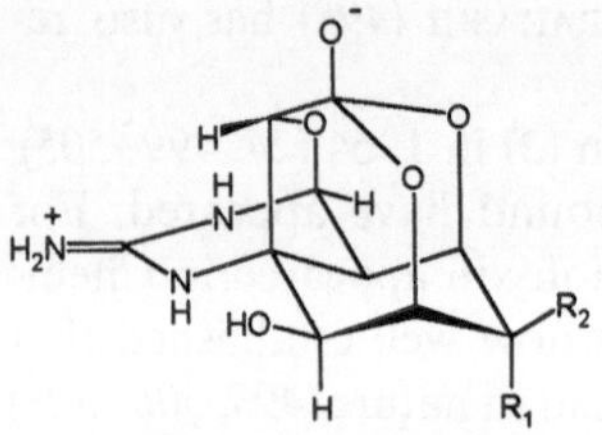

(289a): R$_1$=OH, R$_2$=CH$_2$OH
(289b): R$_1$=OH, R$_2$=CH$_3$
(289c): R$_1$=CH$_2$OH, R$_2$=OH

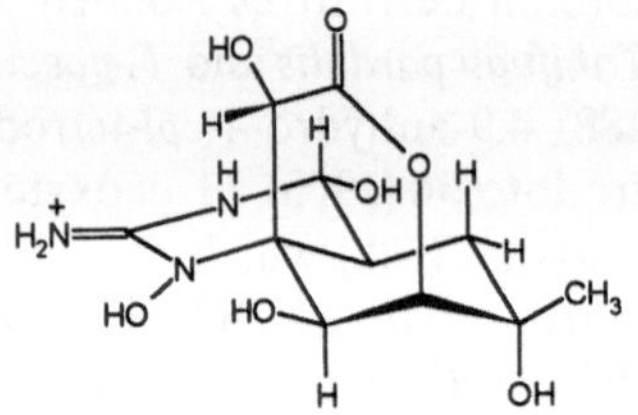

(291)

Considering (291) and the previously reported tetrodotoxin derivatives (288–290h), it seems likely that tetrodotoxin is formed biogenetically
from arginine and an isoprene unit by stepwise oxidation (route a in
Scheme 20) as suggested by YASUMOTO et al. (510), and not by condensation of arginine and a branched sugar (route b in Scheme 20) (525).
Further biogenetic steps leading to tetrodotoxin derivatives proposed by
KOTAKI and SHIMIZU (525) would explain the previously reported un-
successful biosynthetic experiments (491, 526). Efficient fluorescent
HPLC methods have been developed with the aim of detecting small
amounts of tetrodotoxin and its derivatives (527, 528).

The total synthesis of racemic tetrodotoxin (529–531, 542) has been
reviewed in (479) and (485). Many approaches have been developed for
the synthesis of enantiomerically pure tetrodotoxin among them KEANA'S
(532–538), ISOBE's (539–541) and other approaches (543–548). A practical

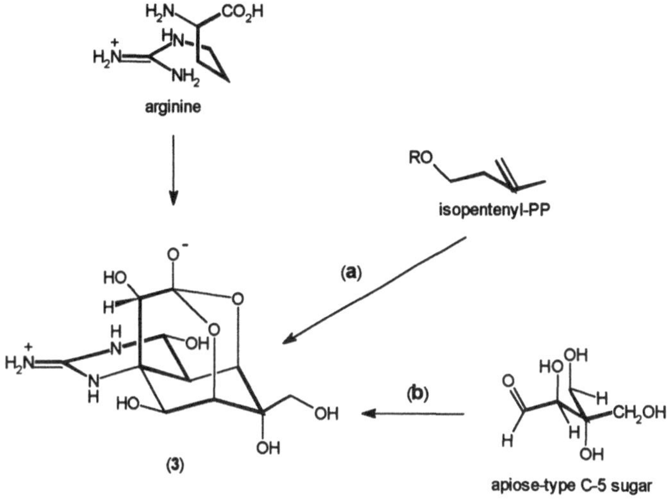

Scheme 20. Proposed biogenetic pathway of tetrodotoxin

synthesis of enantiomerically pure material would make large amounts of this compound and its derivatives available. For synthetic routes to derivatives, see references (*492*) and (*549–554*).

SHIMIZU has discussed most of the work dealing with saxitoxin and its derivatives (*491, 555–563*), and other reviews have also appeared (e.g., *564*). In addition to saxitoxin (**6**) (*565, 566*) a variety of related toxins have been isolated and identified. SHIMIZU listed more than a dozen of saxitoxin derivatives (*560, 563*): gonyautoxin-I (**292a**) (*567, 571, 573, 583–583b*) and gonyautoxin-II (**292b**) (*567, 568*) both isolated from soft-shell clams exposed to *Gonyaulax tamarensis*, the latter being also found in the clam *Mya arenaria*, gonyautoxin-III (**292c**) and gonyautoxin-IV (**292d**) isolated directly from *G. tamarensis* (*567–571*) and gonyautoxin-V (**292e**) isolated from a number of organisms (*572–574, 583*) which is the major toxin of *Protogonyaulax bahamense* var. *compressa* (*575, 576*) whose structure was reported by KOEHN et al. (*577*) and by SHIMIZU et al. (*556*). There are also gonyautoxin-VI (**292f**) (*572–574*) identical with the B2 toxin described by KOEHN and collaborators (*556, 577*), gonyautoxin-VII or decarbamoylneosaxitoxin (**292g**) first isolated from the sea scallop *Placopecten magellanicus* (*578*) and later from the bivalve *Spondylus butleri* (*580*) and from the little neck *Protothaca staminea* (*579*), gonyautoxin-VIII (**292h**), which easily isomerizes to epigonyautoxin-VIII (**292i**) and has an *N*-sulfonyl group and a negative net charge on the molecule (*581, 582*), as well as neosaxitoxin (**292j**), first isolated from the Alaska

butter clam (583) and later found in various other organisms (563), whose
structure was assigned by spectroscopic (mainly ^{13}C-NMR) and chemical
evidence (584) and confirmed by ^{15}N-NMR measurements (558). Finally
there are the C_3 (292k) and C_4 (292l) toxins, from *Protogonyaulax sp.* (574)
whose structure was recently confirmed by X-ray analysis (585). A
detailed discussion of the chemical properties of saxitoxin derivatives has
been presented by SHIMIZU (563).

(6): R_1=CONH$_2$, R_2=R_3=R_4=H

(292a): R_1=CONH$_2$, R_2=OH, R_3=OSO$_3^-$, R_4=H

(292b): R_1=CONH$_2$, R_2=H, R_3=OSO$_3^-$, R_4=H

(292c): R_1=CONH$_2$, R_2=R_3=H, R_4=OSO$_3^-$

(292d): R_1=CONH$_2$, R_2=OH, R_3=H, R_4=OSO$_3^-$

(292e): R_1=CONHSO$_3^-$, R_2=R_3=R_4=H

(292f): R_1=CONHSO$_3^-$, R_2=OH, R_3=R_4=H

(292g): R_1=H, R_2=OH, R_3=R_4=H

(292h): R_1=CONHSO$_3^-$, R_2=R_3=H, R_4=OSO$_3^-$

(292i): R_1=CONHSO$_3^-$, R_2=H, R_3=OSO$_3^-$, R_4=H

(292j): R_1=CONH$_2$, R_2=OH, R_3=R_4=H

(292k): R_1=CONHSO$_3^-$, R_2=OH, R_3=H, R_4=OSO$_3^-$

(292l): R_1=CONHSO$_3^-$, R_2=OH, R_3=OSO$_3^-$, R_4=H

(292m): R_1=R_2=R_3=R_4=H

Almost the complete biogenetic pathway leading to saxitoxin is now
known (558–562). However, as the topic has been reviewed many times
(586–591), will not be discussed here. Saxitoxin and some of its derivatives
occur mainly in microorganisms and molluscs, but also in algae, pufferfish and other organisms (588–592). The FAB and SIMS mass spectra of
tetrodotoxin, saxitoxin and other saxitoxin derivatives have also been
reported (593, 594). Two syntheses of racemic saxitoxin by KISHI (563,
595, 596) and JACOBI (563, 597–600) and an enantioselective synthesis of
decarbamoylsaxitoxin (292m) by the first of these authors (601, 603) were

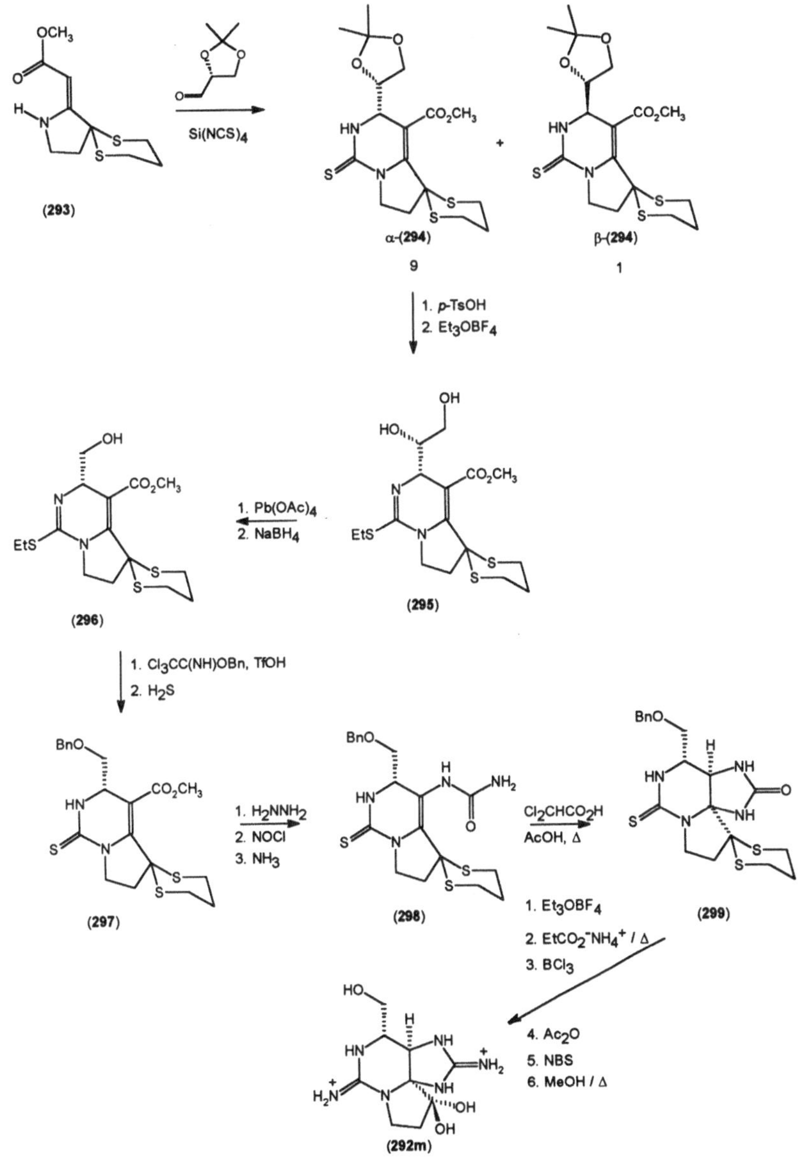

Scheme 21. Enantioselective total synthesis of decarbamoylsaxitoxin

accomplished. Decarbamoylsaxitoxin (**292m**) has been isolated as a naturally occurring derivative of saxitoxin (*580*) and as a biologically active product of saxitoxin acid hydrolysis (*602*).

The enantioselective synthesis of decarbamoylsaxitoxin (Scheme 21) possesses some intriguing features, notably the reaction leading to α-(**294**), whose stereochemistry was determined by X-ray crystallography and the fact that in CDCl$_3$ solution at high concentrations, racemic (**299**) exhibits a molecular recognition-like phenomenon forming a dimeric structure by hydrogen bonding which was verified by ^{1}H-NMR spectroscopy. The tricyclic urea-thiourea (**299**) was converted into (**292m**) by the same procedure employed in the synthesis of racemic saxitoxin. The unnatural antipode of decarbamoylsaxitoxin has no sodium channel blocking activity.

Cypridina luciferin (**300**) is one of the most interesting marine guanidine derivatives. Isolated from sun-dried *Cypridina hilgendorfii* this

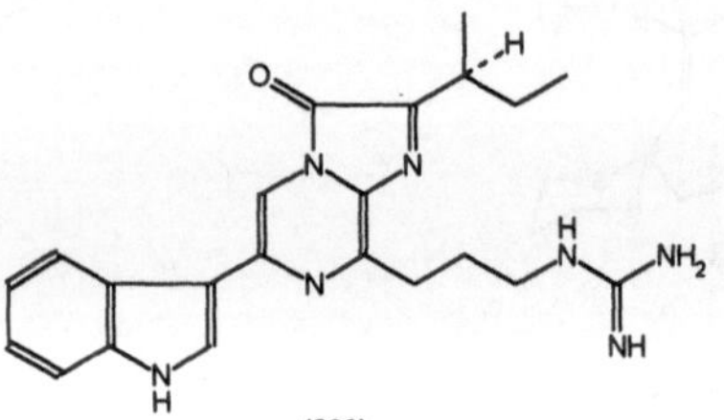

(300)

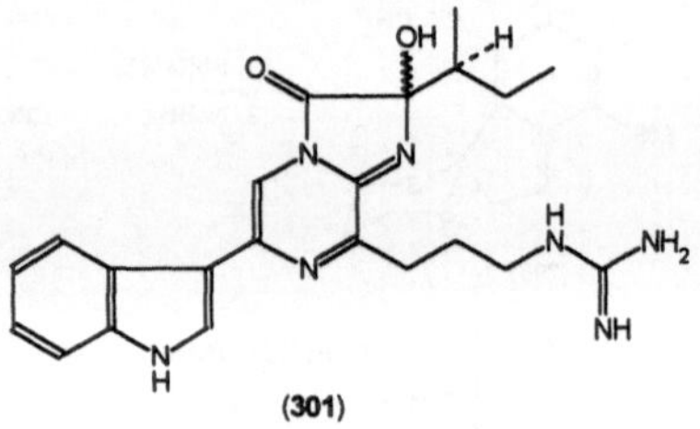

(301)

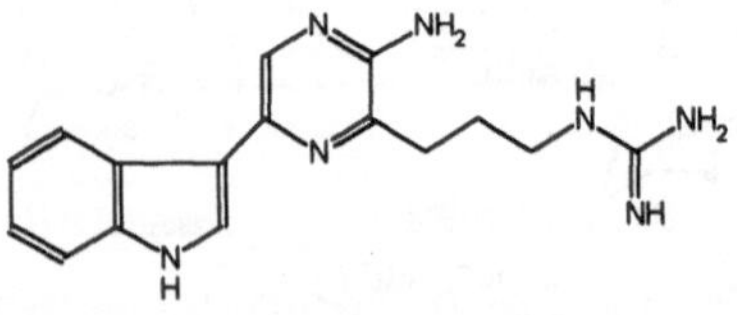

(302)

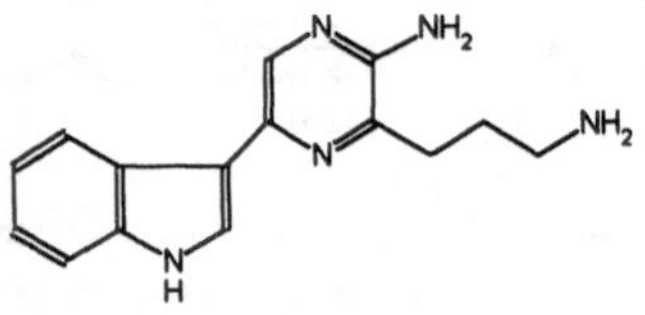

(303)

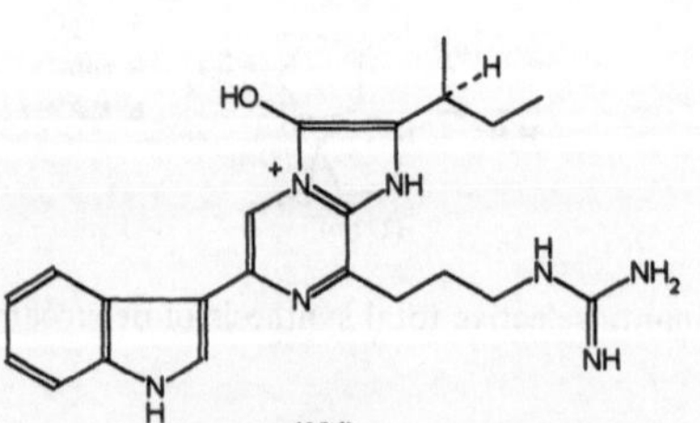

(304)

orange-red crystalline luciferin is one of the simplest in nature requiring only *Cypridina* luciferase in the presence of oxygen to emit light with the formation of *Cypridina* oxyluciferin (**301**) and *Cypridina* etioluciferin (**302**) (*604*). The same products are obtained by treatment of *Cypridina* luciferin with ammonia in the presence of air, but without production of light. Their relatively simple structures and that of etioluciferamine (**303**)

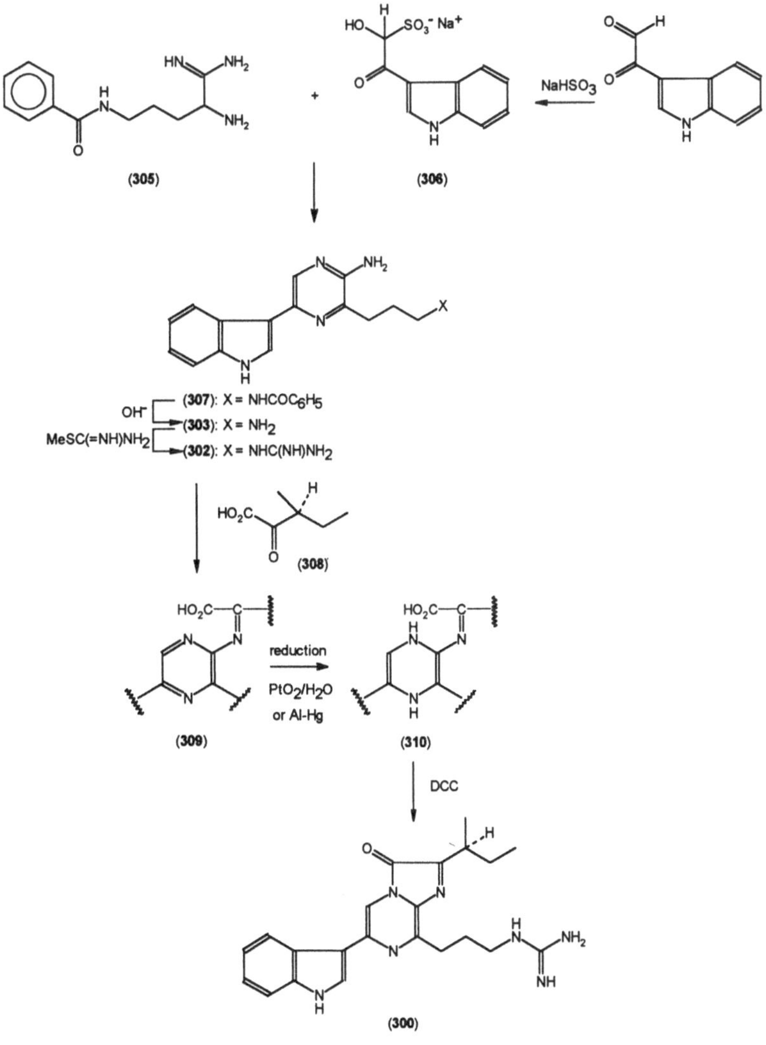

Scheme 22. Synthesis of *Cypridina* etioluciferin, etioluferamine and luciferin

determined by mass spectral and ^{1}H-NMR analysis, chemical degradations and biogenetic considerations allowed GOTO and HIRATA to establish the structure of *Cypridina* luciferin (*605*). The extended dihydropyr-
azine structure (**304**) is responsible for the oxidation state during the
luminescence.

The proposed structures were confirmed by mass spectrometry (*606*),
and total synthesis of etioluciferin and *Cypridina* luciferin itself (*607*)
(Scheme 22). The overall yield of first luciferin synthesis was less than 1%.
This synthesis was later improved by direct condensation between
etioluciferin (**302**) and α-keto-β-methyl-valeraldehyde with dilute hydro-
bromic acid in methanol, giving *Cypridina* luciferin in 30% yield (*607a*).
Marine bioluminescent chemistry has been reviewed by GOTO and KISHI
(*607b*) and also by GOTO (*608*).

The potent toxins microcystin-LR (**311**) and nodularin (**312**) isolated
from *Microcystis aeruginosa* and *Nodularia spumigena*, respectively, are

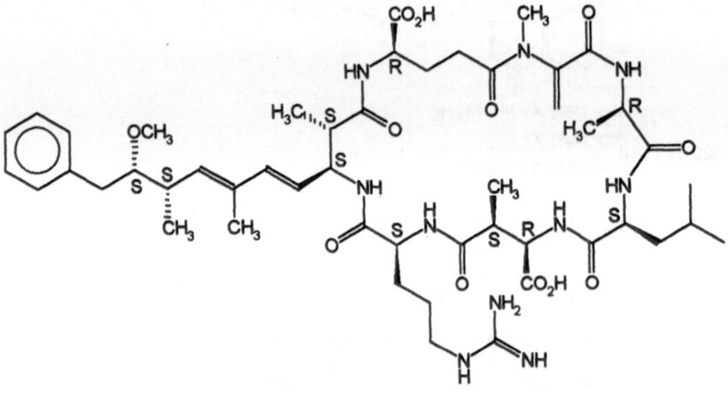

(311)

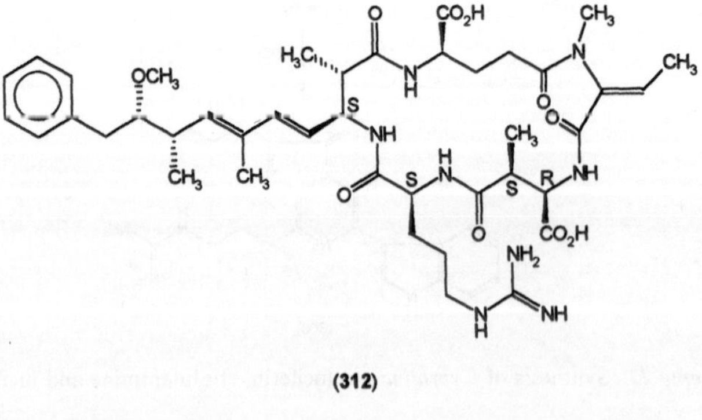

(312)

cyclic peptides which possess the unique amino acid 3-amino-9-methoxy-
-2,6,8-trimethyl-10-phenyl-4,6-decadienoic acid (Adda) (*609*). Their structures were established by chemical degradation and spectroscopic analysis.

Anatoxin-a(s) (**313**) is a neurotoxic alkaloid which was isolated from
the blue-green alga (cyanobacteria) *Anabaena flos-aquae* (*610*). Its potent
toxicity (LD_{50} 20-40 µg/kg mice) is attributed to its exceptional anticholinesterase activity (*611–613*). Anatoxin-a(s) (**313**) decomposes rapidly in
alkaline solution but is stable in neutral or acidic (pH 3–5) media.
Structure determination was achieved by FABMS, ^{1}H-NMR, ^{13}C-NMR,
^{31}P-NMR and ^{15}N-NMR analysis and by degradation to (**314**), (**315**) and
monomethylphosphate.

(313)   (314)   (315)

The absolute configuration of the asymmetric carbon in anatoxin-a(s)
was assigned by synthesis of the *R*- and *S*-derivatives of (**315**) from D- and
L-asparagine, respectively (Scheme 23). The CD spectrum of synthetic
(**315**) derived from L-Asn was identical with that of natural material, thus
establishing the *S* configuration. Anatoxin-a(s) (**313**) has shown good
incorporation of [*methyl*-^{14}C]-L-methionine, [2-^{14}C] glycine and [U-
^{14}C] serine into the methyl substituents during biosynthesis (*614*).
Moreover, the carbon skeleton of anatoxin-a(s) incorporated efficiently
[*amidine*-^{14}C]-L-arginine, [U-^{14}C]-L-arginine, [4,5-^{3}H]-L-arginine,
[U-^{14}C]-L-ornithine and sodium [1,2-^{14}C] acetate. These results suggested two possible biogenetic precursors for anatoxin-a(s), *erythro*-4-
hydroxy-L-arginine or enduracididine (**174**). The former was isolated in
minute quantities from *A. flos-aquae* (*614*). Recently, CARMICHAEL has
reviewed chemical, pharmacological and biological aspects of anatox-
in-a(s) and other cyanotoxins (including anatoxin-a, saxitoxin, the micro-
cystins and nodularin) (*615, 616*).

More recently, OHTANI et al. (*617*) isolated the potent hepatotoxin
cylindrospermosin (**320**) from the blue-green alga (cyanobacteria) *Cylindrospermopsis raciborskii*. The structure was determined by extensive
spectroscopic analysis, such as CIDMS and HMQC, HMBC, NOESY
and COSY NMR experiments. MOORE et al. (*591*) have proposed a
biogenetic pathway for the biosynthesis of cylindrospermosin.

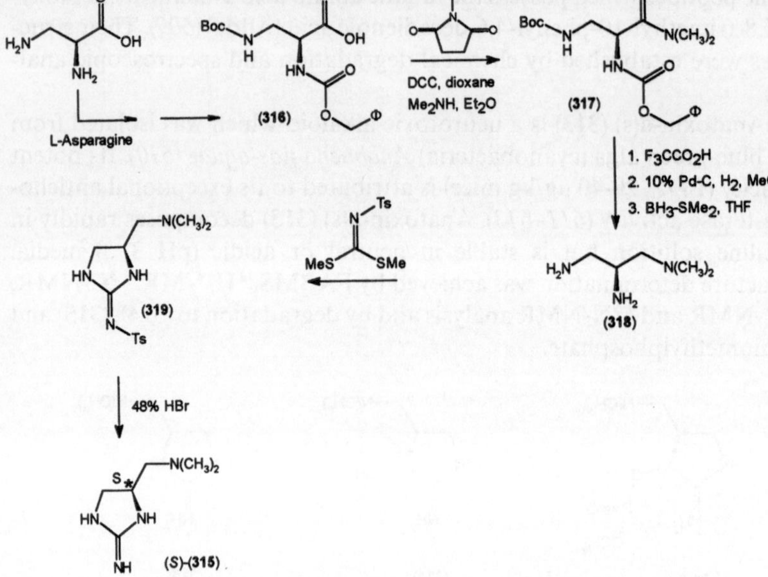

Scheme 23. Synthesis of (S)-de-(methylphosphate)anatoxin-a(s)

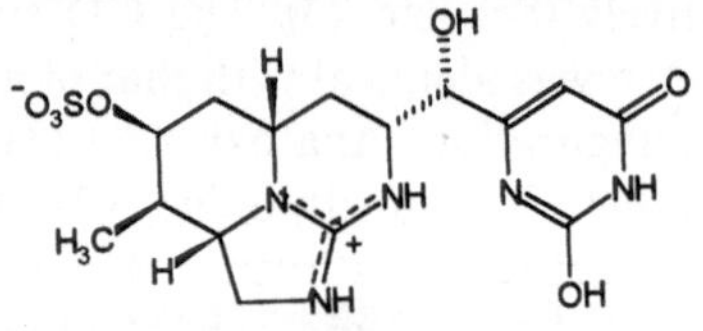

(320)

4.2. Marine Algae

A cyclopropyl amino acid named carnosadine (321) was isolated from a red alga *Grateloupia carnosa* by WAKAMIYA *et al.* (*618*). The structure of carnosadine (321) was fully elucidated by spectroscopic analysis (*618*) and confirmed by two syntheses (Schemes 24 and 25) (*619–621*). In the first,

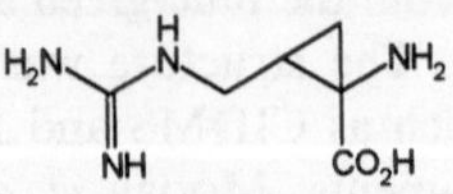

(321)

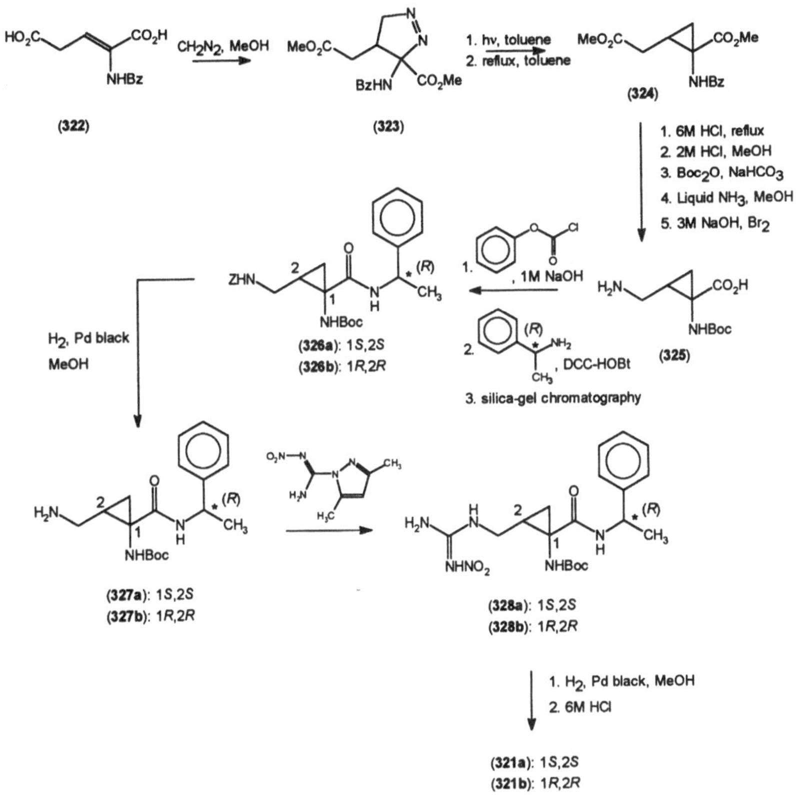

Scheme 24. Synthesis of carnosadine

the absolute stereochemistry of the final product was established by X-ray analysis of the intermediate (1*S*, 2*S*)-(**327a**).

From aqueous extracts of the red alga *Schottera nicaeensis* CHILLEMI *et al.* (*48*) isolated nicaeensin (**339**) structurally characterized by spectroscopic analysis and chemical degradation. Drastic acid hydrolysis afforded *N*-methyl-1,4-butanediamine (**340**) together with guanidine. Partial acid hydrolysis of nicaeensin gave in addition deacetyl nicaeensin (**341**) and *N*-methyl-*N*-acetyl-1,4-butanediamine (**342**). Finally, treatment of nicaeensin with pentane-2,4-dione in alkaline medium gave the expected 1-[3-(2-amino-4,6-dimethylpyrimidinyl)ureido]-4-(*N*-methylacetamido) butane derivative (**343**). Nicaeensin (**339**) is probably biogenetically related to gongrine (**344**) and gigartinine (**345**), whose chemistry has been reviewed (*622*). Some new arginine and gigartinine derivatives have been reviewed by IRELAND *et al.* (*623*), such as (*S*)-citryllinyl-(*S*)-arginine (**346**),

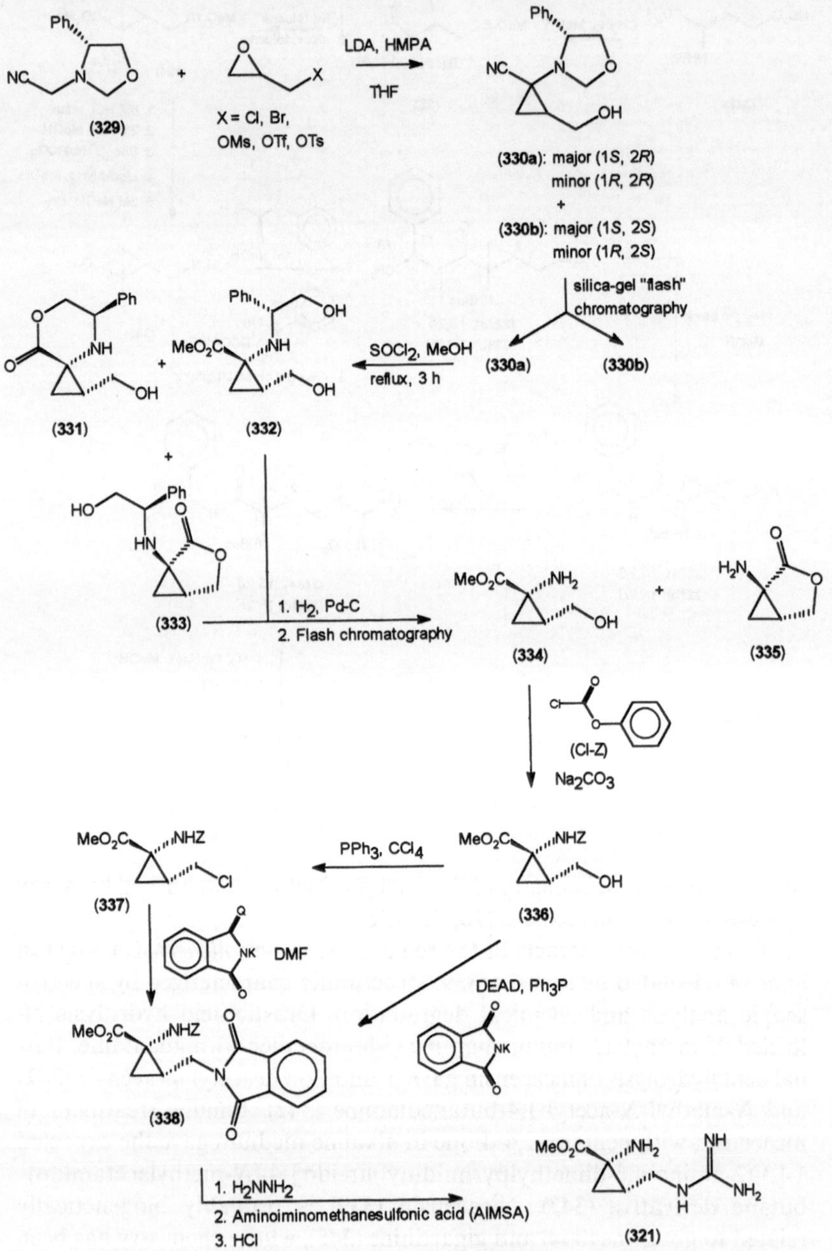

Scheme 25. Asymmetric synthesis of carnosadine

isolated from the marine algae *Grateloupia turutura* (S)-gigartinyl-(S)-gigartinine (347) from *G. livida* and (S)-arginyl-(S)-glutamine (348) from *Enteromorpha linza*.

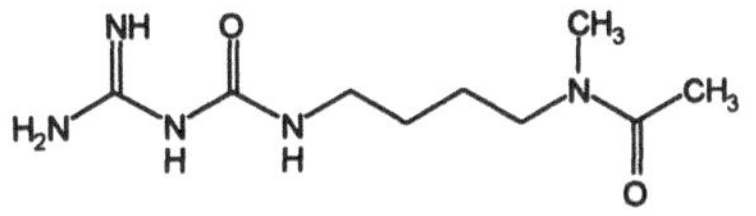

(339)

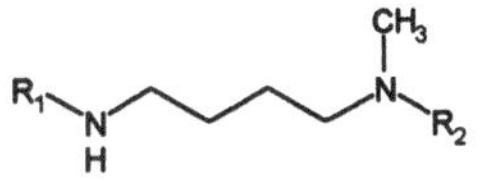

(340): R$_1$=R$_2$=H
(341): R1=H$_2$N-C(-NH)-NH-CO- , R$_2$=H
(342): R1=H, R$_2$=H$_3$CCO-

(343): R$_2$=H$_3$CCO- , R$_1$= -CO-N

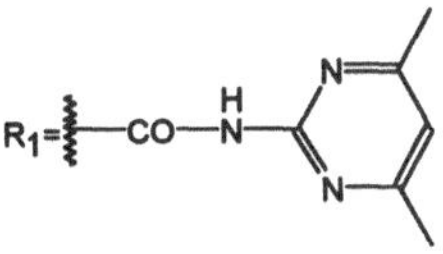

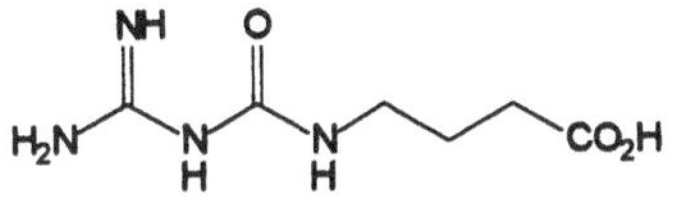

(344)

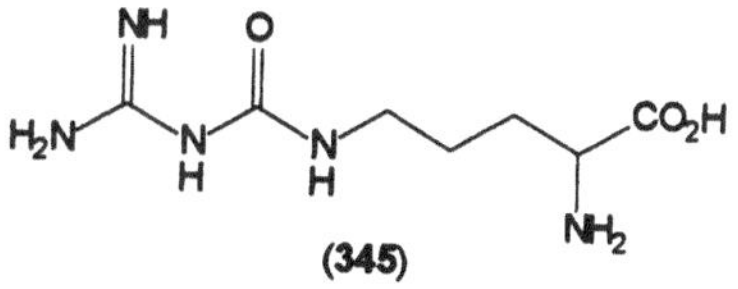

(345)

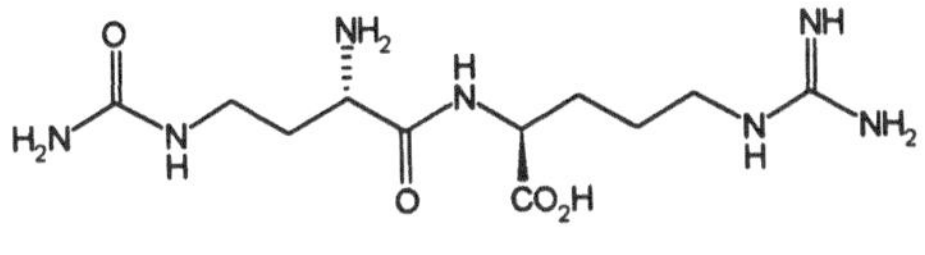

(346)

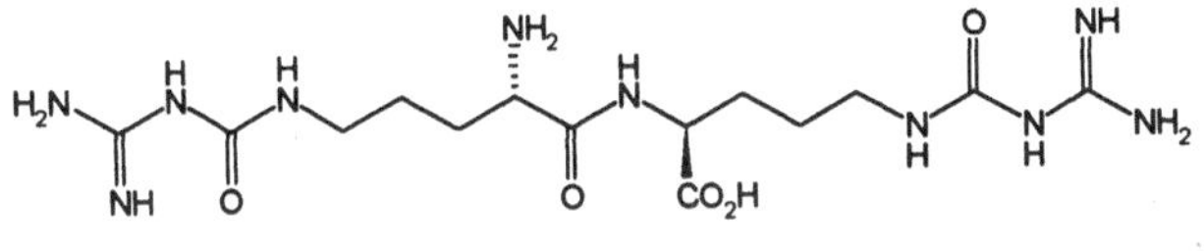

(347)

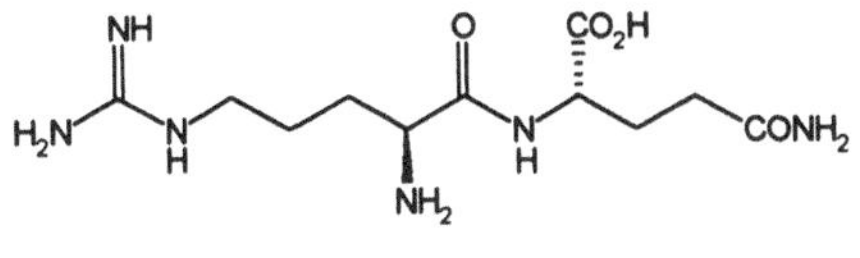

(348)

4.3. Marine Sponges

Marine sponges have provided an array of polycyclic guanidine metabolites. From *Phakellia flabellata*, SHARMA *et al.* (*25, 624, 625*) isolated two weakly basic guanidine compounds named monobromophakellin (**4**) and (−)-dibromophakellin (**5**). These compounds had pKa values below 8, which are rather low when compared with other guanidines. Their structures were deduced by UV, IR, mass spectra, ^{1}H-NMR and ^{13}C-NMR analysis (*25*), and confirmed by X-ray diffraction. The low basicity of the guanidine in phakellins is partially explained by the angle strain in the polycyclic system, the loss of planarity inhibiting charge stabilization in the salt (*25*). The chemistry of the phakellins has been reviewed by CHEVOLOT (*10*).

(**4**): R$_1$=H, R$_2$=Br
(**5**): R$_1$=R$_2$=Br

A biomimetic synthesis of dibromophakellin was achieved (*626*) *via* synthesis of dihydrooroidin (**353**) (Scheme 26). Dihydrooroidin hydrochloride (**353**) was cyclized to dibromophakellin *via* the proposed intermediate (**354**). Alternatively, *N*-methylated dihydrooroidin (**355**) gave the supposed analogous intermediate (**356**). The intermediate (**354**) was exposed to potassium tert-butoxide/2-butanol, giving a quantitative yield of racemic dibromophakellin (**5**).

Compounds biogenetically related to the phakellins have been isolated from other sponges. A "yellow compound", 2-debromohymenialdisine (**357**), has been isolated from *Phakellia flabellata* extracts (*627*) and from *Hymeniacidon aldis* (*629*). Acylation of (**357**) free base gave the diacetyl derivative (**358**). Potassium permanganate oxydation of (**357**) free base yielded guanidine and (**359**), the latter on reduction with sodium borohydride giving (**360**). These results corroborated the proposed structure (**357**) (*627*). Intramolecular hydrogen bonding was verified by the pronounced downfield shift of the pyrrole N-H resonance in the ^{1}H-NMR spectrum of (**357**), (**357**) free base, (**359**) and (**360**).

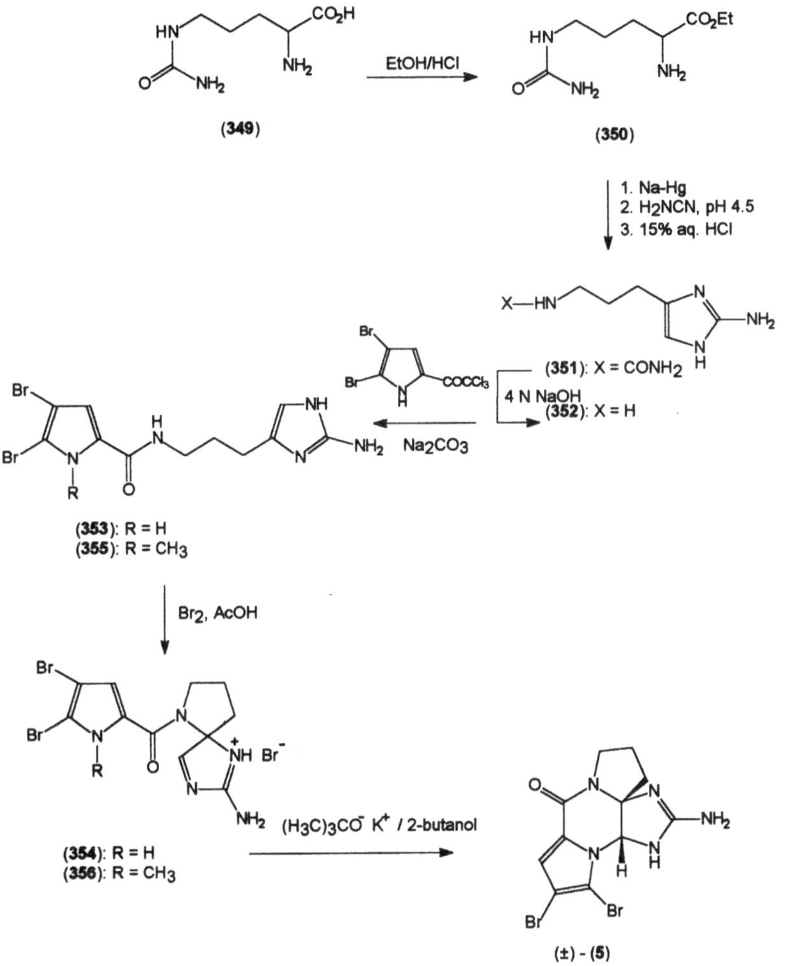

Scheme 26. Synthesis of racemic dibromophakellin

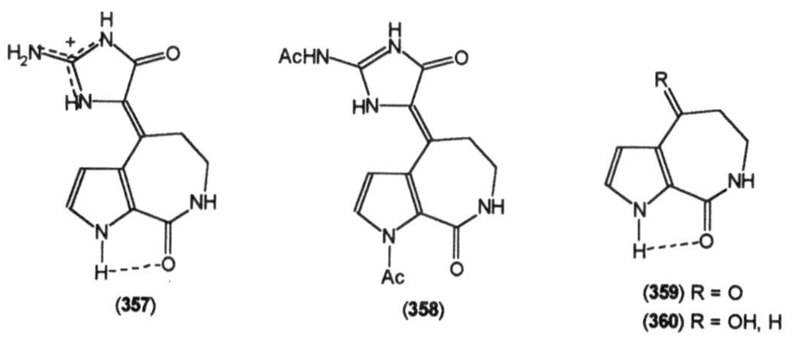

The 2-bromo derivative hymenialdisine (**361**) was found in *H. aldis,* *Axinella verrucosa* and *Acanthella aurantiaca* polar extracts (*628, 629*). Hymenialdisine (**361**) was identified by analysis of spectral data and X-ray diffraction, as well as by analysis of the spectral properties of its diacetyl derivative.

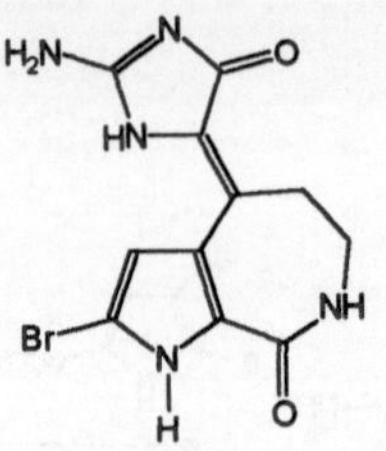

(361)

A biogenetic pathway was proposed for monobromophakellin (**4**), dibromophakellin (**5**), 2-debromohymenialdisine (**357**) and hymenialdisine (**361**), with proline and α-guanidinoornithine as their precursors. 2-Debromohymenialdisine (**357**) and hymenialdisine (**361**) were subsequently isolated from Palau (Caroline Islands), Truk (Caroline Islands) and Papua New Guinea *Axinella* and *Hymeniacidon* sponge specimens (*630*). SCHMITZ et al. (*631*) also isolated 2-debromohymenialdisine (**357**) and the bicyclopyrrololactam moieties aldisin (**359**) and 2-bromoaldisin (**362**) from *Hymeniacidon aldis* and an unidentified sponge, all being identified by analysis of their spectral data.

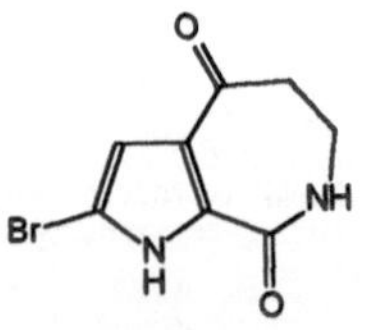

(362)

Pseudaxinyssa canthanella from a New Caledonian lagoon yielded three related compounds, (9R),(10S)-dibromocantharelline (**363**), odiline (**364**) and the (+)-enantiomer of dibromophakellin (**5**), together with 2-debromohymenialdisine (**357**) and hymenialdisine (**361**) (*632*). (+)-Dibromophakellin (**5**) was identified by comparison of its spectral data with those reported for (−)-dibromophakellin earlier, and independent

X-ray analysis. Dibromocantharelline (**363**) and its di-debromo deriva-tive were identified by ^{1}H-NMR, ^{13}C-NMR and mass spectral analysis. Odiline (**364**) and its mono N-acetyl derivative (at the primary nitrogen of guanidine) were identified by analysis of spectral data.

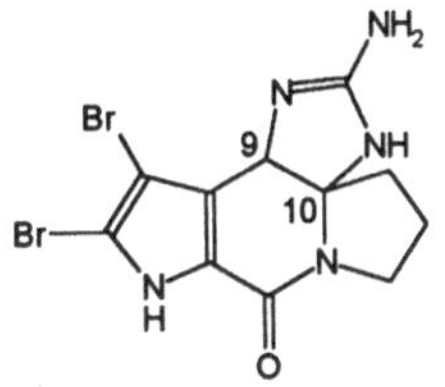

(363) 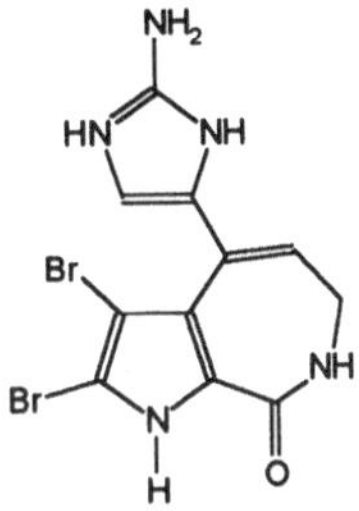   (364)

Dibromoisophakellin, the (9S),(10R)-enantiomer of dibromocanthar-elline (**363**), was isolated from the sponge *Acanthella carteri* (*633*) and was identified by spectral data and X-ray analysis. The same Russian researchers also isolated dibromoagelaspongin (**365**) from a sponge of the genus *Agelas* (*634*). These compounds, as well as didebromo-(**365**) formed by catalytic hydrogenation were identified by their spectroscopic proper-ties and by X-ray analysis. A biogenetic pathway was suggested for the formation of (**365**). The chemotaxonomy of the genus *Agelas* has been reviewed on the basis of the biogenetic relationships of "phakellin" related compounds (*635*).

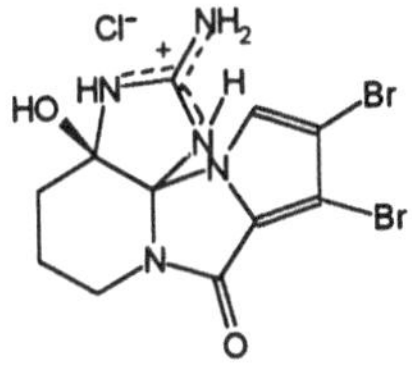

(365)

KOBAYASHI *et al.* (*636*) have reported the isolation of oxysceptrin (**366**), an oxidized form of sceptrin (**367**), from *Agelas* cf. *nemoechinata*. Oxyscep-trin (**366**) was identified by comparison of its spectroscopic properties with those of sceptrin and has the same absolute stereochemistry. Both oxysceptrin (**366**) and sceptrin (**367**) are considered to be 2 + 2 cycloaddi-tion products of hymenidin (**368**); the oxidation of C-15 of oxysceptrin may occur after the cycloaddition. Oxysceptrin (**366**) and its 3-debromo

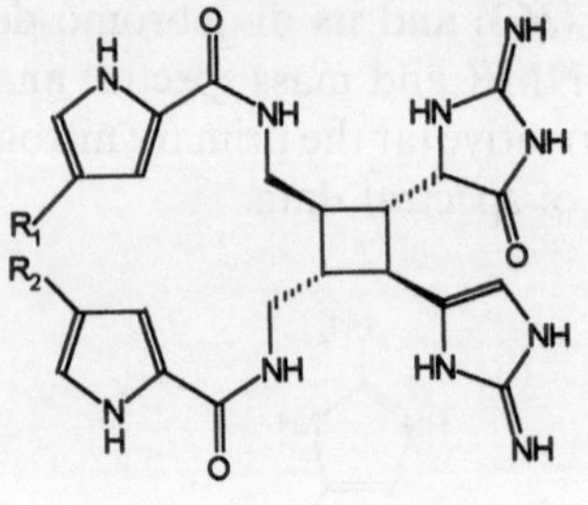

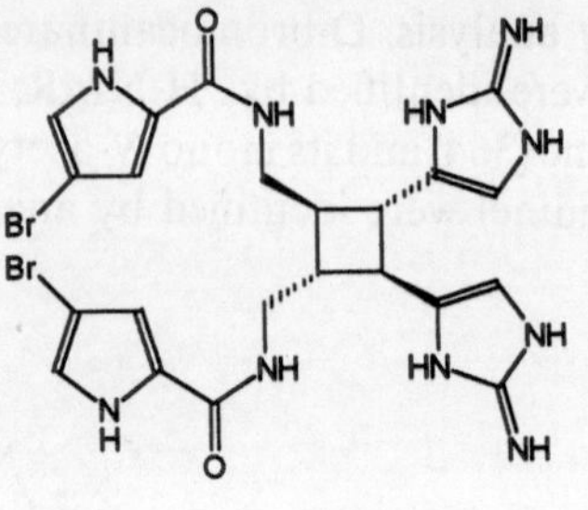

(366): R_1=R_2=Br
(369): R_1=H, R_2=Br

(367)

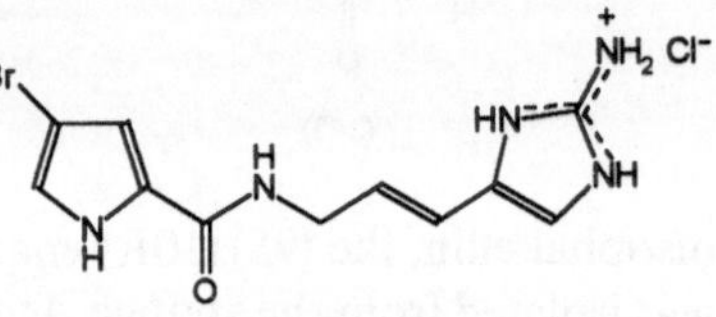

(368)

derivative (**369**) were also isolated from a Caribbean specimen of *Agelas conifera* (*637*).

Kinnel *et al.* (*638*) have reported the isolation from the sponge *Stylotella agminata* of a substance named palau'amine (**370**) with an unprecedent polycyclic skeleton incorporating a *cis*-fused bicyclo[3.3.0] azaoctane ring. The structure of palau'amine (**370**) was deduced by different NMR techniques such as HMBC, COSY, HMQC and ROESY. Together with palau'amine, the authors also isolated many other related compounds among them sceptrin (**367**), hymenidin (**368**), dibromophakellin (**5**), hymenialdisine (**361**) and 2-debromohymenialdisine (**357**). Palau'amine possesses remarkable immunosuppressive activity.

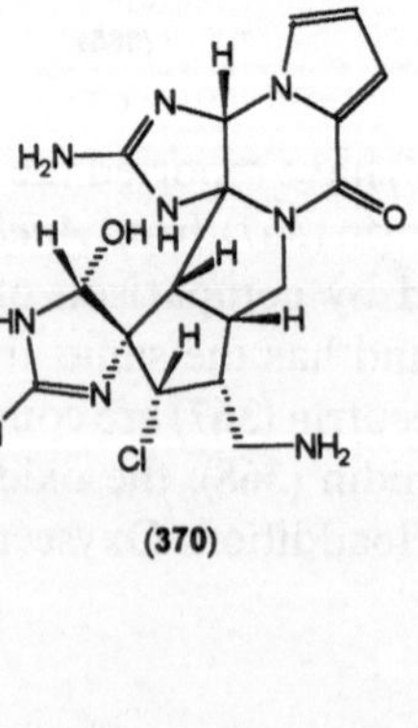

(370)

References, pp. 248–295

Aplysinopsin (**373**) was simultaneously isolated in two different laboratories (*639, 640*). KAZLAUSKAS *et al.* (*639*) isolated it from the sponge *Thorecta sp.* (more probably *Aplysinopsis* sp.) and identified it by spectroscopic means and by synthesis: condensation of indole-3-aldehyde (**371a**) with N^3-methylcreatinine (**372a**) gave aplysinopsin. HOLLENBEAK and SCHMITZ (*640*) isolated aplysinopsin from another sponge *Verongia spengelli*. Comparison of the spectral data with those reported by KAZLAUSKAS *et al.* (*639*) established the identity. While the stereochemistry of the double bond remained unresolved by KAZLAUSKAS *et al.* (*639*), the *E* configuration was proposed by HOLLENBEAK and SCHMITZ (*640*) based on NOE experiments. A third species, *Dercitus sp.* also yielded aplysinopsin (**373**) together with 2'-*N*-demethylaplysinopsin (**374**) and 6-bromo-2'-*N*-demethylaplysinopsin (**375**) (*641*) similarly identified by spectroscopic properties and synthesis (Scheme 27). Aplysinopsin (**373**) and 6-bromoaplysinopsin (**376**) also occur in the sponge *Smenospongia aurea* (*642*). DALKAFOUKI *et al.* (*643*) describe a similar synthetic route for these and related compounds (*644*).

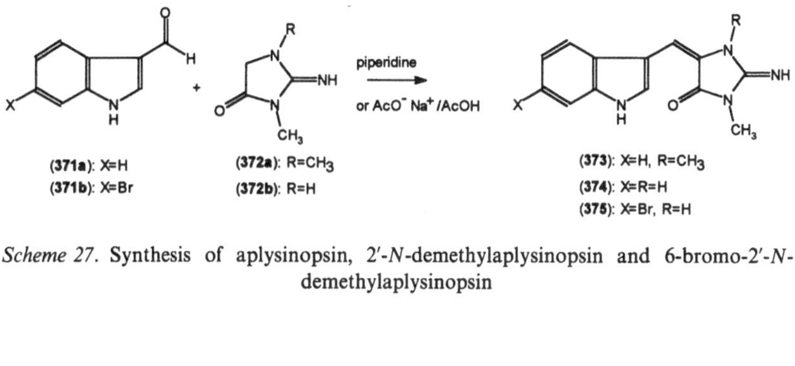

(**371a**): X=H
(**371b**): X=Br

(**372a**): R=CH$_3$
(**372b**): R=H

(**373**): X=H, R=CH$_3$
(**374**): X=R=H
(**375**): X=Br, R=H

Scheme 27. Synthesis of aplysinopsin, 2'-*N*-demethylaplysinopsin and 6-bromo-2'-*N*-demethylaplysinopsin

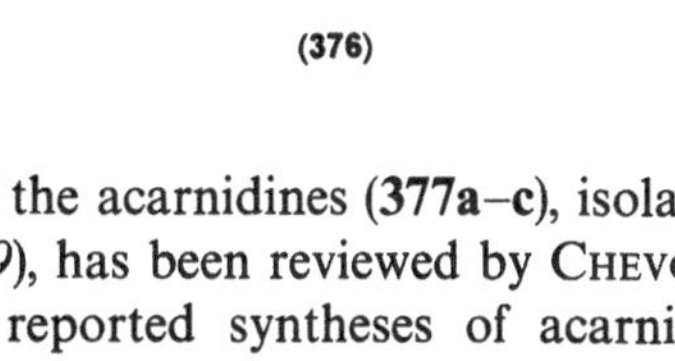

(**376**)

The chemistry of the acarnidines (**377a–c**), isolated from the sponge *Acarnus erithacus* (*49*), has been reviewed by CHEVOLOT (*10*). In view of this, only the two reported syntheses of acarnidine (**377a**) will be mentioned.

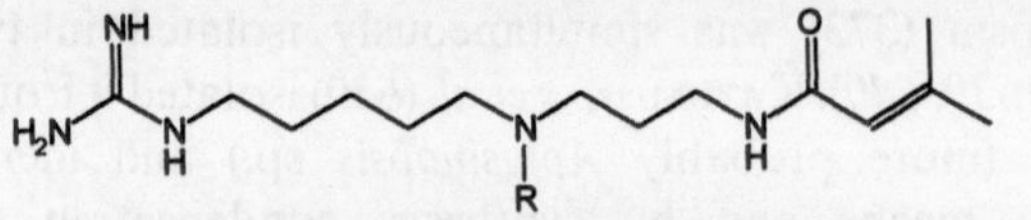

(377a): R = CO(CH$_2$)$_{10}$CH$_3$
(377b): R = CO(CH$_2$)$_3$CH=CH(CH$_2$)$_5$CH$_3$
(377c): R = CO(CH$_2$)$_{12}$CH$_3$

In the first synthesis (645, 646) (Scheme 28), SWERN oxidation of alcohol (380) gave aldehyde (381) which was condensed with mono-protected 1,5-diaminopentane (382, cadaverine) leading through several more steps via the unstable primary amine (385) to 3,5-acarnidine (377a).

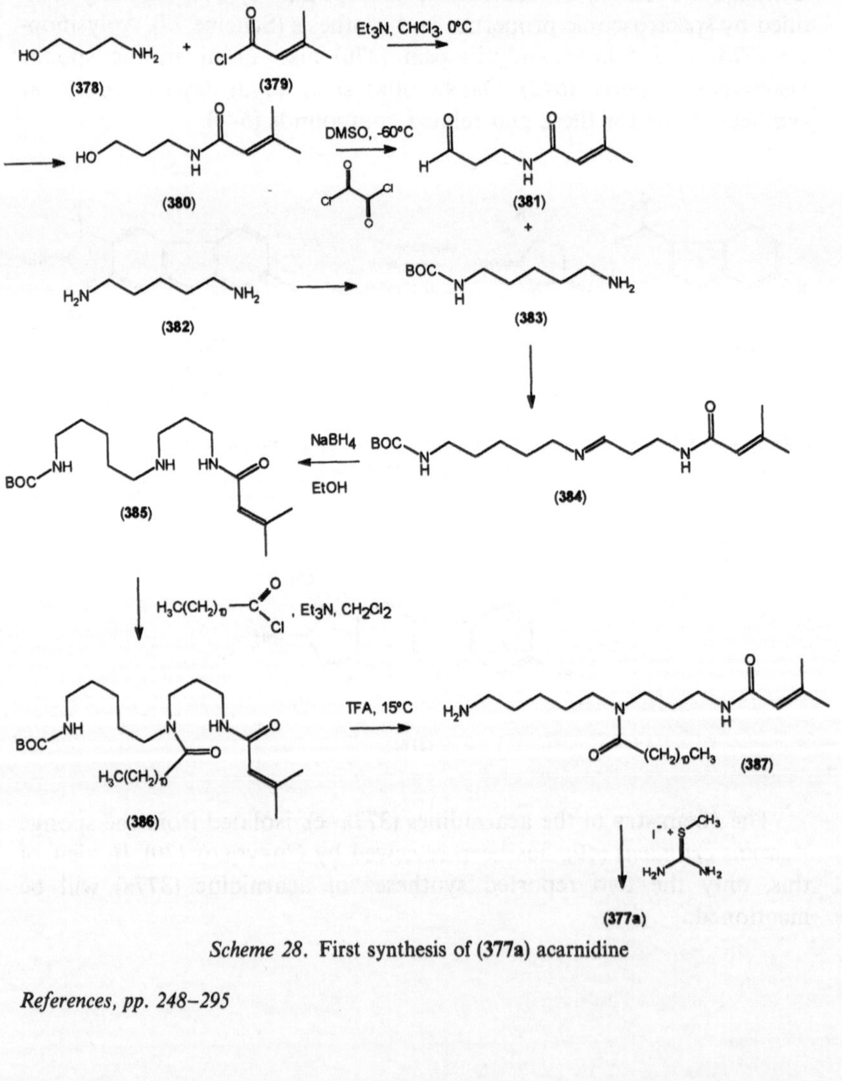

Scheme 28. First synthesis of (377a) acarnidine

In a second synthesis reported for (377a) (647) (Scheme 29), synthon (381) was condensed with amine (381) to give imine (382). After its transformation into nitroguanidine (384), the nitro group could only be removed by reduction with a mercury cathode (− 0.9 V *versus* standard

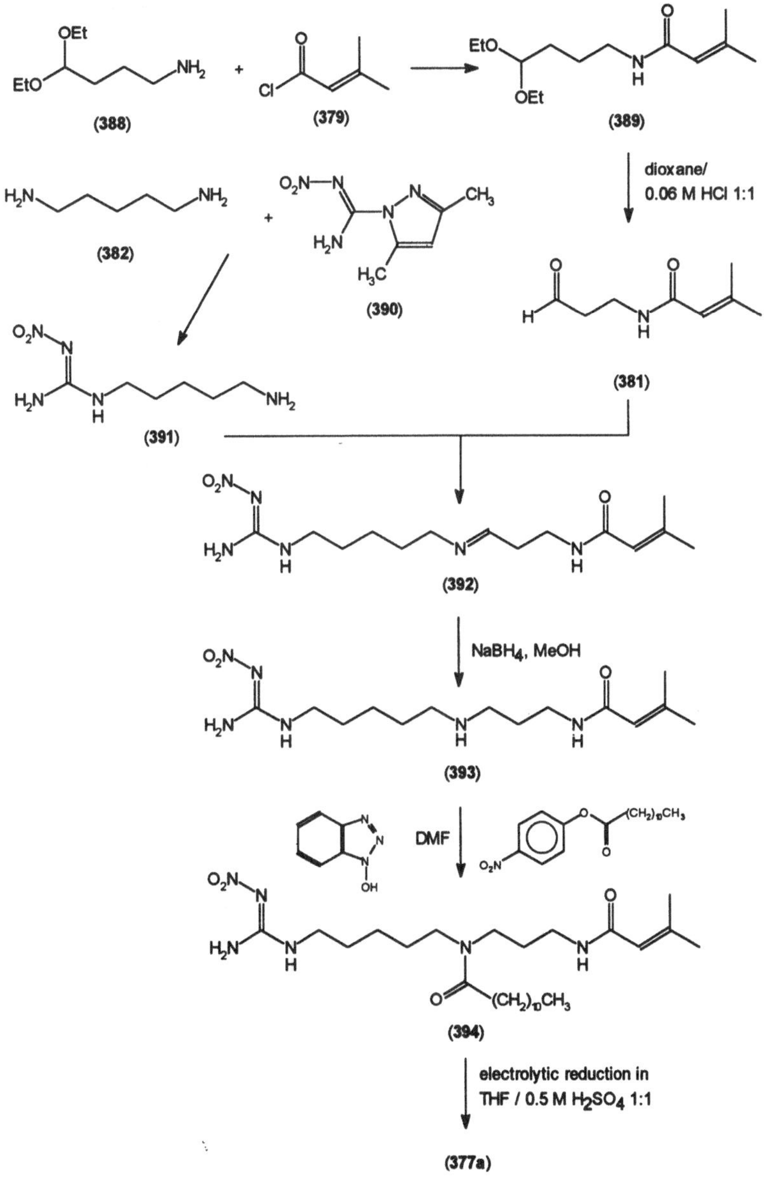

Scheme 29. Second synthesis of (377a) acarnidine

calomel electrode) in 1:1 THF/0.5 M aq. H_2SO_4 yielding (**377a**) isolated as its sulfate.

Ptilocaulin (**395**) and isoptilocaulin (**396**) are antimicrobial cytotoxic tricyclic guanidine derivatives isolated from the sponge *Ptilocaulis* aff. *Ptilocaulis spiculifer* (*648*) and were identified by ^{1}H-, ^{13}C-NMR studies, and by X-ray diffraction. The biogenetic pathway leading to these metabolites seems to involve addition of a guanidine to a polyketide chain. The interesting polycyclic framework of ptilocaulin has led to several syntheses (Schemes 30–34) (*649–656*).

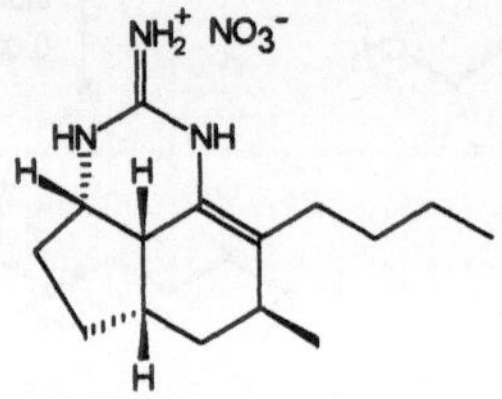

(+)-(**395**) 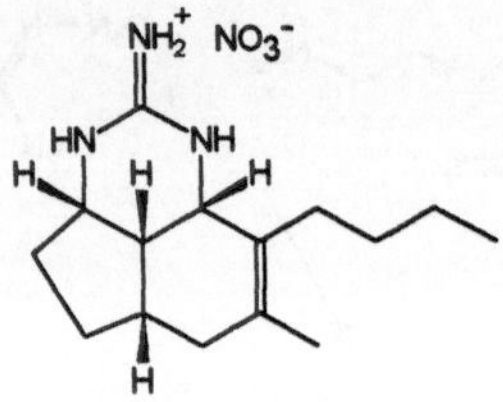 (**396**)

The first synthesis of (±)-ptilocaulin (**395**) (*649, 650*) was the simplest one (Scheme 30); a somewhat similar synthetic path starting with R-(+)-5-methylcyclohexanone (**405**) led to (−)-ptilocaulin (Scheme 31). The fact that this (−)-ptilocaulin exhibited a sign of rotation and CD spectrum opposite to those of the naturally occurring product established the absolute stereochemistry (+)-(**395**) for ptilocaulin.

The other syntheses comprise those of ROUSH and WALTS which also gave (−)-ptilocaulin, the enantiomer of (+)-(**395**) (Scheme 32) (*651, 652*), the elegant synthesis of (±)-ptilocaulin by UYEHARA *et al.* (Scheme 33) which involved a rearrangement of a bicyclo[3.2.2]non-6-en-2-one system (**422**) (*653*) and a synthesis by HASSNER (*654, 655*) (Scheme 34) involving an unusual annelation via an intramolecular nitrile oxide-olefin dipolar cycloaddition (INOC) and a rare example of an aldol condensation between a ketone and a protected aldehyde enolate. Finally, natural (+)-ptilocaulin has been synthesized (*656*) from (*S*)-(+)-5-trimethylsilyl-2-cyclohexenone (**444**) in manner analogous employed to that of SNIDER and FAITH.

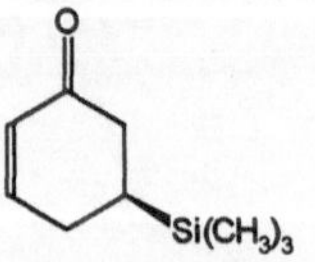

-(+)-(**444**)

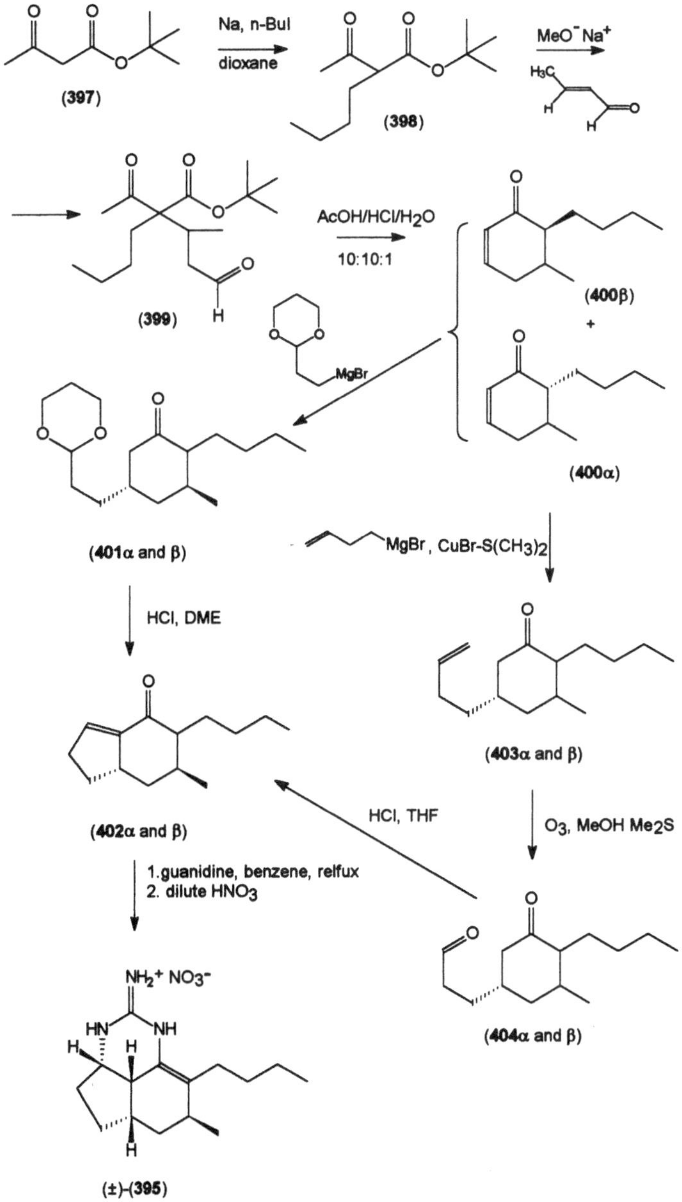

Scheme 30. SNIDER and FAITH synthesis of racemic ptilocaulin

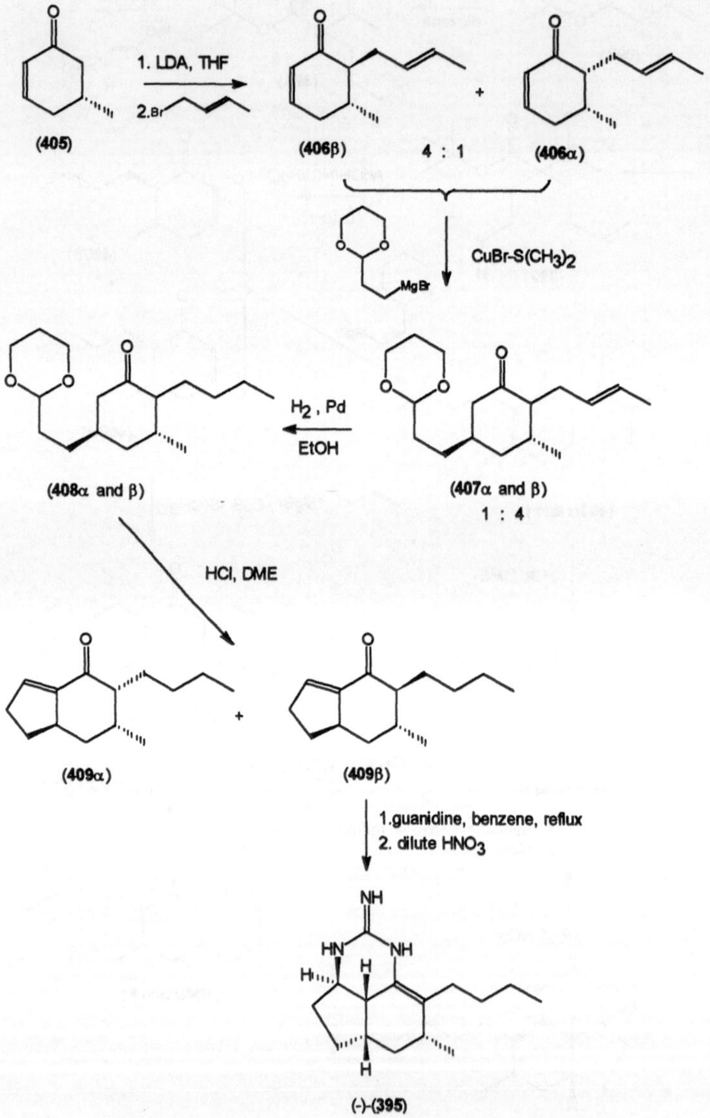

Scheme 31. SNIDER and FAITH synthesis of (−)-ptilocaulin

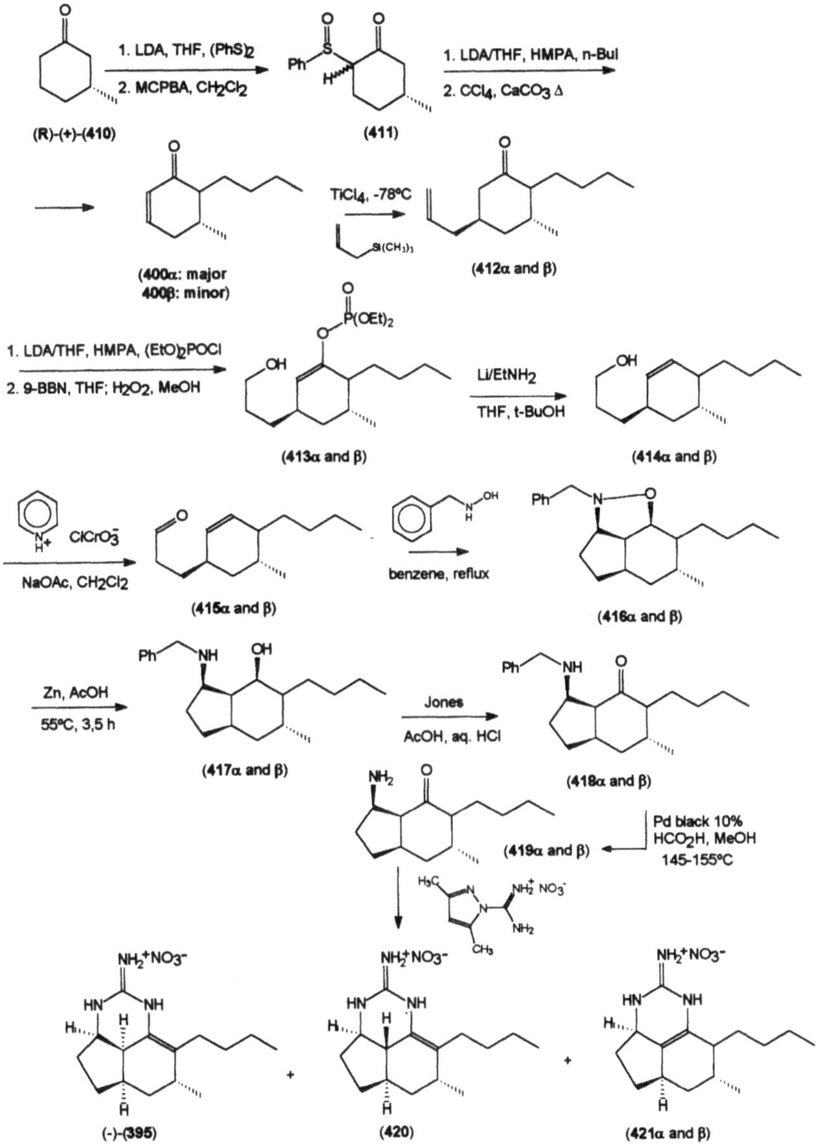

Scheme 32. ROUSH and WALTS synthesis of (−)-ptilocaulin

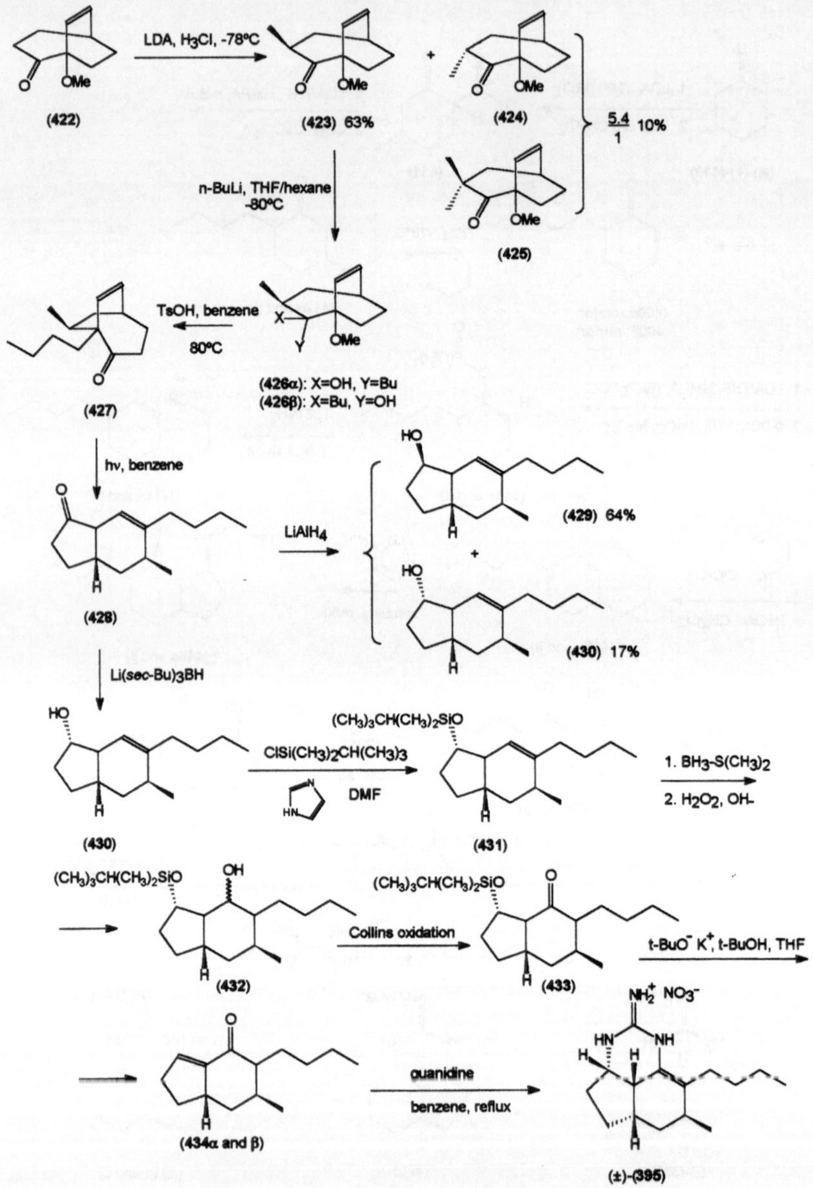

Scheme 33. Uyehara *et al.* synthesis of racemic ptilocaulin

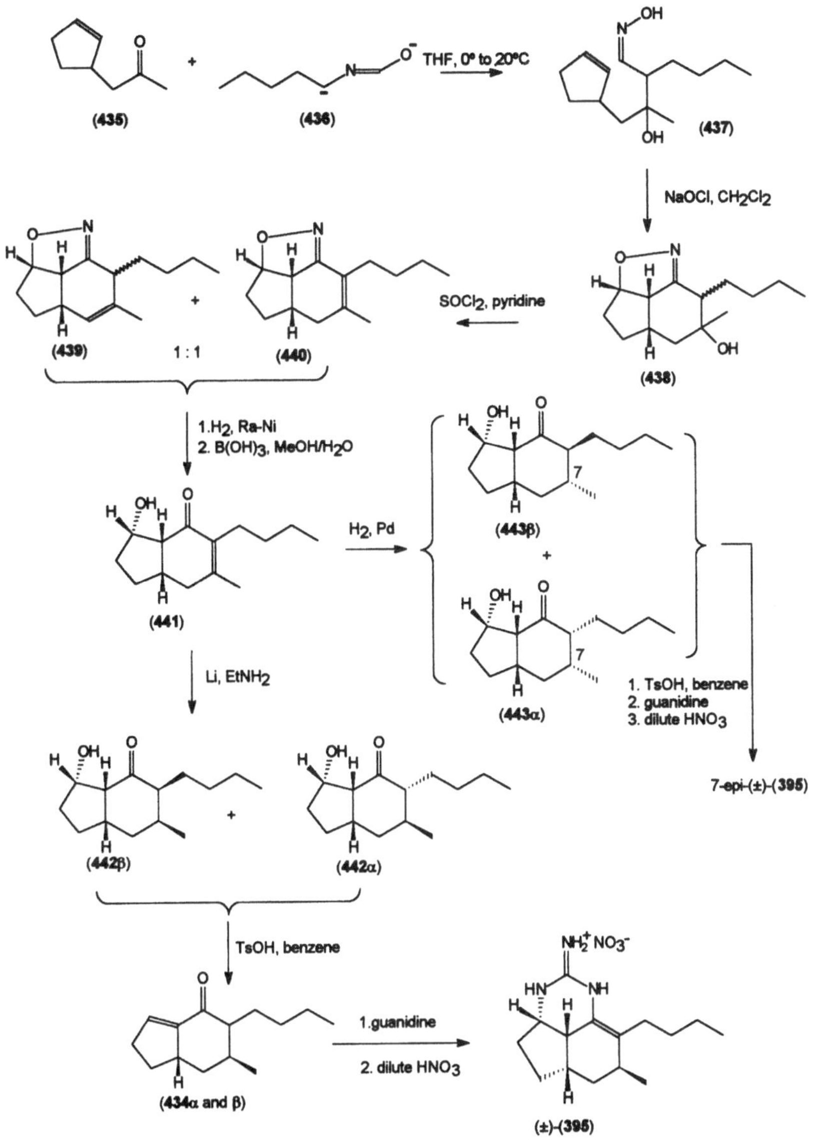

Scheme 34. HASSNER and MURTHY synthesis of racemic ptilocaulin

Siphonodictidine (**445**) is a guanidine furanosesquiterpene found in the mucus secreted by the sponge *Siphonodictyon* sp. (*657*). This compound is responsible for the inhibition of coral growth in the vicinity of the sponge. Siphonodictidine was identified by analysis of spectral data of

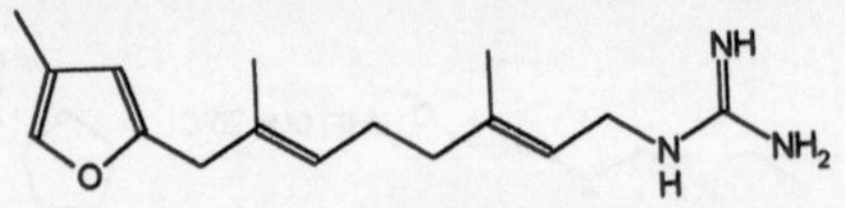

(445)

the natural product and of the 4,6-dimethylpyrimid-2-yl derivative. A simple but efficient synthesis of siphonodictidine has been reported (Scheme 35) (*658, 659*).

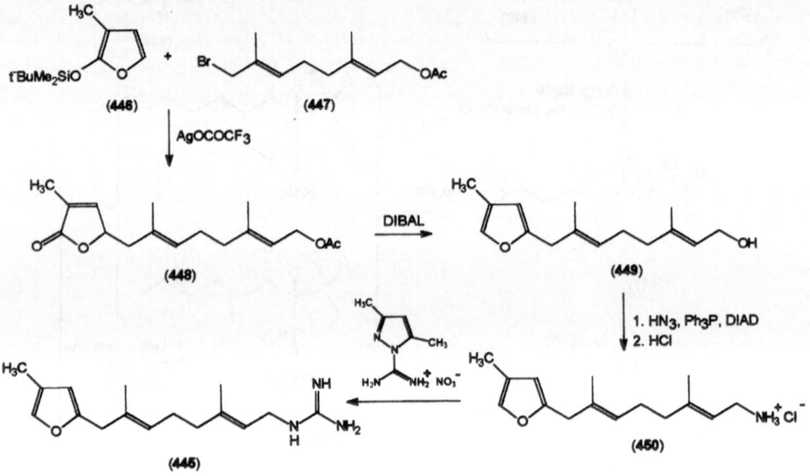

Scheme 35. Synthesis of siphonodictidine

Agelasidine A (**451**) as well as its congeners agelasidine B (**452**) and agelasidine C (**453**) have been isolated from sponges of the genus *Agelas* (*660–662*). They were identified by spectroscopic analysis of the natural products and of their 2-amino-4,6-dimethylpyrimidyl derivatives, and also by chemical degradation (*660, 662*): ozonolysis of agelasidine A, followed by reduction and acetylation gave a diacetate (**454**) and a tetraacetate (**455**). The absolute configurations of the asymmetric carbons in the cyclohexane ring moiety of agelasidines B and C were defined by chemical degradation and circular dichroism measurements as 1*R*, 4*R* by comparison with those of ageline A (**456**) (*662*). The absolute configurations of C-10 in agelasidine A (**451**) and of C-13 in agelasidine B (**452**) are opposite as shown by the opposite rotations of their respective (**455**) derivatives and tentatively assigned as *S* and *R* based on BREWSTER'S empirical rules (*662*). The suggested biogenetic pathway leading to the agelasidines may proceed from the corresponding terpene alcohols through direct replacement by hypotaurocyamine or by sigmatropic rearrangement of their hypotaurocyamine sulfenic esters (*662*). More recently, (−)-agelasidine C (**457**) and (−)-agelasidine D (**458**) were isolated from the Carribean marine sponge *Agelas clathrodes* (*663*).

Structures of these compounds were elucidated by spectroscopic analysis and by comparison of their physical and chemical properties with those of (+)-agelasidine C (**453**).

Agelasidine A (**451**) has been synthesized by ICHIKAWA (*664*) *via* a thermal rearrangement of (**460**) to the dithiocarbonate (**461**) (Scheme 36).

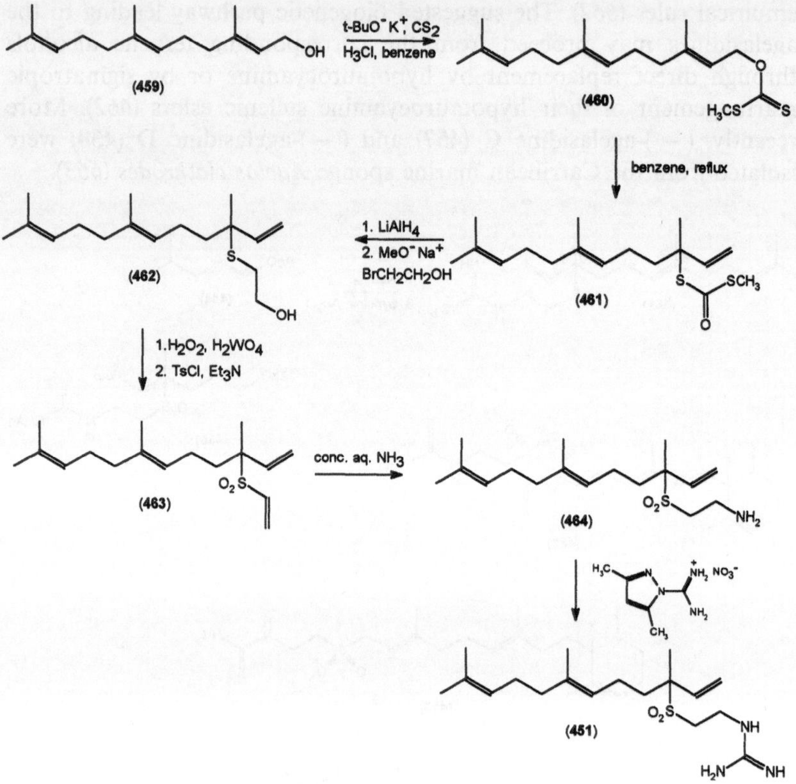

Scheme 36. Synthesis of agelasidine A

Suvanine (**465**), isolated from the sponges *Ircinia sp.* (*665*) and from *Coscinoderma matthewsi* (*666*) is a sesterpene derivative with an *N,N*-dimethylguanidinium counter-ion. The initial structure proposal was (**466**) (*665*), but was later revised (*666*) due to the fact that other cations could replace the *N,N*-dimethylguanidinium counter-ion. The suvanine ozonolysis/reduction product (**467**) was subjected to by X-ray diffraction, thus establishing the stereochemical framework. The *N,N*-dimethylguanidinium ion does not appear to be responsible for the biological activities reported for suvanine (*666*).

Sulfircin (**468**), from the sponge *Ircinia* sp., is the second sesterpenoid marine metabolite which possess a *N,N*-dimethylguanidinium counter ion of an organic sulfate (*667*). Its structure was elucidated by analysis of ¹H-NMR, ¹³C-NMR, FABMS spectra and X-ray diffraction.

Onnamide A (**469**) from a *Theonella sp.* possesses an unusual poly-functionalized skeleton (*668*). Its structure was completely elucidated by

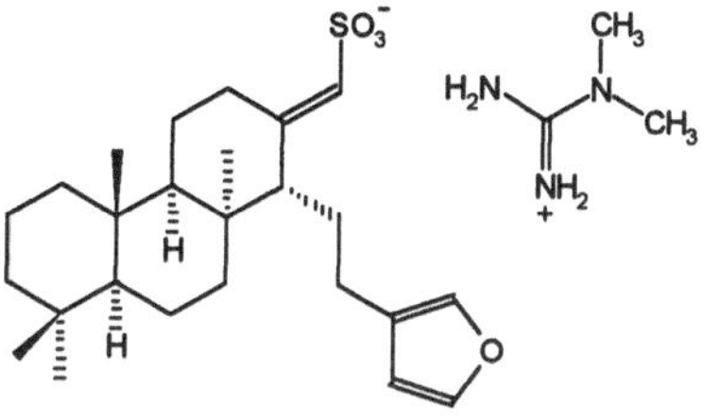

(465)

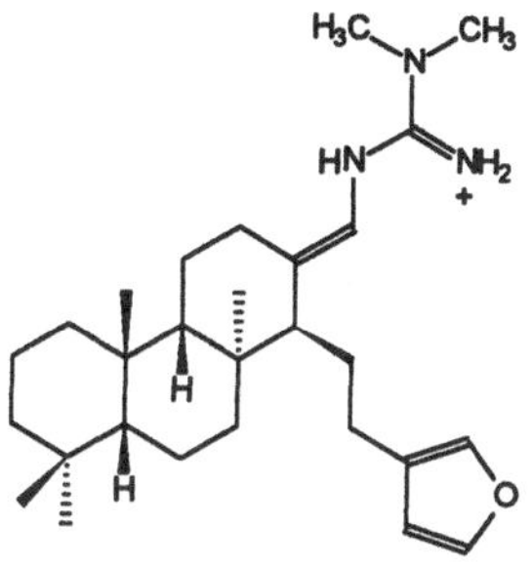

(466)

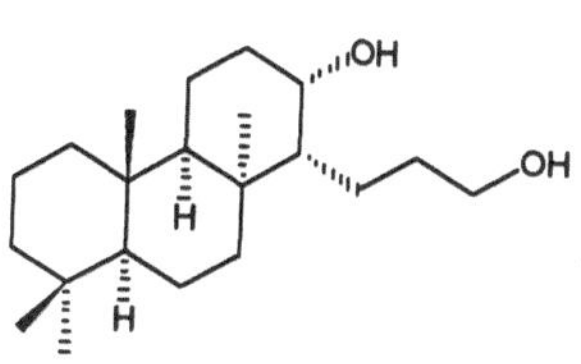

(467)

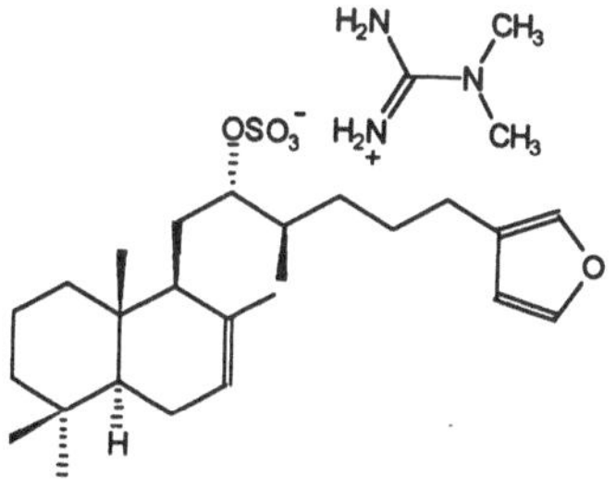

(468)

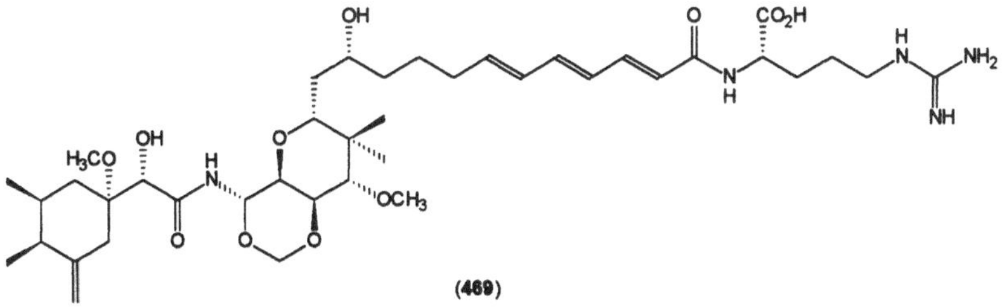

(469)

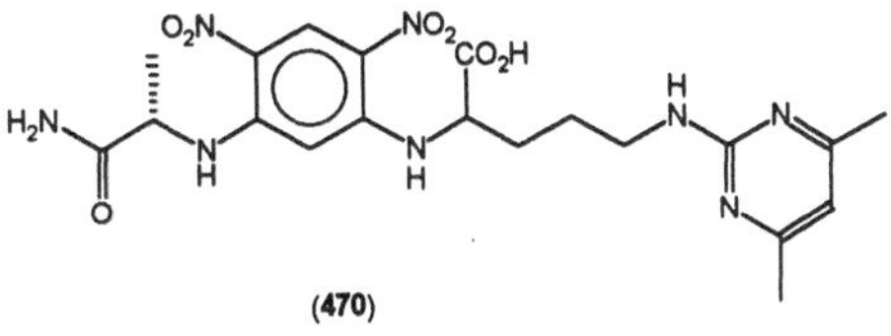

(470)

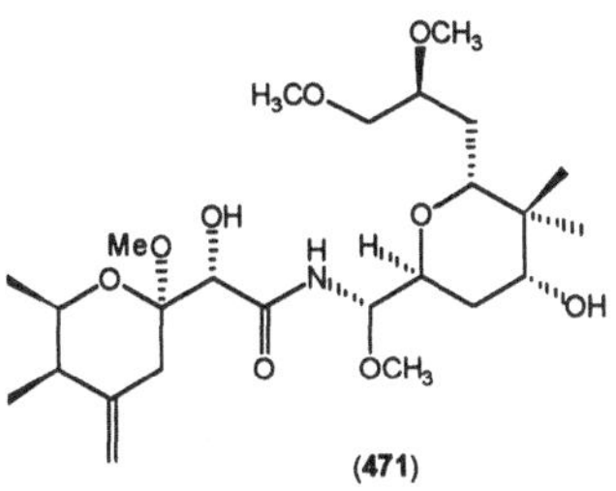

(471)

^{1}H- and ^{13}C-NMR spectroscopy, various 2D NMR techniques, FABMS and UV analysis of the natural product, of its methyl ester and of the degradation product (470). The relative configurations of the asymmetric centers were assigned on the basis of NOE difference spectroscopy and by comparison of the spectral data with that of pederin (471), an insect toxin (668 and references therein). More recently, a series of related "onnamides" was also isolated from sponges of the *Theonella* genus, and identified by their spectroscopic properties (669). The total synthesis of onnamide A has been reported (670, 671) (Scheme 37).

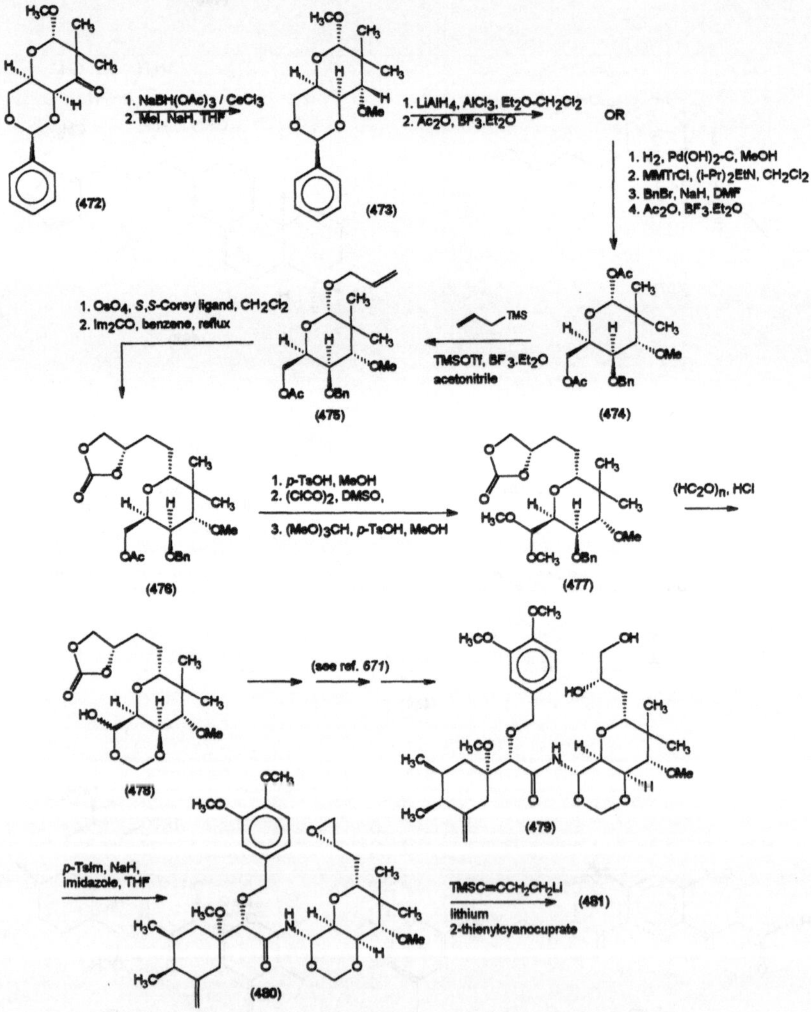

Scheme 37a. Synthesis of onnamide A (first part)

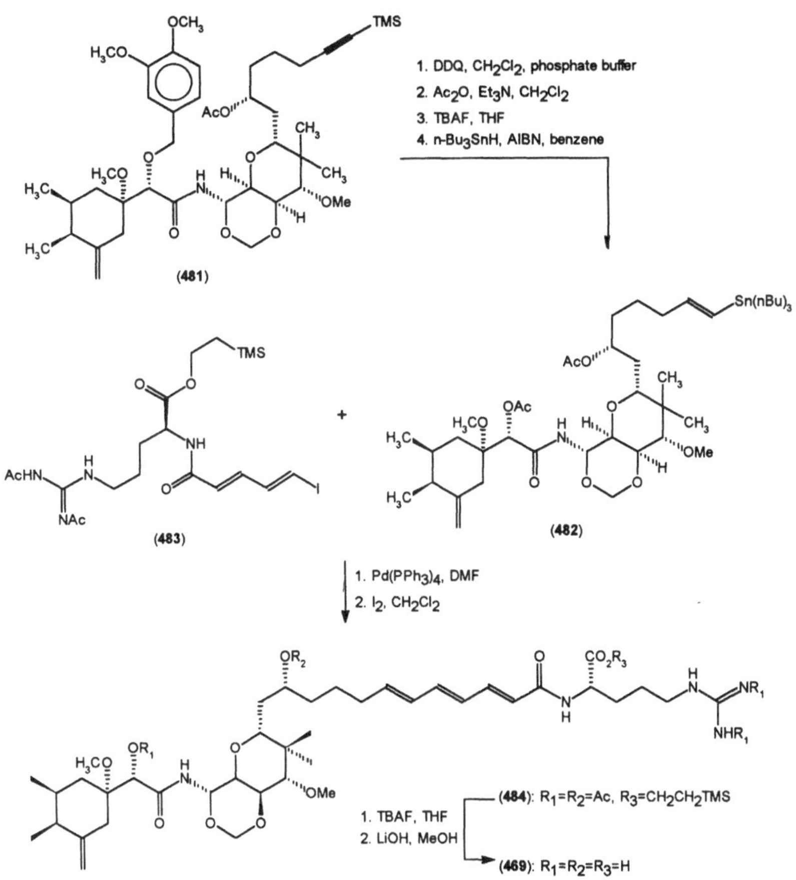

Scheme 37b. Synthesis of onnamide A (second part)

DEBITUS et al. (672) have isolated corallistine (485) from the sponge Corallistes fulvodesmus. The structure of corallistine (485) was elucidated by ¹H- and ¹³C-NMR spectroscopy and by X-ray diffraction of the acyl derivative (486).

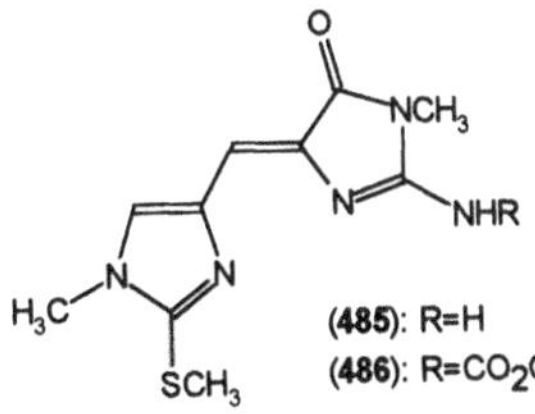

(485): R=H
(486): R=CO₂CH₂C(CH₃)₃

From two marine sponges, *Ptilocaulis spiculifer* and *Hemimycale* sp., KASHMAN *et al.* (*673–675*) have isolated the polycyclic guanidine alkaloid ptilomycalin A (**487**). The structure of the bis(trifluoroacetyl) derivative (**488**) was elucidated by various 2D-NMR techniques and conformation studies were also performed. The presence of a polysubstituted guanidine function was elegantly demonstrated by deuterium/proton exchange experiments.

(**487**): R=H, X=unspecified
(**488**): R=COCF$_3$, X=F$_3$CCOO

Biogenetically related to ptilomycalin A are the crambines A (**489**), B (**490a**), C1 (**491a**) and C2 (**492**) isolated as a mixture of homologues from the Mediterranean sponge *Crambe crambe* (*676, 677*). Their isolation was achieved by a complex chromatographic procedure (*678, 682*) and the structures were established by extensive NMR studies in one and two dimensions, as well as by FAB mass spectra analysis. Seasonal and geographical content variations in the amounts of these compounds were observed. BERLINCK *et al.* (*679*) have suggested a biogenetic pathway for these compounds. Very recently, JARES-ERIJMAN *et al.* (*680*) and also SNIDER and SHI (*687*) revised the structures of the major homologues crambine B to (**490b**) and crambine C1 to (**491b**) by tandem mass spectrometry and total synthesis, respectively. It was also suggested (*680*) that the name crambescins instead of crambines be used for these compounds, in order to avoid nomenclatural confusion.

JARES-ERIJMAN *et al.* (*681*) isolated polycyclic guanidine alkaloids closely related to ptilomycalin A from the more polar fraction of *Crambe crambe* extracts: crambescidin 800 (**493**), crambescidin 816 (**494**), crambescidin 830 (**495**) and crambescidin 844 (**496**). Besides the first two of these compounds BERLINCK *et al.* (*682*) also isolated iso-crambescidin 800 (**497**)

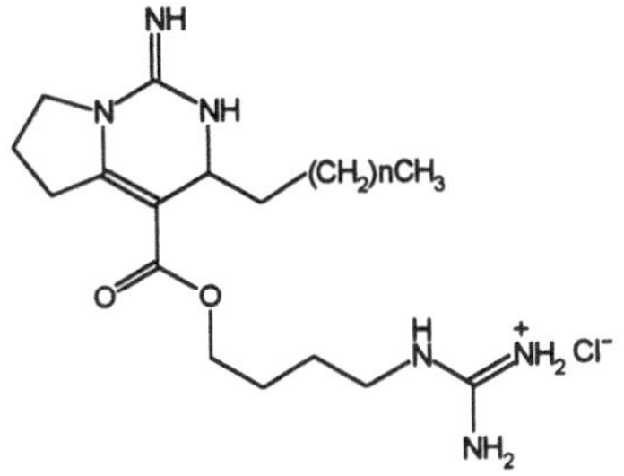

(489): n=8,9,10,11

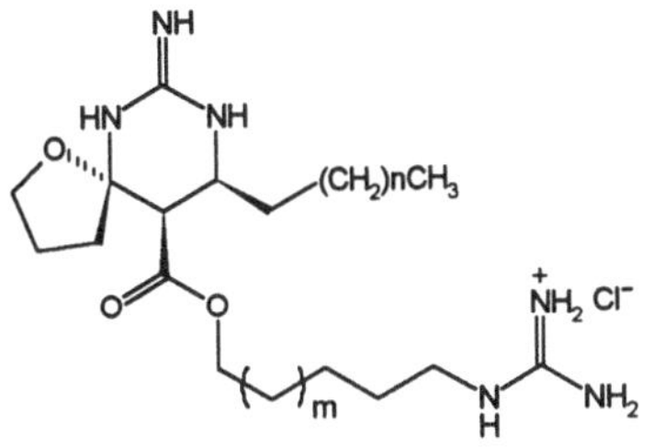

(490a) n=8,9,10,11; m=1
(490b) n=7; m=3

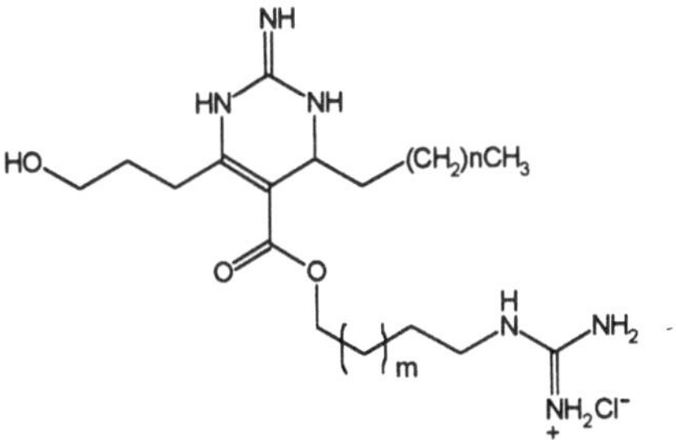

(491a): n=8,9,10,11; m=2
(491b): n=7; m=4
(492): n=9, m=1

and crambidine (498). Iso-crambescidin 800 (497) was isolated subsequently from the same sponge by JARES-ERIJMAN *et al.* (*683*) who also determined the absolute stereochemistry of the polycyclic moiety of these alkaloids by chemical correlations. Structures of these complex polycyclic guanidine alkaloids were elucidated by various 2D-NMR techniques, their assigned stereochemistry being based on NOE experiments (differential NOE spectra, NOESY and ROESY). It is not known whether crambidine (498), a 2-aminopyrimidine rather than a true guanidine, is a precursor or a derivative of the crambescidins.

SNIDER and SHI (*684*) proposed that the polycyclic skeleton of these ptilomycalin A-like guanidine alkaloids may be formed by addition of guanidine to a polyketide chain, in analogy to ptilocaulin and isoptilocaulin, and based on these assumptions, they synthesized the bicyclic moieties of crambescin A and crambescin B major homologues in a biomimetic fashion (Scheme 38). After determining the relative stereochemistry of the different synthetic stereoisomers of crambescin B derivatives (505), (506) and (507) by NMR analysis it was concluded that

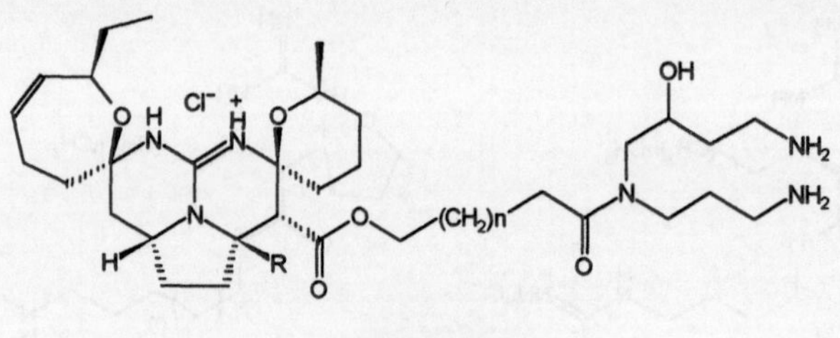

(493): R=H, n=13
(494): R=OH, n=13
(495): R=OH, n=14
(496): R=OH, n=15

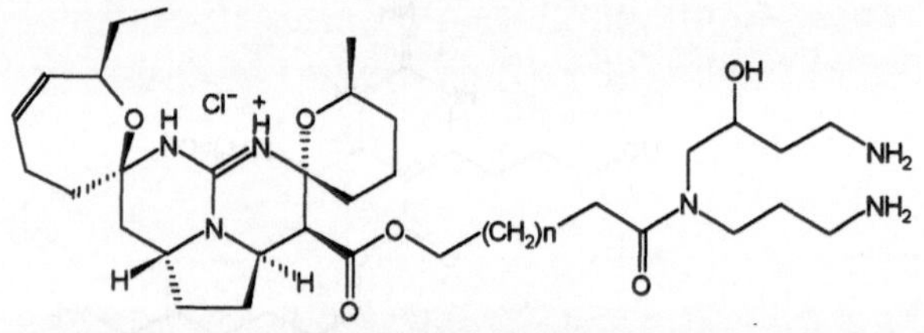

(497): n=13

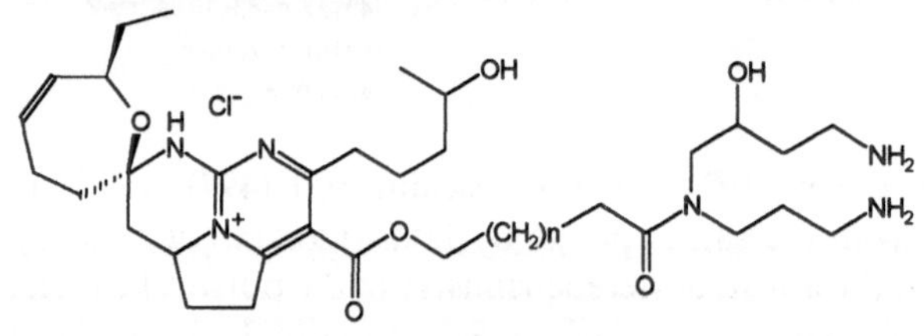

(498): n=13

(505) had the actual stereochemistry of natural crambescin B at C-8 [erroneously assigned in the paper dealing with its isolation (676)]. Hence, SNIDER and SHI (684) propose the stereochemistry of crambescin B depicted in (490b). The 61:22:13:2 mixture of (505), (506), (507) and (508) obtained by synthesis appears to be an equilibrium mixture that forms in methanol either under acidic or basic conditions. The yield and selectivity for the desired spiroaminal (505) could be improved by heating (508) with triethylamine in $CHCl_3$ leading to 94% 20:2:1 mixture of (505), (506) and (507), respectively. The spiroaminal (505) which possesses the natural configuration thus seems to be the major product under kinetically controlled conditions as a result of steric and stereoelectronic effects. The

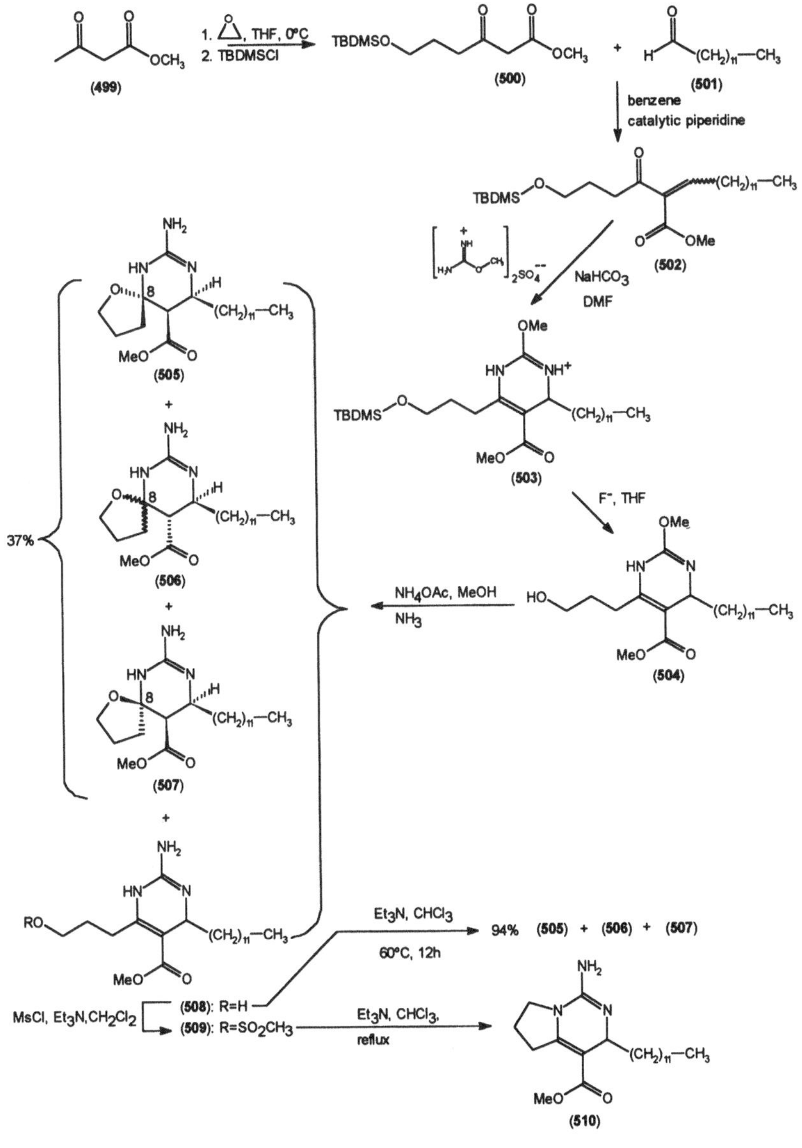

Scheme 38. Synthesis of the bicyclic moieties of crambescin A and crambescin B major homologues

synthesis of the bicyclic portion of crambescin A (**510**) was completed by reaction of (**508**) with mesyl chloride followed by cyclization in refluxing Et$_3$N/CHCl$_3$. Syntheses of the central pentacyclic moiety of ptilomycalin A (**487**) (*685, 686*) and of crambescins A (**489**), B (**490b**), C1 (**491b**) and C2 (**492**) (*687*) were subsequently achieved by the same approach.

FUSETANI *et al.* (*688*) have isolated the cyclotheonamides A (**511**) and B (**512**) from a marine sponge of the genus *Theonella*. These incorporate two previously undescribed amino acids (**513**) and (**514**). The structures of the polypeptides were determined by FABMS and mono- and bi-dimensional NMR analysis. The sterochemistries of the known amino acids Pro, Phe and Dpr were determined by chiral gas chromatography as L, D and L, respectively. Cyclotheonamide A has been synthesized (*689*) and the complex formation between cyclotheonamide A and bovine β-trypsin was studied by X-ray analysis (*690*).

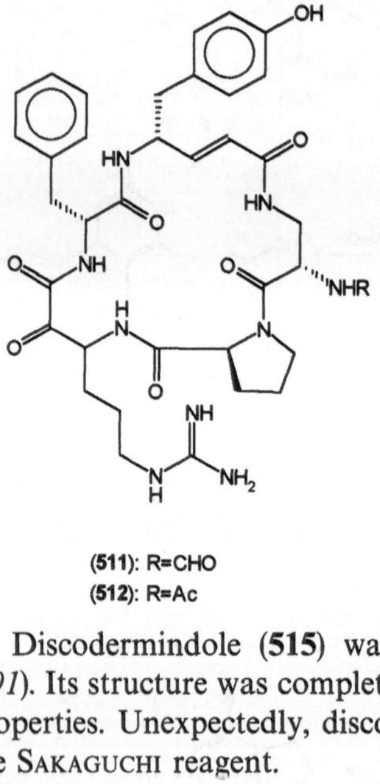

(**511**): R=CHO
(**512**): R=Ac

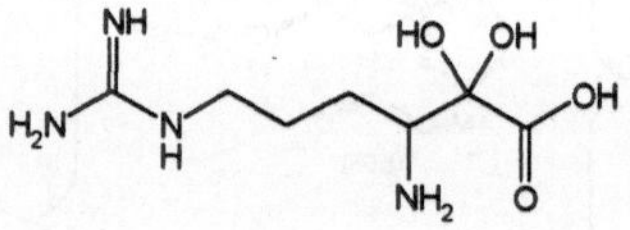

(**513**)

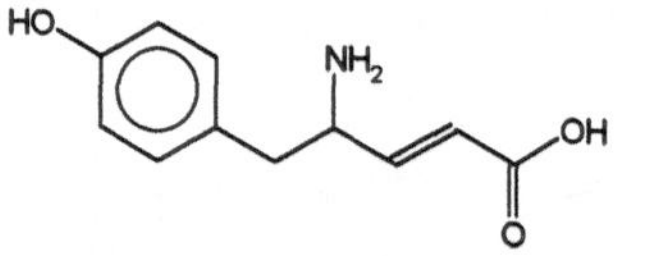

(**514**)

Discodermindole (**515**) was isolated from *Discodermia polydiscus* (*691*). Its structure was completely determined by analysis of its spectral properties. Unexpectedly, discodermindole gave a positive response to the SAKAGUCHI reagent.

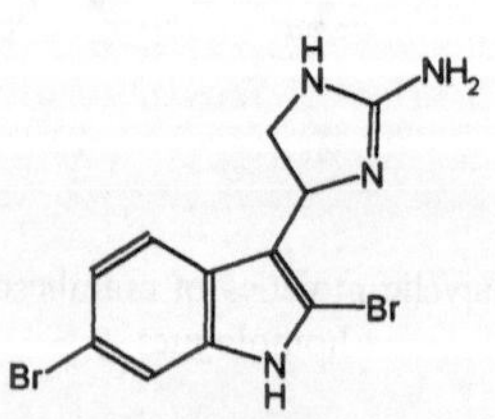

(**515**)

From the New Caledonian sponge *Phloeodictyon* sp. KOURANY-LEFOLL *et al. (692)* isolated two antibacterial and cytotoxic bicyclic guanidinium derivatives named phloeodictines A (**516**) and B (**517**). These compounds were identified by FAB mass spectra and 2D-NMR analysis as well as by catalytic hydrogenation followed by derivatization with pentane-2,4-dione.

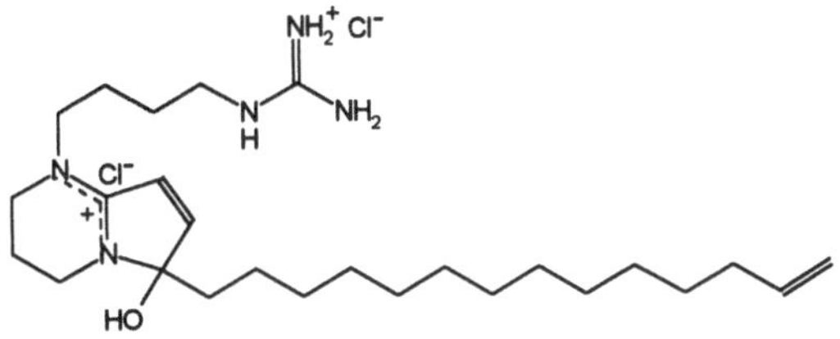

(516)

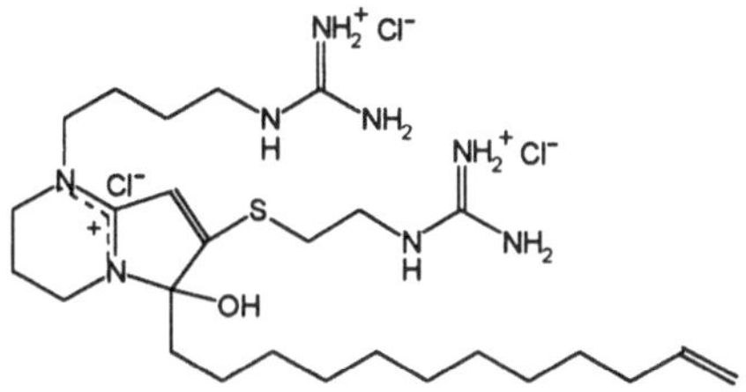

(517)

Nazumamide A (**518**) has been isolated from the marine sponge *Theonella* sp. *(693)* and identified by spectroscopic analysis, mainly NMR techniques. Nazumamide A appears to be the first naturally-occurring peptide containing the *N*-2,5-dihydroxybenzoate terminus. It was recently obtained by standard peptide synthesis *(694)*.

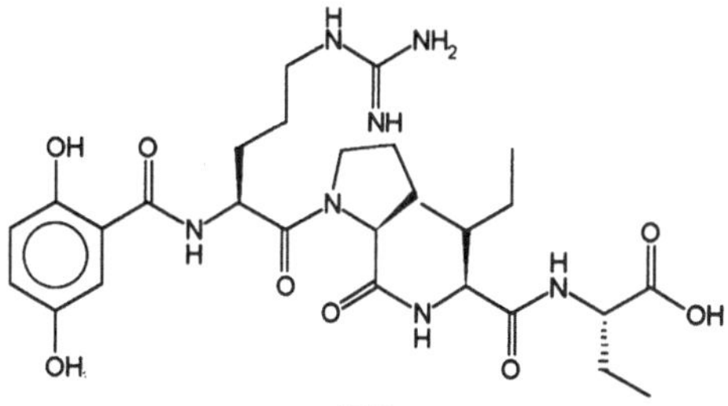

(518)

More recently, O,O-dimethyl-N^2-creatinylphosphate (519) and ulo-santoin (520) were isolated from the sponge *Ulosa ruetzleri* (*695*). Compound (519) appears to be an artifact of isolation. Moreover, while (519) is devoid of any insecticidal activity, (520) shows very impressive toxicity against the tobacco hornworm, the Mexican bean beetle, the southern armyworm and against the American cockroach *Periplaneta americana*. Both (519) and (520) were identified by X-ray crystallographic analysis.

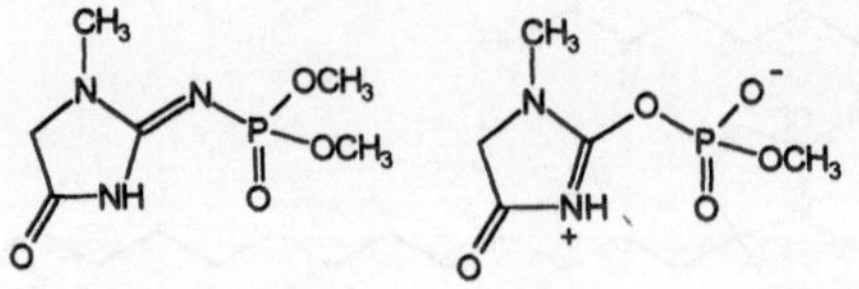

(519)   (520)

Also recently anchinopeptolide A (521) was isolated from the sponge *Anchinoe tenacior* (*696*), together with unidentified related compounds. Its structure was determined by various spectroscopic techniques.

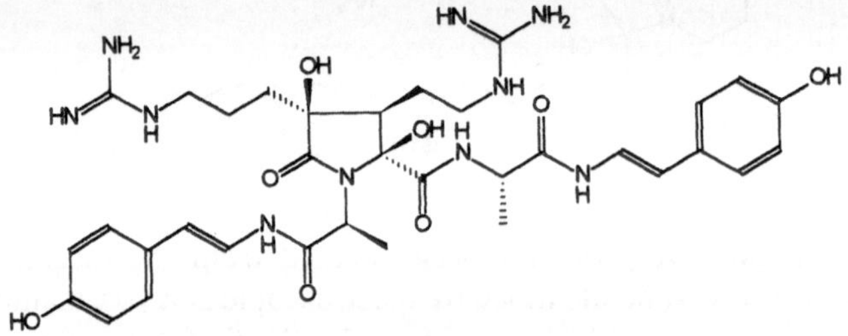

(521)

Leucettamine B (522) was isolated from the sponge *Leucetta micro-raphis* together with related imidazole metabolites (*697*). It was identified by analysis of spectral data. Leucettamine B was essentially inactive in a membrane receptor leukotrine B, LTB$_4$ binding assay.

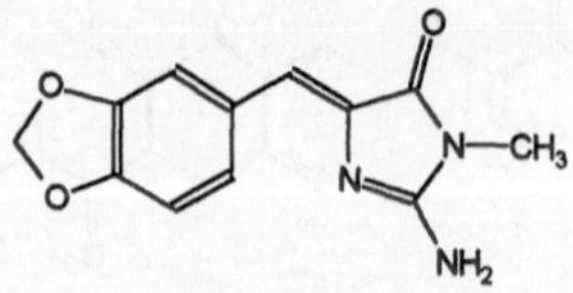

(522)

4.4. Other Marine Invertebrates

6-Bromoaplysinopsin (376) was isolated from the Mediterranean anthozoan *Astroides calycularis* (698), together with aplysinopsin (373) and their 1-N-propionyl derivatives (523) and (524). The derivatives are probably artifacts of isolation. All compounds were identified by their spectral properties and by synthesis in a fashion analogous to the syntheses outlined in Scheme 26. Aplysinopsin (373) was also isolated from the scleratinian coral *Tubastrea aurea* (699). Five indole derivatives (374), (375) and (376), together with the dihydro analogues (525) and (526), were isolated from the coral *Tubastrea coccinea* and from its predator, the nudibranch *Prestilla melanobranchia* (700).

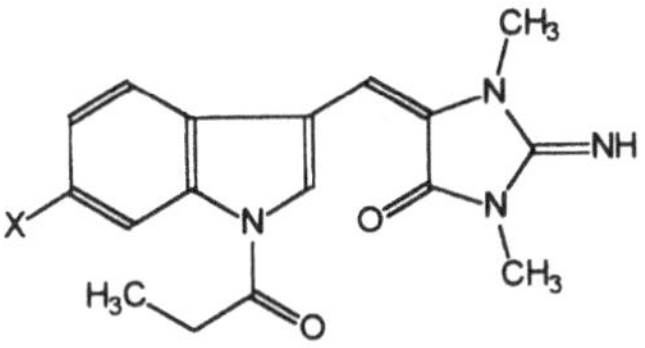

(523): X = H
(524): X = Br

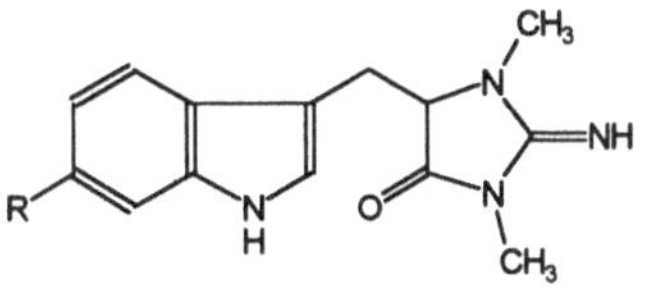

(525): R=H
(526): R=Br

An interesting study of aplysinopsin derivatives has appeared (701). 2'-N-demethylaplysinopsin (374), 6-bromo-2'-N-demethylaplysinopsin (375), 2'-N-demethyl-3'-N-methylaplysinopsin (527) and 6-bromo-2'-N-demethyl-3'-N-methylaplysinopsin (528) were isolated from the scleratinian coral *Dendrophyllia* sp. as a mixture of Z and E isomers. The configuration assignments were made by measuring the ^{1}H-^{13}C heteronuclear coupling constants. Synthesis of the 2'-N-demethyl derivatives gave mixtures in which the Z isomer predominates, whereas the inverse was observed for compounds bearing a methyl group at N(2). Compound (528), with the natural composition $Z/E > 95:15$ undergoes facile photoisomerization giving a mixture richer in the E isomer.

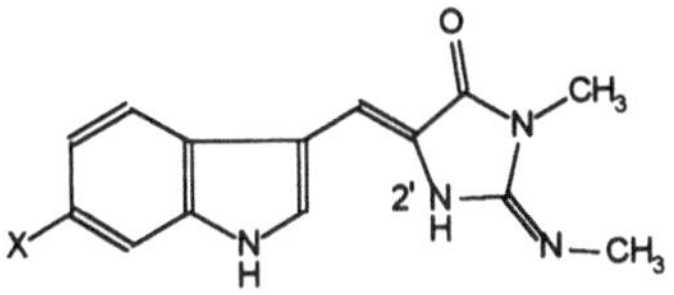

(527): X = H
(528): X = Br

Tubastrine (529) has been isolated from the coral *Tubastrea aurea* (702). Tubastrine gave a positive response to SAKAGUCHI reagent but did not react with pentane-2,4-dione, although the substance formed by catalytic reduction of the conjugated double bond did, giving (531). The structure of tubastrine was deduced by analysis of its spectral data and of its derivatives (530) and (531).

Two acylguanidine derivatives were isolated from the sea anemones *Actinia equina* and *Actinia fragacea* (*703*). The first was identified as *N*-acetylagmatine (532); the second, actiniamine (533), was identified by means of chemical degradation and spectroscopic analysis. HCl hydrolysis of actiniamine gave two guanidine derivatives, (534) and a second, for which structure (535) was suggested.

Cyclopropene structures previously assigned to the polyandrocarpidines I and II isolated from the tunicate *Polyandrocarpa sp.* (*704*) have been revised by CARTÉ and FAULKNER on chemical and spectral evidence (*705*). On reisolating these compounds, they obtained four compounds named polyandrocarpidines A (formerly I) (536), B (formerly II) (537), C (538) and D (539). The stereochemistry assigned to the polyinsaturated side chains was based on analysis of spectral data. Ozonolysis of the polyandrocarpidine A–D mixture gave the *N*-guanidinoalkyl succinimides (540) and (541), which confirmed the presence of the *N*-guanidinoalkyl-5-methylene-γ-lactam moiety. One of the possible hexa-

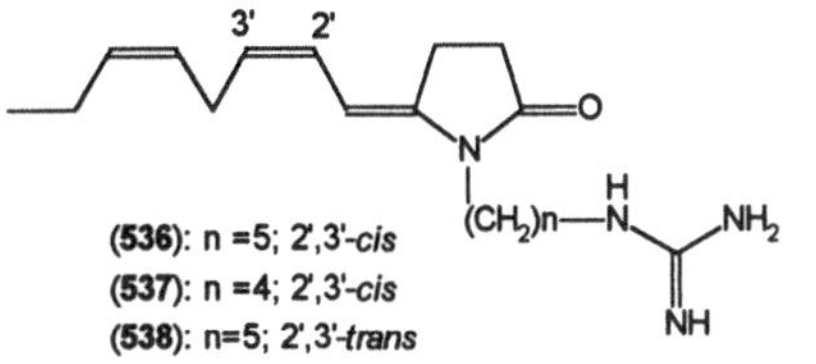

(536): n =5; 2',3'-cis
(537): n =4; 2',3'-cis
(538): n=5; 2',3'-trans
(539): n=4; 2',3'-trans

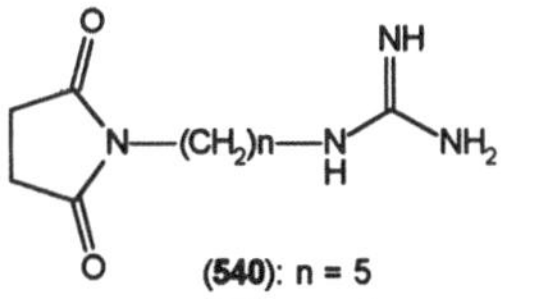

(540): n = 5
(541): n = 4

hydropolyandrocarpidine was synthesized (706) in order to confirm the revised structures (Scheme 39).

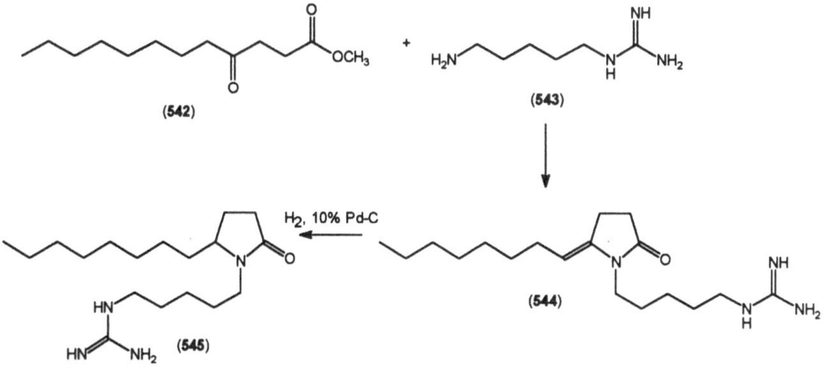

Scheme 39. Synthesis of hexahydropolyandrocarpidine

GUSTAFSON and ANDERSEN (707) have isolated triophamine (546) from the dorid nudibranch *Triopha catalinae*. This symmetric diacylguanidine derivative was identified by means of ^{1}H-, ^{13}C-NMR, IR, UV and mass spectral analysis. Alkaline hydrolysis yielded guanidine and the carboxylic acid (547). The fact that triophamine was also isolated by the same authors (708) from another nudibranch (*Polycera tricolor*) suggests that triophamine is probably incorporated through the food chain. The stereochemistry of the double bonds was assigned as *E* by the synthesis of triophamine (546) outlined in Scheme 40 (709). Acid (557) was prepared first, but the signals of the double bond carbons and hydrogens were clearly different from those of (547). It was therefore presumed that the

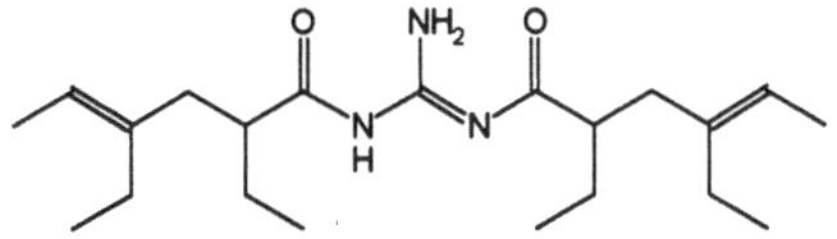

(546) 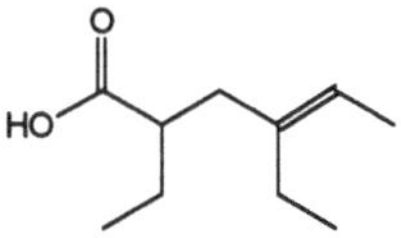   (547)

double bonds in the two acyl residues of triophamine were (*E*) rather than (*Z*). Synthesis of the required carboxylic acid (**547**) was accomplished in an analogous fashion as described for (**557**) and the synthesis completed to give (±)-triophamine (**546**) and a stereoisomer whose structure was not reported.

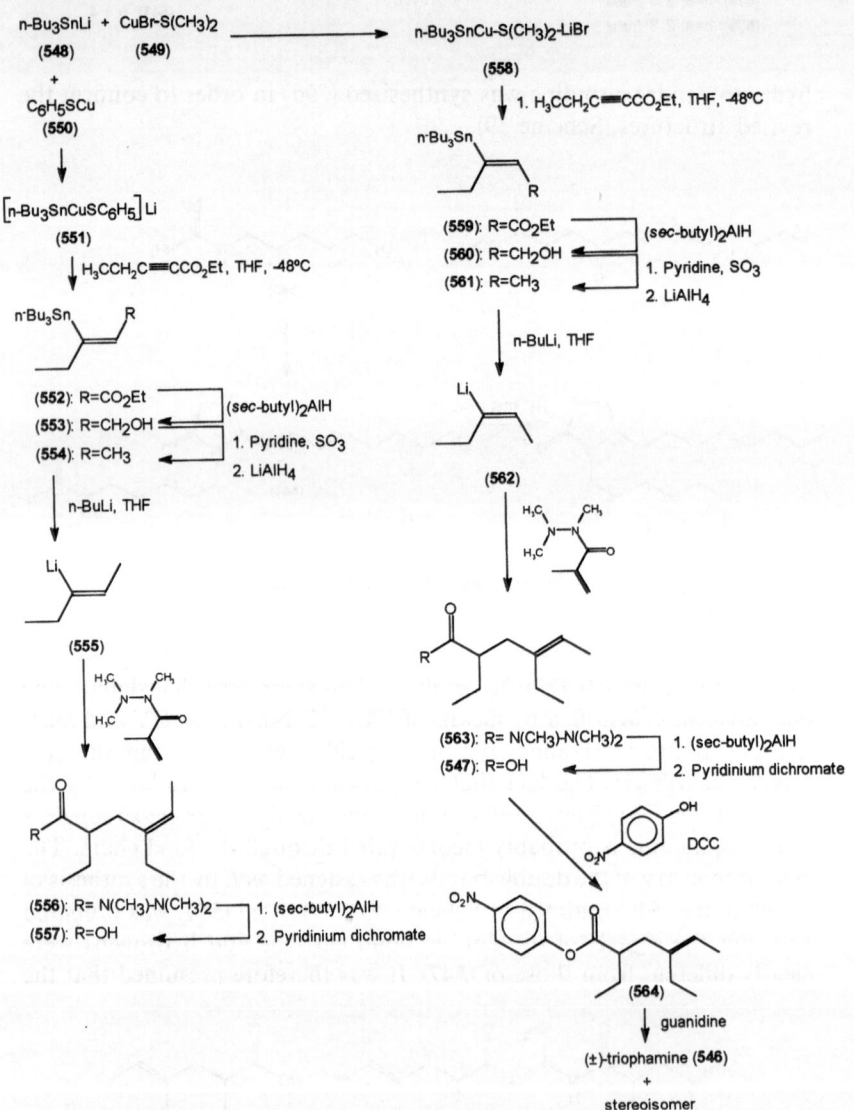

Scheme 40. Synthesis of triophamine

5. Guanidine Secondary Metabolites from Higher Plants

Higher plants are not common sources of guanidine alkaloids. Those that do occur are largely prenylated guanidines. Commonly important biological activities are reported.

Galega officinalis (Leguminosae) has been known since antiquity of its galactogenic and hypoglycemic properties as well as for being an antidote for venomous bites (*710*). As long ago as 1914 TANRET isolated galegine from the seeds of *G. officinalis* (*711–713*) and showed that it was an optically inactive base of molecular formula $C_6H_{13}N_3$, but misinterpreted chemical data and suggested erroneous structural formulae. After a hiatus brought on by the first World War, work by BARGER and WHITE (*714, 715*) and SPÄTH and PROKOPP (*716*) led to the correct structure (**565**),

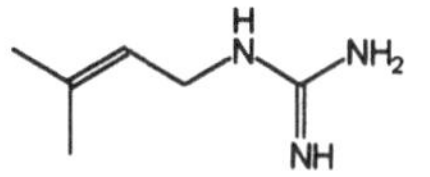

(565)

which was subsequently confirmed by total synthesis (Scheme 41) (*717*). Galegine has hypoglycemic properties with a slight margin between the hypoglycemic and the lethal dose, which indicates high toxicity (*718*). Galegine showed good incorporation of $^{14}CO_2$ and [*amidine*-^{14}C] arginine during its biosynthesis, suggesting that galegine formation from arginine might involve a transamidination reaction (*3*). The chemistry of galegine was reviewed by BRAUN (*719*).

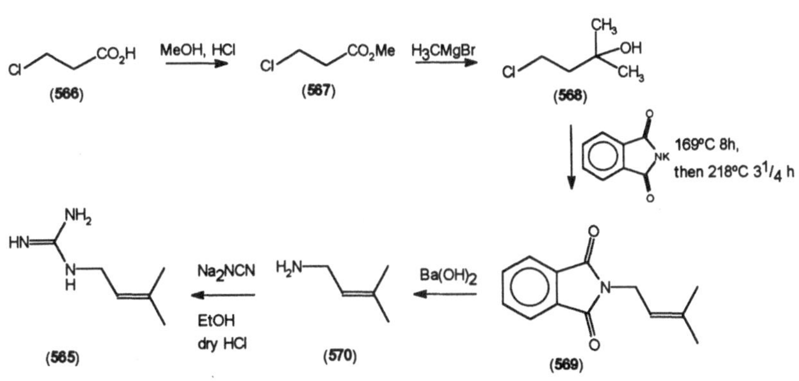

Scheme 41. Synthesis of galegine

4-Hydroxygalegine (**571**) has also been isolated from *Galega officinalis* (*720–722*) and identified by chemical degradation (catalytic reduction, acetylation and permanganate oxidation) and spectroscopic analysis (IR). The structure was also confirmed by synthesis (*723*).

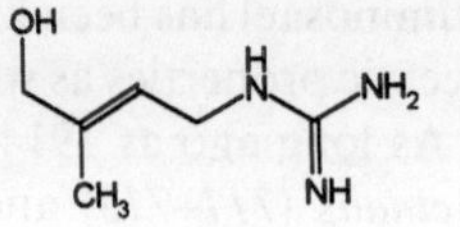

(**571**)

Spherophysine (**572**), a galegine homologue, was isolated from *Sphaerophysia salsula* (*724a, b*). Hydrogenation of spherophysine gave a dihydroderivative, while alkaline hydrolysis of spherophysine and dihydrospherophysine yielded urea and the corresponding amines $C_9H_{20}N_2$ and $C_9H_{22}N_2$, respectively. The latter was identified as isopentyltetramethylenediamine (**573**). Of the two possible formulas (**574**) and (**575**) of $C_9H_{20}N_2$, (**574**) was preferred because permanganate oxidation under acid conditions of $C_9H_{20}N_2$ furnished the glyoxal derivative (**576**). Somewhat later, BIRCH et al. (*724c*) argued that spherophysine and the amine $C_9H_{20}N_2$ did not exhibit the characteristics of vinylamines and showed that synthetic (**575**) when subjected to acid permanganate oxidation formed (**576**), presumably *via* further oxidation of an intermediate 1,2,3-triol. Consequently spherophysine was (**572**).

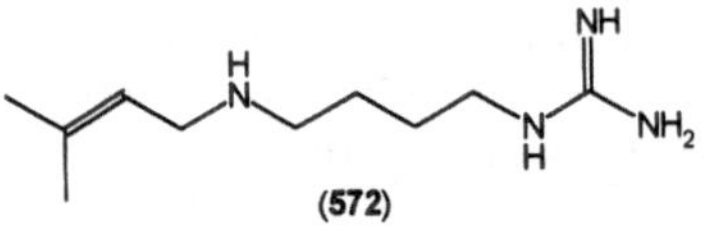

(**572**)

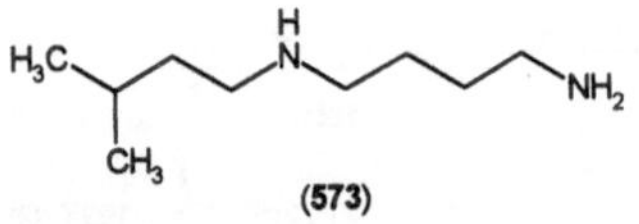

(**573**)

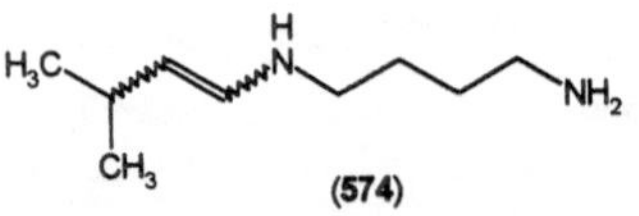

(**574**)

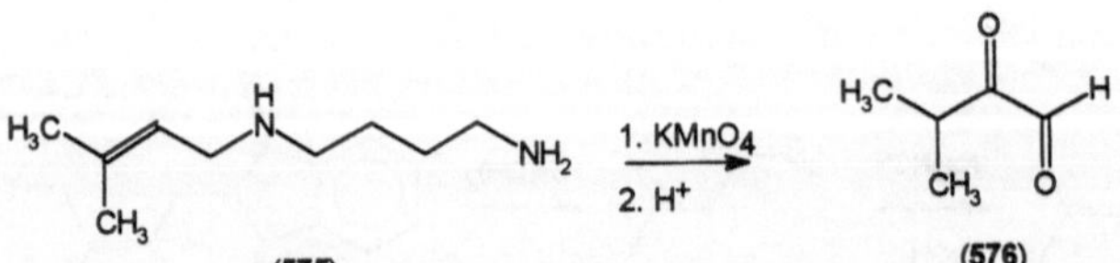

(**575**)

1. KMnO$_4$
2. H$^+$

(**576**)

Pterogynine (**577**) was isolated from the bark of *Pterogyne nitens* (Leguminosae) (*725*). Mass spectral analyses of the hydrochloride and of the hydrogenation product (578) were informative. The structure of pterogynine (577) was established by ^{1}H-NMR analysis of both (577) and (578) and by alkaline hydrolysis of (578), which gave diisoamylamine.

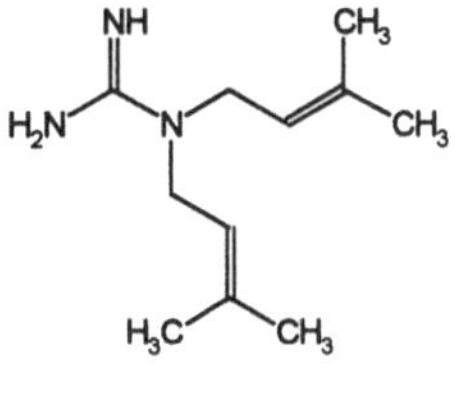

(577) 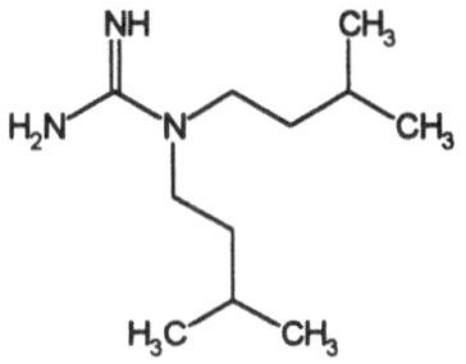   (578)

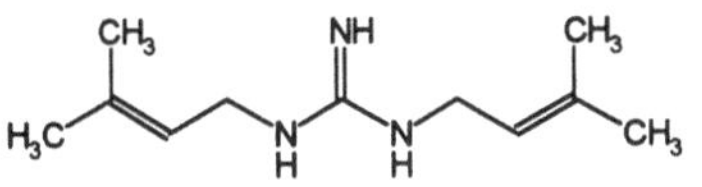

(579)

Pterogynidine (579) occurs in *Pterogyne nitens* (726), and it could be differentiated from pterogynine (577) by total synthesis of the two compounds as depicted in Scheme 42.

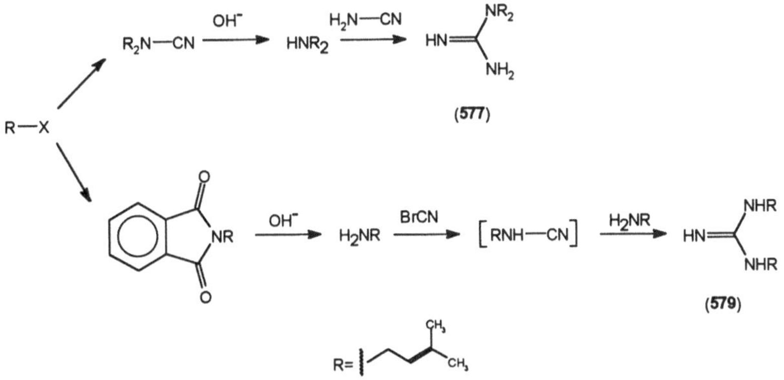

Scheme 42. Synthesis of pterogynine and pterogynidine

Various monoterpene guanidine alkaloids have been isolated from *Alchornea javanensis* (Euphorbiaceae) (40, 727). The first, alchornine (580), was identified by spectral analysis and chemical degradation. The second, alchornidine (581), was also identified by analysis of the spectral data. Hydrolysis of alchornidine in aqueous ethanolic KOH afforded 3,3-dimethylacrylic acid and alchornine (580), while aqueous acetic acid hydrolysis of alchornidine (581) gave an isomer of alchornine, isoalchornine (582). The latter was converted into alchornine by treatment with hot dilute ethanolic NaOH, a reaction which involves cleavage of the lactam and recyclization to the alternate nitrogen of the five-membered ring. The other guanidine constituents were a substance later identified as pterogynidine (579) (40) and N,N',N'''-triisopentenylguanidine (583). ^{1}H-NMR signals of the bicyclic compounds (580), (581) and the derivatives obtained by chemical degradation were completely assigned. Proposed mass spectral fragmentation pathways for alcornine and isoalchornine were shown to be of value in determining the position of the isoprenyl substituent.

KHUONG-HUU et al. (728) isolated alchornéine (584) from the leaves, roots and bark of *Alchornea floribunda* (Euphorbiaceae) and from the leaves of *Alchornea nintella*. These plants are utilised in the Congo and Cameroun as intoxicants and aphrodisiacs. Alchornéine (584) was identified by its spectroscopic properties (UV, ^{1}H-NMR at 60 MHz, IR and MS) and by chemical degradation. Alkaline treatment of the iodomethylate (585) yielded the corresponding substituted urea (586) and alchornéine (584). Hydrolysis of (586) gave the imidazolidinone (587). The structure of alchornéine (584) was confirmed by X-ray diffraction (729, 730).

Two additional related alkaloids, isoalchornéine (**588**) and alchor-néinone (**589**), were subsequently isolated from *Alchornea floribunda* (*731*). These compounds were fully characterized by their respective spectral data, including the stereochemical assignments. That the isopropenyl substituent was attached to C-2 in these alkaloids was confirmed by synthesis of racemic 4-isopropyl-3-methoxyimidazo-2-one, the dihydro derivative of (**587**). The C-2 (*R*) configuration of alchornéine (**584**) was established by comparison of the rotation of (+)-(**590**), a degradation product of alchornéine, with that of synthetic (−)-(**593**) obtained from (*S*)-valinamide (**591**). The C-2 absolute configuration of isoalchornéine (**588**) was established as (*R*) in the same manner. Interestingly, isoalchornéine itself has a negligible rotation although it was optically active. The (*S*)

absolute configuration was suggested for C-6 based on symmetry considerations; the (R) configuration would probably produce a higher value for
the optical rotation due to the ensuing more pronounced dissymmetry.
The chemistry of *Alchornea* alkaloids has been reviewed in the series "The
Alkaloids" (*732*).

The alkaloid chaksine was isolated from *Cassia absus* (Leguminosae)
in 1935 (*733*), and studied chemically by Guha and Ray (*734*), Singh *et al.*
(*735*) and Siddiqui *et al.* (*736, 737*). Subsequently, Wiesner *et al.* (*738*)
proposed a monoterpenoid structure (**594**) for chaksine after further
degradative studies. This proposal conformed with the ^{1}H-NMR studies
reported later by the same group (*739*) and with X-ray diffraction of the
ureido acid (**595**), a degradation product of chaksine (*741*).

(**594**)   (**595**)

Considerably later, in 1985, Voelter and Winter reisolated chaksine
as the iodide salt from *C. absus* (*740*). Although the ^{1}H- and ^{13}C-NMR
data agreed with those reported earlier for the common (**594**), field
desorption mass spectrometry showed a quasi-molecular ion peak
$(M^{2+}I^-)^+$ at m/z 579 corresponding to a "dimer" whose structure was
established by X-ray diffraction as (**596**). Chaksine possesses the (R)
absolute configuration at all asymmetric carbons. The early literature on
chaksine has also been reviewed in the series "The Alkaloids" (*741*).

(**596**)

Leonurine (**2**) was isolated from *Leonurus sibiricus* (Labiatae) (*742,
747*) and from *L. artemisia* (*748*). The original structure proposal (**597**) was
based on degradative and spectroscopic studies (*23, 743*). For example,

hydrolysis of leonurine gave syringic acid and 4-guanidinobutanol (*742*). Three possibilities, (**2**), (**597**) and (**598**), exist for linking the two hydrolysis products. Of these, (**2**) was initially discarded in the basis of the pKa (7.9), erroneously attributed to an acyl derivative of guanidine. However, synthesis of (**598**) and then of (**2**) (*744, 745*) (Scheme 43) shown that (**2**) was, in fact, identical with natural leonurine. CHENG *et al.* (*746*) have reported an improved synthesis of leonurine. Leonurine showed potent uterotonic activity (*748*).

Young barley (*Hordeum* sp., Gramineae) seedlings possess effective resistance against the fungal parasite *Heminthosporum sativum* (*749*) and a relationship exists between antifungal activity and the strongly basic fractions obtained by chromatographic separation of barley seedling extracts (*750*). STOESSL (*751–754*) identified two antifungal basic sub-stances in these fractions and showed that they were α-D-glucosides (**606**) and (**607**) of the two aglycones hordatine A (**608**) and hordatine B (**609**), with (**606**) predominating. Presence of the respective Z-isomers could not be excluded. Subsequent countercurrent distribution of the mixture obtained by hydrogenation led to partial separation of the dihydro derivatives, which were partially hydrolyzed to (**610**) and (**611**). Further hydrolysis gave (**612**) and (**613**) which on heating at the melting points were converted to lactones (**614**) and (**615**). Structures of (**612**)–(**615**) were confirmed by synthesis. That of (**612**) is shown in Scheme 44.

Small amounts of free hordatine A and B as well as *p*-coumaroyllag-matine (**620**) were also present in the basic fractions. The latter served as a relay in the total synthesis of racemic hordatine A. Oxydative coupling between two units of (**620**) with dilute H_2O_2 in the presence of catalytic amounts of horseradish peroxidase gave ($\pm$)-hordatine A in 35% yield.

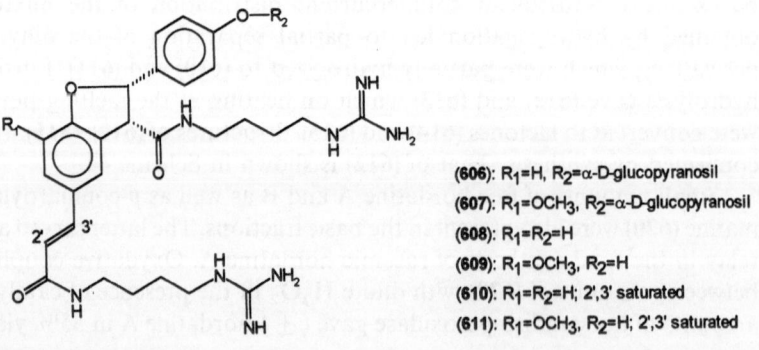

Scheme 43. Synthesis of leonurine

(606): R₁=H, R₂=α-D-glucopyranosil
(607): R₁=OCH₃, R₂=α-D-glucopyranosil
(608): R₁=R₂=H
(609): R₁=OCH₃, R₂=H
(610): R₁=R₂=H; 2',3' saturated
(611): R₁=OCH₃, R₂=H; 2',3' saturated

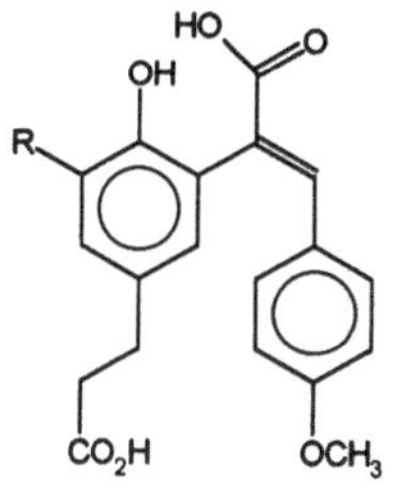

(612): R=H
(613): R=OMe

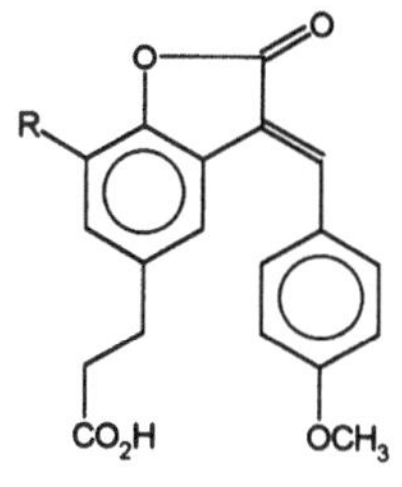

(614): R=H
(615): R=OMe

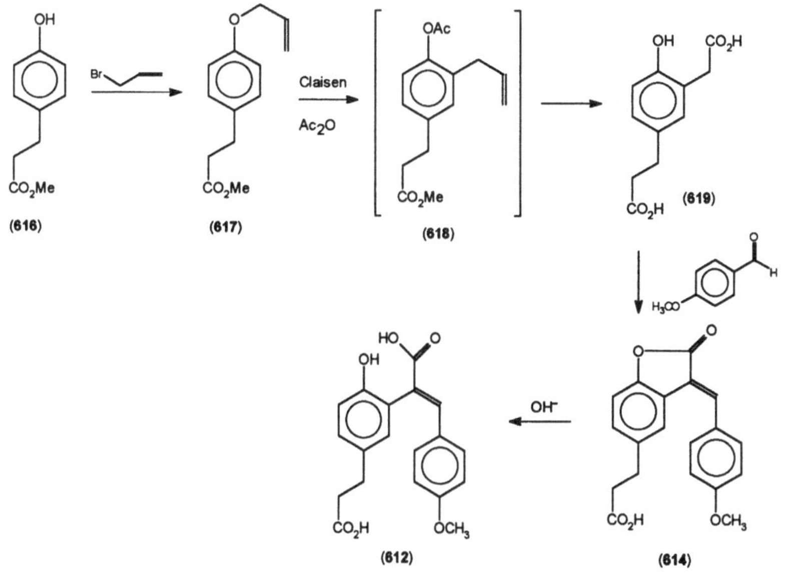

(616)

(617)

(618)

(619)

Claisen
Ac₂O

OH⁻

(612)

(614)

Scheme 44. Synthesis of the aromatic moiety of hordatine A

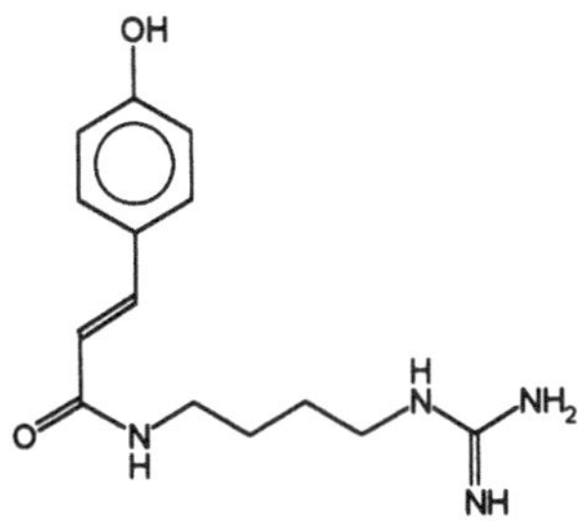

(620)

Presumably this resembles the last step in the biosynthesis of the hordatines (Scheme 45). SMITH and coworkers (755–757) have shown that barley seedlings incubated with coumaroyl-CoA and [U-^{14}C]agmatine incorporated label in both coumaroylagmatine and the hordatines.

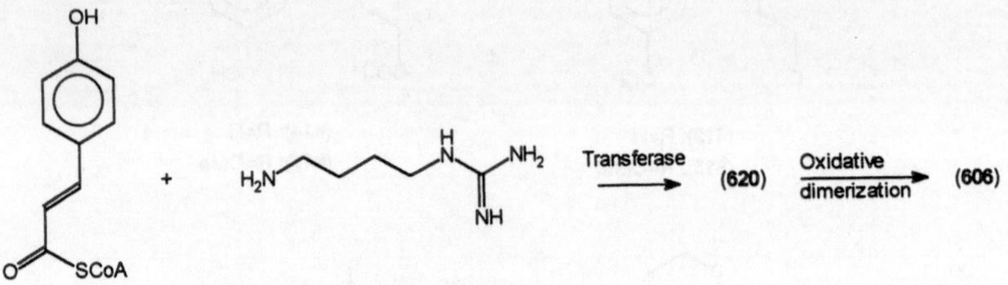

Scheme 45. Biosynthesis of hordatines

Structure determination of the guanidine alkaloid arenaïne (621) from *Plantago arenaria* (Plantaginaceae) exemplifies an early use of ^{13}C-NMR spectroscopy (758). Reaction of the linalool-derived acid (622) or its equivalent with guanidine is probably involved in the biogenesis of arenaïne.

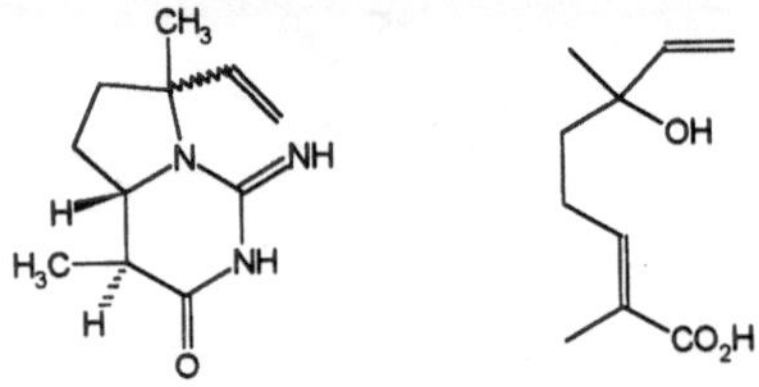

(621)   (622)

YOSHIDA (759) has isolated stizolamine (623) from the seeds of *Stizolobium hassjoo* (Leguminosae) and determined its structure by analysis of its spectral properties and chemical degradation. Hydrolysis of (623) with 6N HCl gave guanidine, N-methylalanine, oxalic acid, alanine and glycine. Hydrolysis of the KMNO$_4$ oxidation product (624) gave guanidine, methylamine and 2 mols of oxalic acid. Heating of stizolamine in H$_2$O at 95 °C for 48 h produced two pyrazines (625) and (626). From these results the structure 1-methyl-3-guanidino-6-hydroxymethyl-pyrazin-2-one (623) was proposed for stizolamine and this was verified by X-ray analysis. Stizolamine showed good incorporation of guanosine triphosphate during its biosynthesis (760). It is found in the seeds and pericarp, but not in the mature leaves, stems and roots and is widely

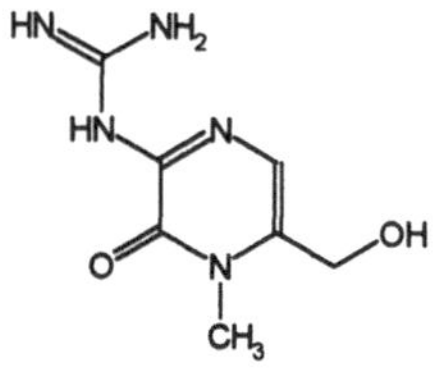

(623)

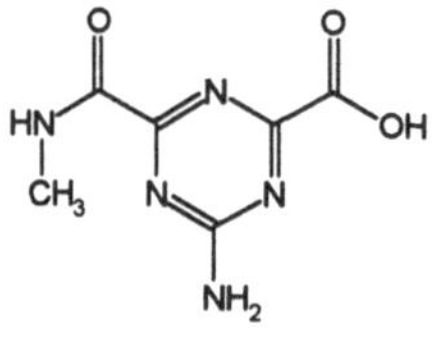

(624)

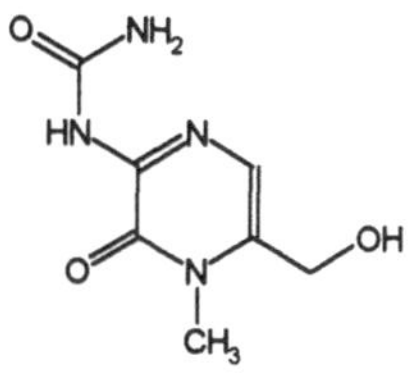

(625)

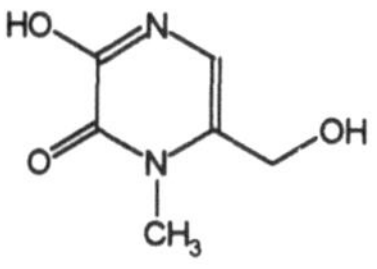

(626)

distributed in seeds of Leguminosae, especially in subfamily Lotoidae (760).

Infection of ryegrass (*Lolium perenne*) by an endophytic fungus (*Acremonium loliae*) confers resistance against the Argentine stem weevil [*Listronotus bonariensis* (Kushel), Coleoptera: Curculionideae] (761). Peramine (627) isolated from infected ryegrass clones is the antifeedant responsible for this activity. It as well as its diacetyl derivative (628) were identified by spectroscopic analysis (51). Biogenetically peramine may be derived from proline and arginine. A synthesis of peramine is shown in Scheme 46 (762).

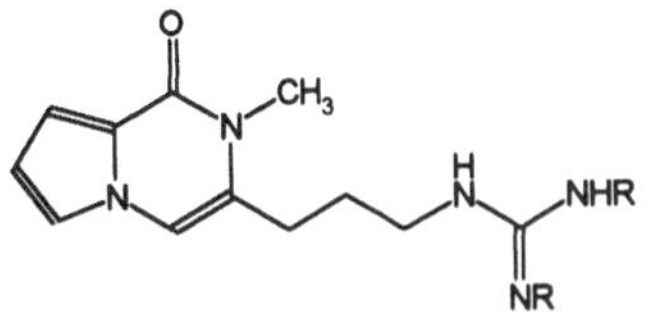

(627): R=H
(628): R=Ac

A simpler method for the isolation of peramine has been reported (763), and several peramine analogues have been synthesized with the aim of studying structure-activity relationships (764). It appears that the pyrrolopyrazinone ring system rather than the propylguanidyl chain is the important moiety for antifeedant activity.

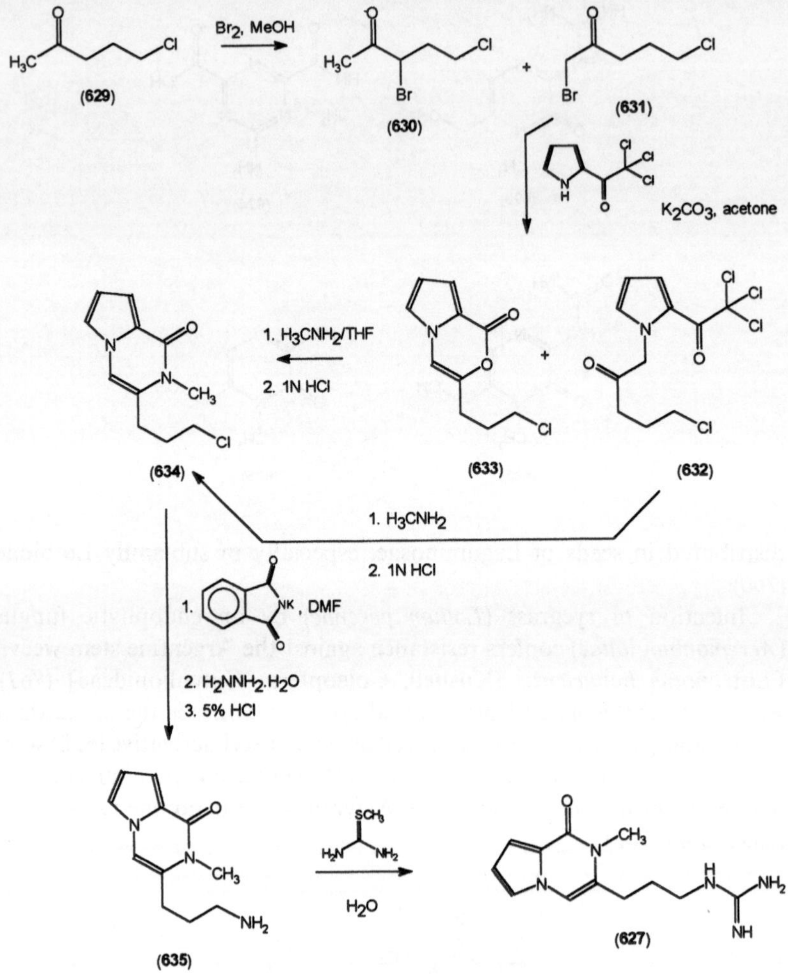

Scheme 46. Synthesis of peramine

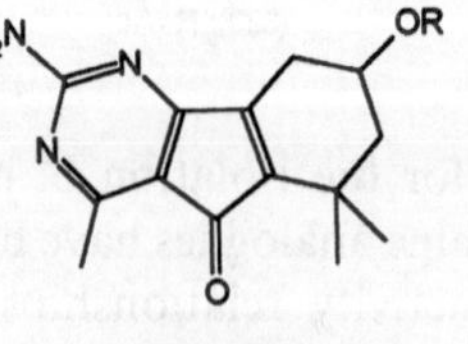

(636): R=H
(637): R=Ac,

Millaurine (**636**) and acetylmillaurine (**637**) have been isolated from the seeds of *Millettia laurentii* (Leguminosae) (*765*). Although these are 2-aminopyrimidine derivatives, these compounds were described as guanidine alkaloids. Their structures were established by spectroscopic and X-ray analysis.

6. Guanidine Secondary Metabolites from Terrestrial Invertebrates and Vertebrates

6.1. Spiders

It has been known for a long time that many spiders possess glands which produce neurotoxins that paralyse or kill their prey. As early as 1957 FISCHER and BOHN (*766*) isolated strongly basic fractions from the venom glands of seven species of spiders of suborder Orthognata, family Aviculariidae. These fractions were composed of a combination of spermine or triethylenediamine with *p*-hydroxyphenylpyruvic acid and other acids resulting from degradation of tyrosine. However, it was only in the early 1980's that thorough studies were initiated on the isolation and constitution of the gland venom of orb-web spiders. While few accidents to humans have been reported due to bites from *Argiopidae* (*767*), pharmacological studies dealing with spiders of various genera (*Argiope*, *Nephila* and *Araneus*, among others) showed that their toxins are potent antagonists with high specificity of the glutamate receptor (*768–785*).

In 1986, GRISHIN *et al.* (*786*) reported the isolation and structure determination of argiopine (**638**), from the venom gland of *Argiope lobata*. The compound was identified by amino acid analysis and by spectroscopic means (¹H, ¹³C-NMR and MS). It has an unusual polyamine framework which is linked to a 2,4-dihydroxyphenylacetic acid terminus.

(**638**)

Related toxins were isolated from other spider species. ARAMAKI *et al.* (*787*) obtained four toxins [Joro spider toxins (JSTX)-1, -2, -3 and -4] from *Nephila clavata* and three toxins [Nephila spider toxins (NSTX)-1, -2 and -3] from *Nephila maculata* venom glands. All these toxins possess an argiopine-like structure whose common feature is the presence of a 2,4-dihydroxyphenylacetylasparaginyl-cadaverinocarboxy (or putrescininyl) ethylamino group. The structure of NSTX-3, proposed as 2,4-dihydroxyphenylacetyl-asparaginyl-(arginyl-cadaverino-β-alanyl)-cadaverine (**639**), was established by spectroscopic analysis (*787, 788*). JSTX-3 (**640**), the analogue with an amine rather than a guanidine terminus (*789*), has been synthesised (*790*).

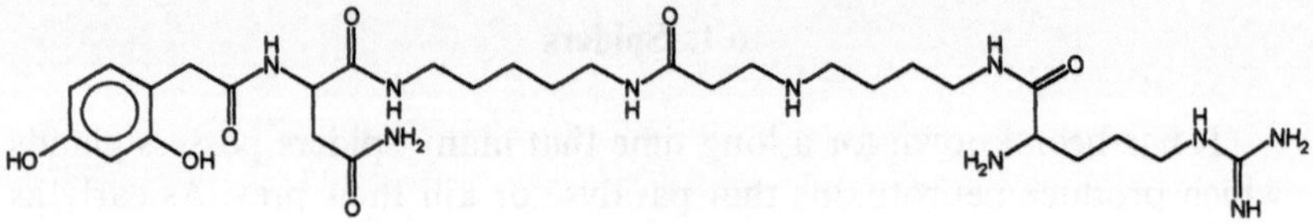

(639)

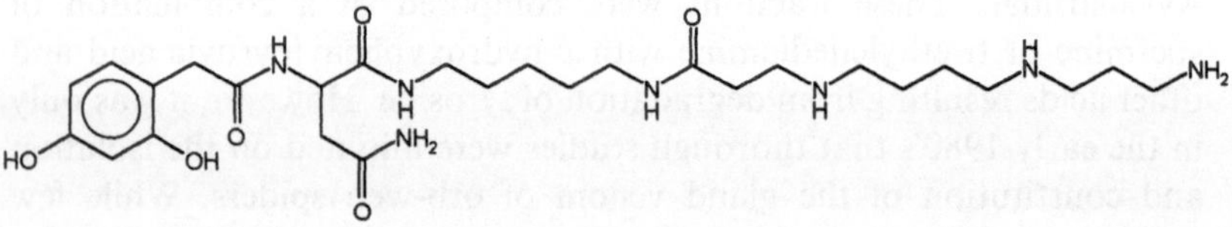

(640)

From *Argiope aurantia*, ADAMS et al. (*791*) isolated the argiotoxins AR$_{659}$ (**641**) and AR$_{673}$ (**642**), in which the 2,4-dihydroxyphenylacetyl acid residue is replaced by a 4-hydroxyindole-3-acetyl group. They also isolated a substance named AR$_{636}$ which is identical with argiopine (**638**). Compounds (**641**) and (**642**) were identified by spectroscopic analysis, by gas-phase EDMAN degradation and synthesis.

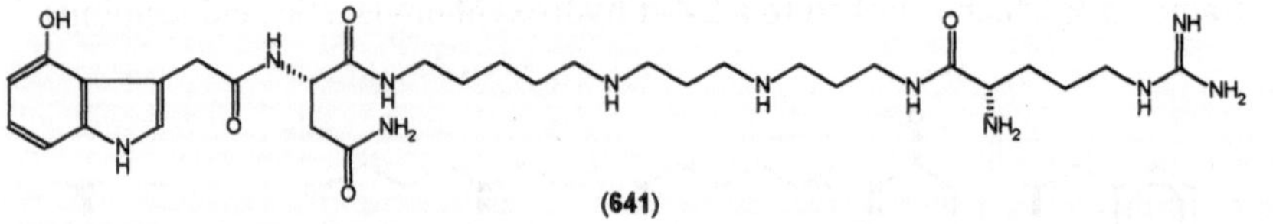

(641)

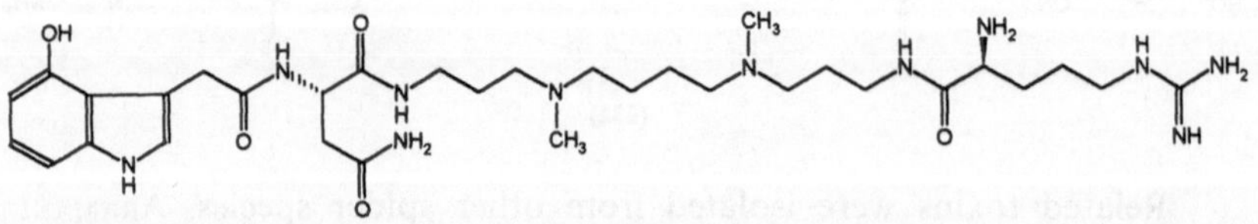

(642)

References, pp. 248–295

Toki *et al.* (*792*) have isolated and identified compounds from *Nephila clavata* named nephilatoxins-7 (**643**), -8 (**644**) and -9 (**645**). Other members of this group of compounds isolated by the same authors (*793*) were the nephilatoxins NPTX-1 through NPTX-12 (**646**)–(**654**).

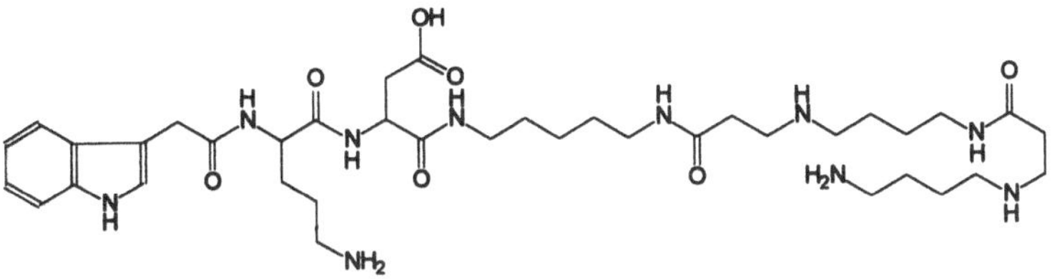

(643)

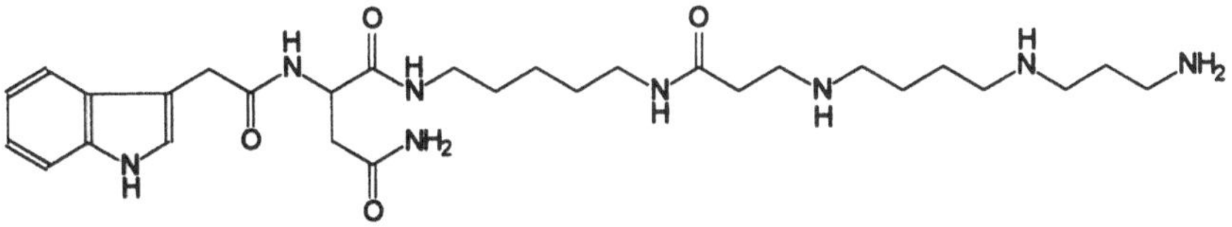

(644)

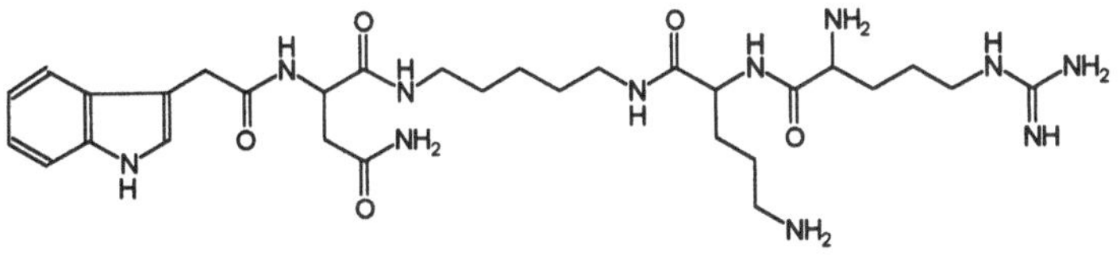

(645)

Further members of this group of compounds isolated from *Argiope lobata* (*794, 795*) are: argiopinin I (**655**), argiopinin II (**656**), argiopinin III [identical with argiotoxin AR_{659}, (**641**)], argiopinin IV (**657**), argiopinin V (**658**), pseudoargiopinin I (**659**), pseudoargiopinin II (**660**) and pseudoargiopinin III (**661**). Argiopinin I (**655**) and pseudoargiopinin I (**659**) are the only toxins which contain a tetraalkyl ammonium group in the polyamine chain, while argiopinin IV (**657**) and argiopinin V (**658**) contain a *N*-methyllysine and a *N,N*-dimethyllysine residue in place of the usual asparagine residue.

Toki *et al.* (*796, 797*) have reported structures of the Jorotoxins-1 (**662**), -2 (**663**) and -4 (**664**), from *Nephila clavata* venom glands.

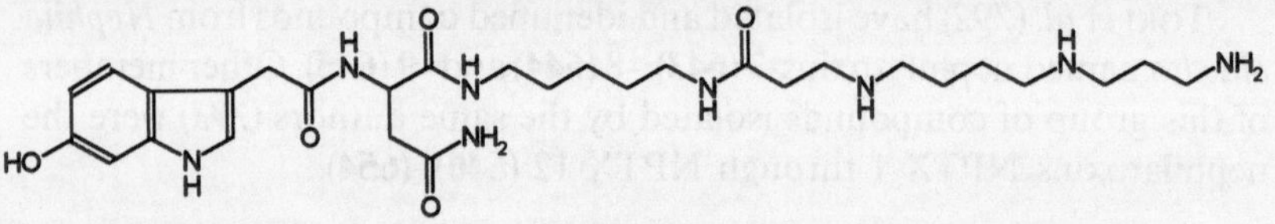

(646): NPTX-1

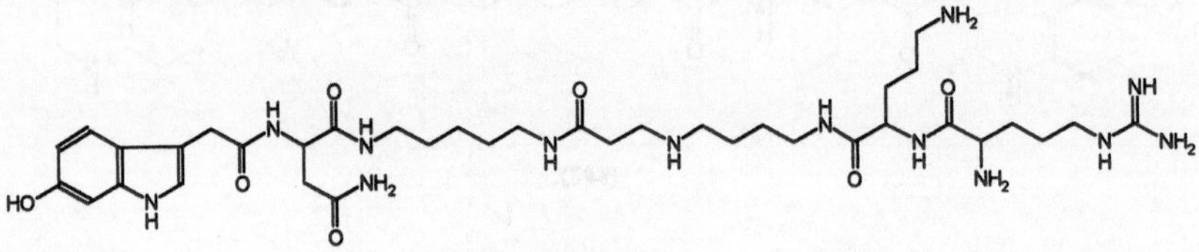

(647): NPTX-2

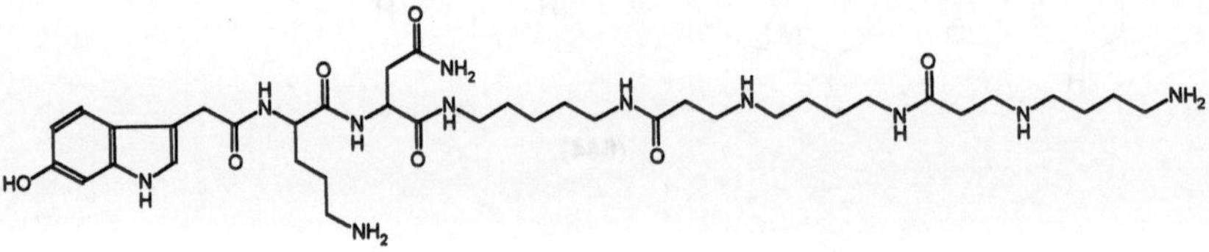

(648): NPTX-3

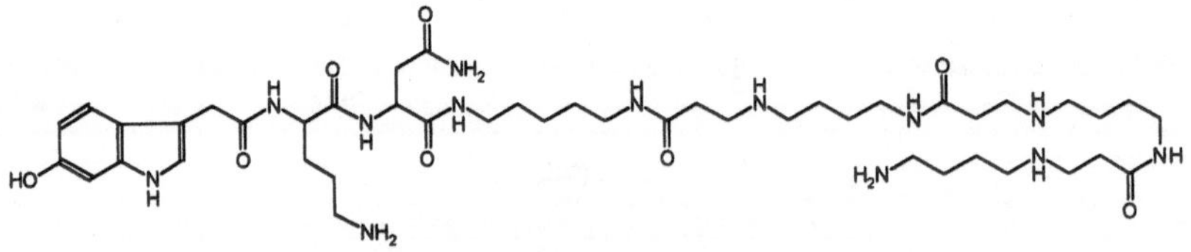

(649): NPTX-4

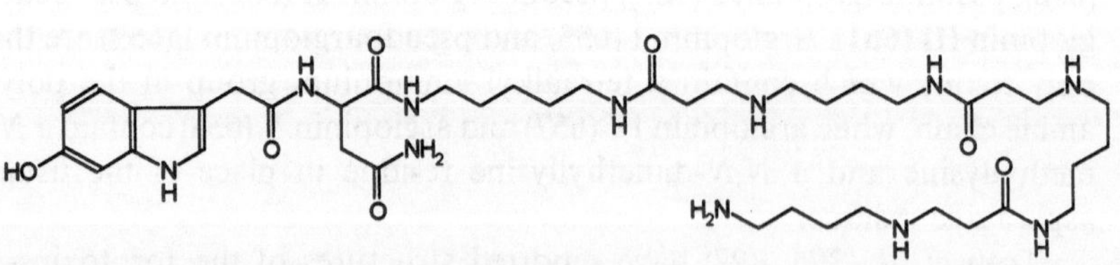

(650): NPTX-5

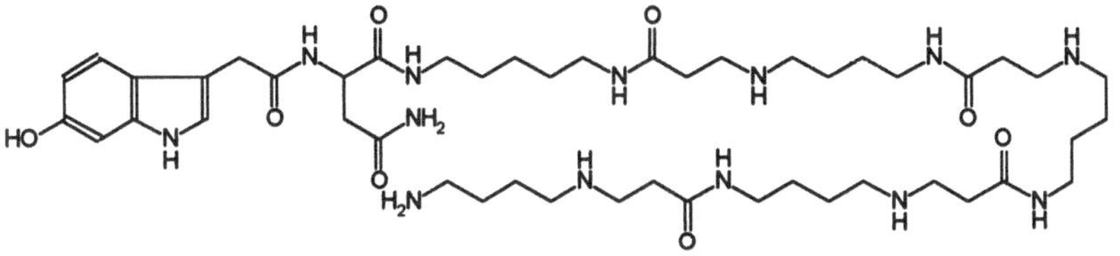

(651): NPTX-6

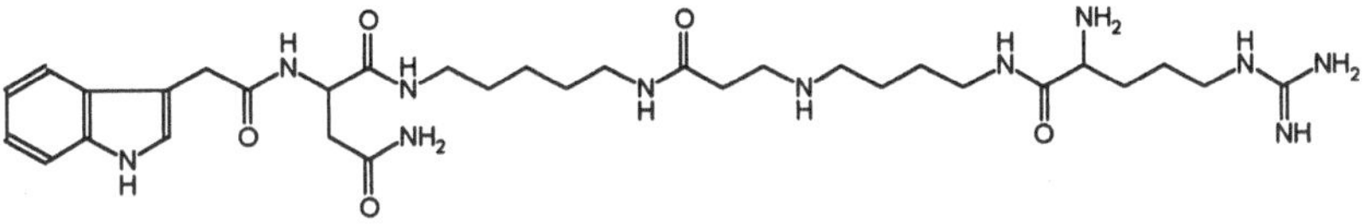

(652): NPTX-10

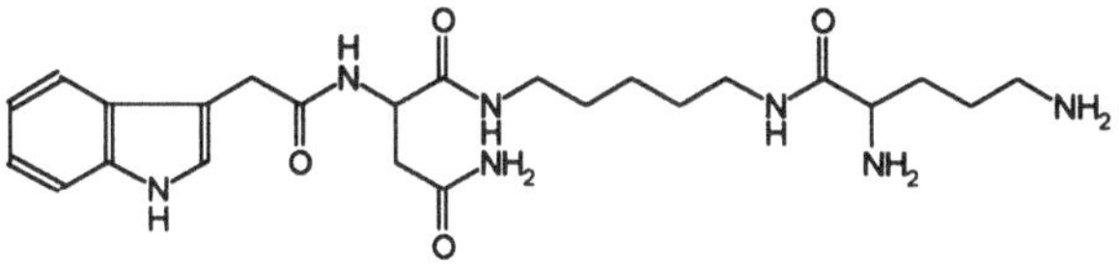

(653): NPTX-11

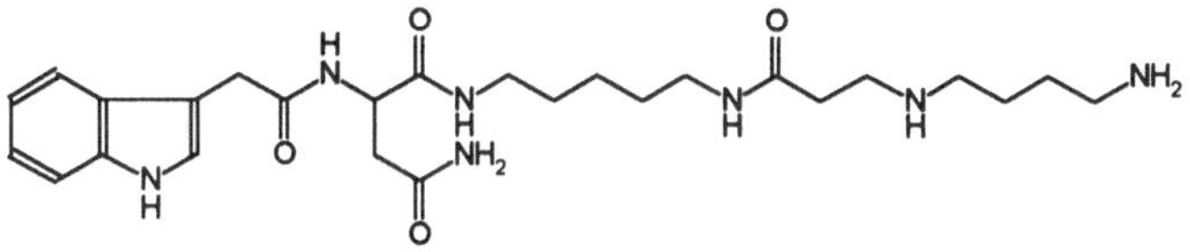

(654): NPTX-12

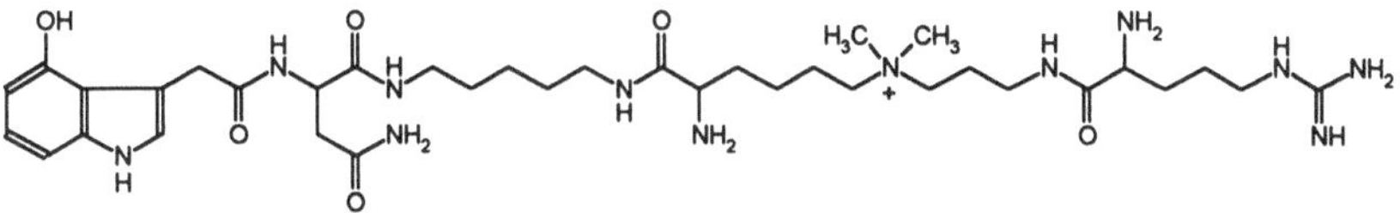

(655)

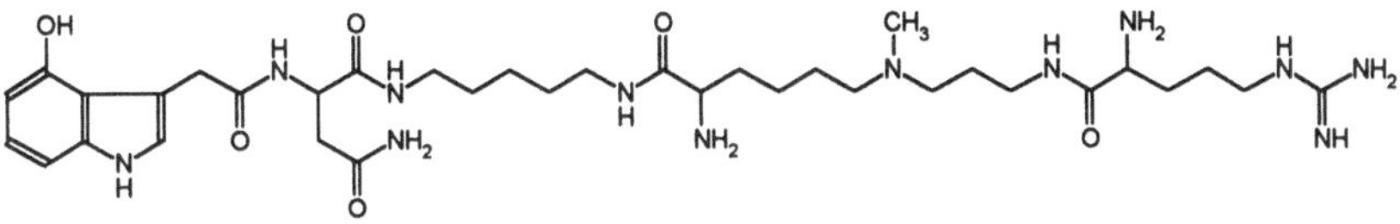

(656)

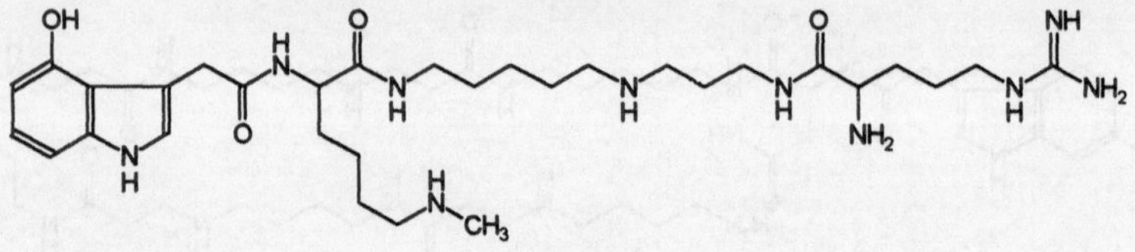

(657)

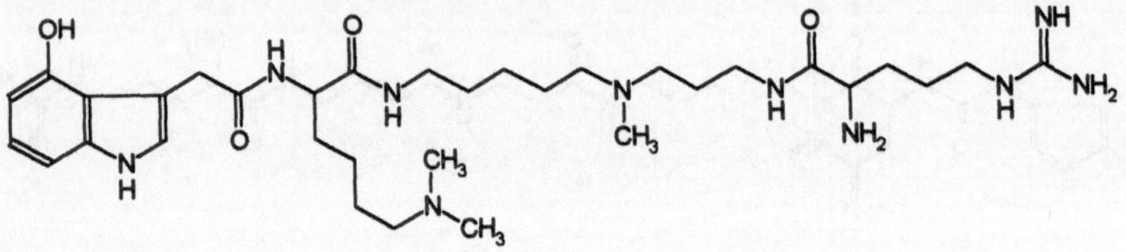

(658)

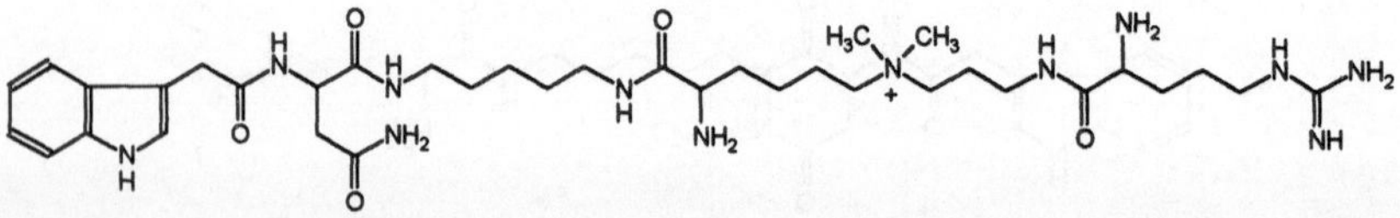

(659)

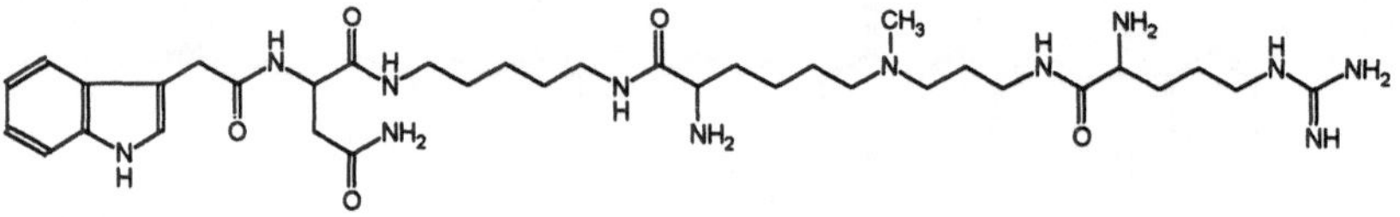

(660)

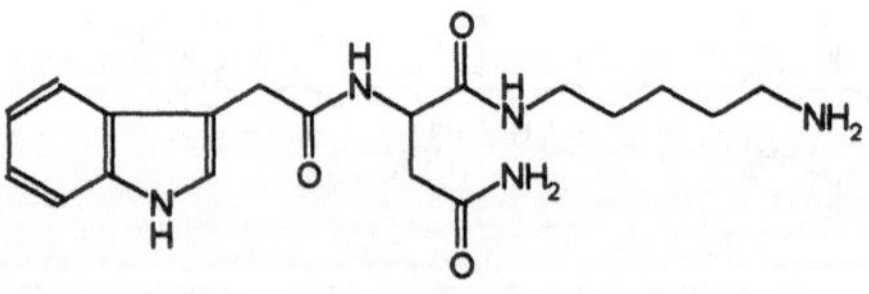

(661)

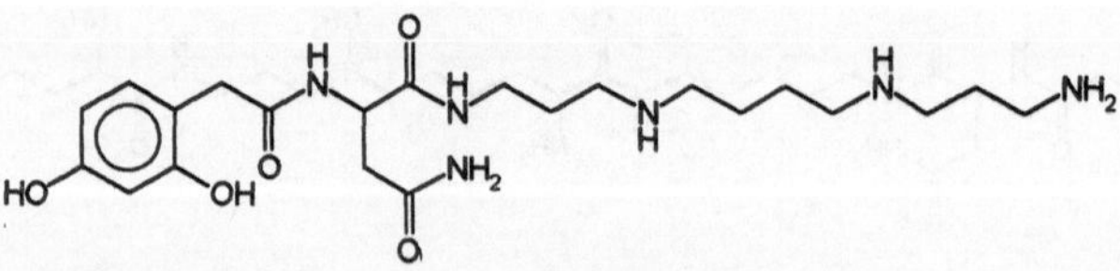

(662)

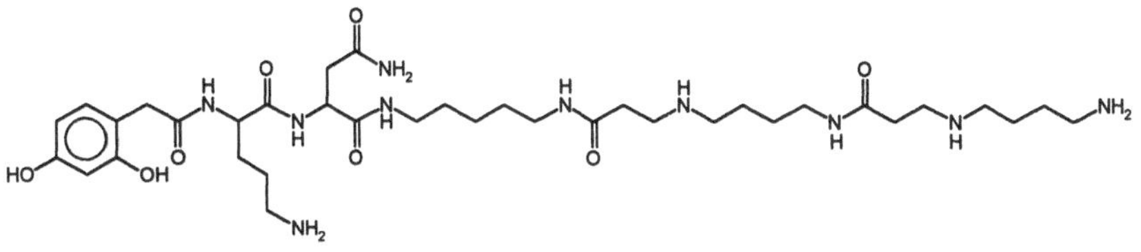

(663)

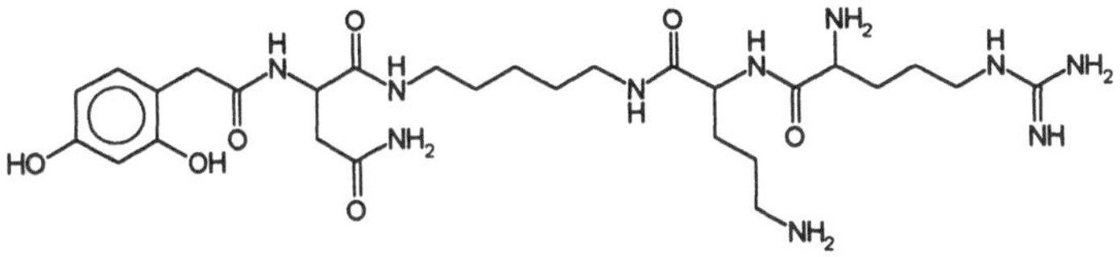

(664)

BUDD *et al.* (*798*) have reported isolation of similar toxins from the venoms of *Argiope trifasciata* and *A. florida* and a series of related hydroxylamine toxins (**665**)–(**669**) and oxyguanidyl derivatives (**670a**) and (**671a**) were recently isolated from the spider *Agelenopsis aperta* (*799–801*). The structure of these two later compounds have been revised to (**670b**) and (**671b**), by spectroscopic analysis and total synthesis (*801a*).

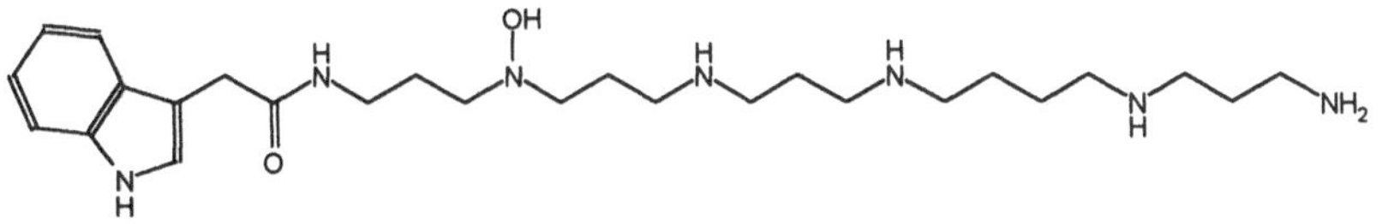

(665)

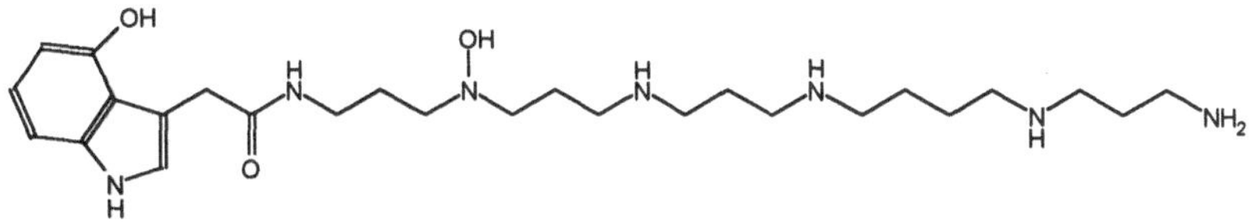

(666)

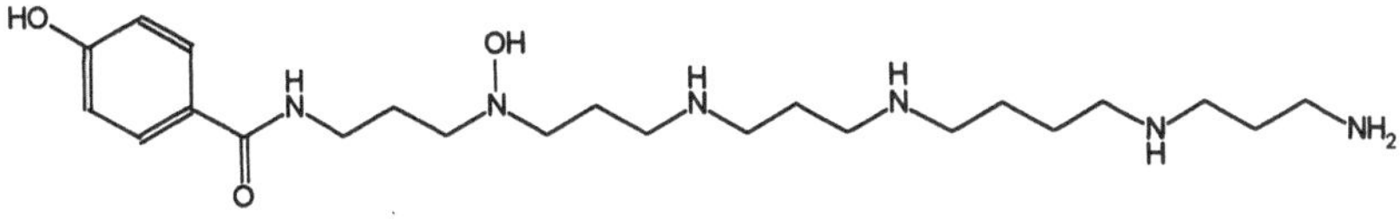

(667)

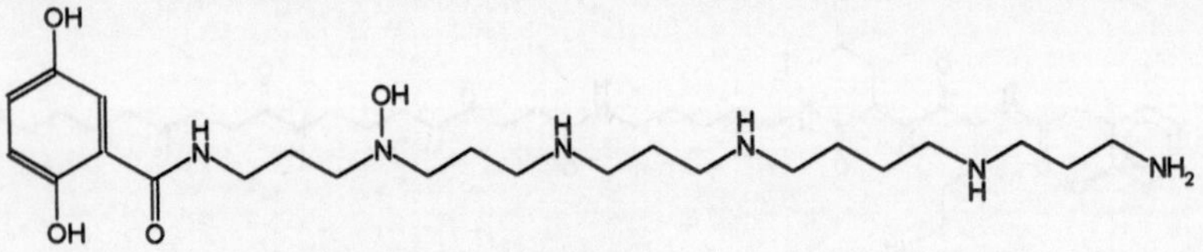

(668)

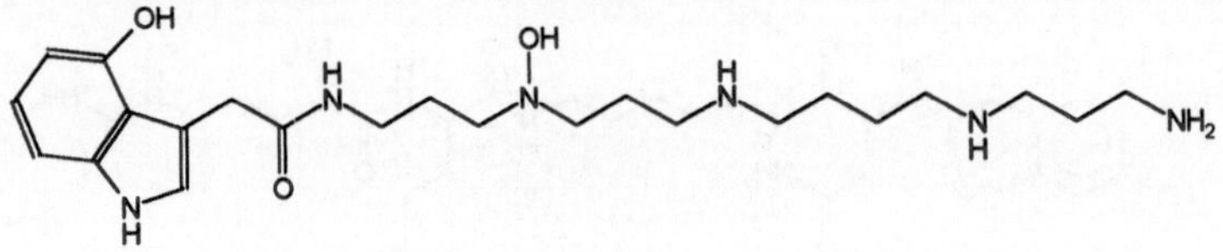

(669)

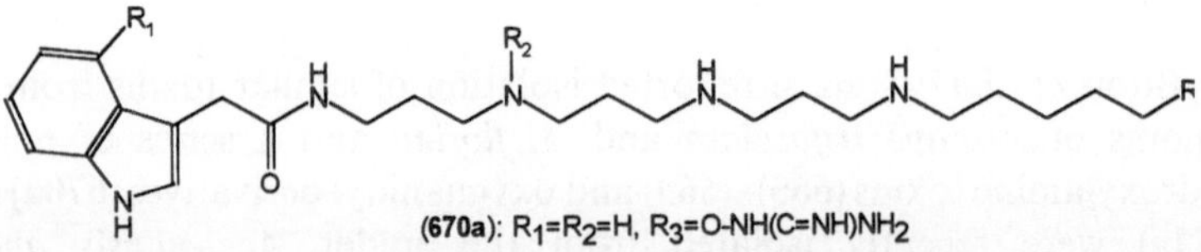

(670a): R₁=R₂=H, R₃=O-NH(C=NH)NH₂

(670a): $R_1=R_2=H$, $R_3=O\text{-NH(C=NH)NH}_2$

(671a): $R_1=OH$, $R_2=H$, $R_3=O\text{-NH(C=NH)NH}_2$

(670b): $R_1=H$, $R_2=OH$, $R_3=N(CH_3)_3{}^+$

(671b): $R_1=R_2=OH$, $R_3=N(CH_3)_3{}^+$

Many authors reviewed different aspects of these toxins, including the chemistry, synthesis, biological activities and pharmacology (*802–803*). Many toxins have been synthesised (*790, 800, 804–813*) as well as some analogues (*814–817*). As stated by SACCOMANO *et al.* (*803*), the polyamine backbones are generally assembled in a stepwise fashion using amine alkylations, reductive aminations and acrylonitrile additions. Standard peptide coupling reactions are usually utilized for formation of the amide bond. Pharmacology is the primary objective for eventual use of the toxins and work along this line continues to provide important knowl-edge about the toxins' mode of action and its implications. Some discussions of structure-activity have appeared (*815, 818*).

The digger wasp *Philanthus triangulum* produces a somewhat related neurotoxin, δ-philanthotoxin (**672**) (*819–830*) which has been synthesised (*830*), as well as a series of analogues (*831–835a*), in order to enhance the activity as a quisqualate antagonist. It appears that a guanidine terminus within an arginine residue enhances the activity of δ-philanthotoxin analogues about four times. Other structural modifications also en-hanced the glutamate antagonist activity of δ-philanthotoxin (*832*).

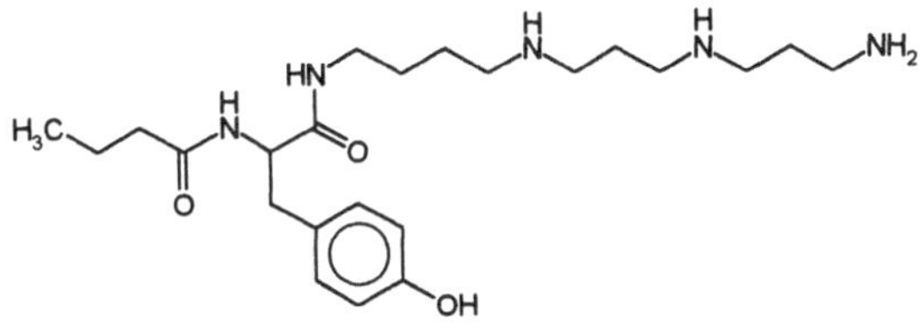

(672)

6.2. Toads

Since the isolation of bufotalin and bufotoxin by WIELAND and WEIL in 1913 (836) an array of bufotoxins have been isolated from toads of the family Atelopidae, genus *Bufo*. Pharmacological studies revealed the cardiotonic properties of these toxins (837–844) and their structures were elucidated by extensive chemical degradations (845–864). The structure of bufotoxin itself, also named vulgarobufotoxin (846, 854, 856, 858), was finally shown to be (673) by chemical (865, 866) and enzymatic (867) degradations, and also by synthesis (868–870). PETTIT and KAMANO (868, 869) have synthesized vulgarobufotoxin (673) as outlined in Scheme 47. An alternative route was developed by SHIMADA *et al.* (870) by condensing the 3-suberoyl-*p*-nitrophenol ester of digitoxigenin (871) with L-arginine hydrochloride.

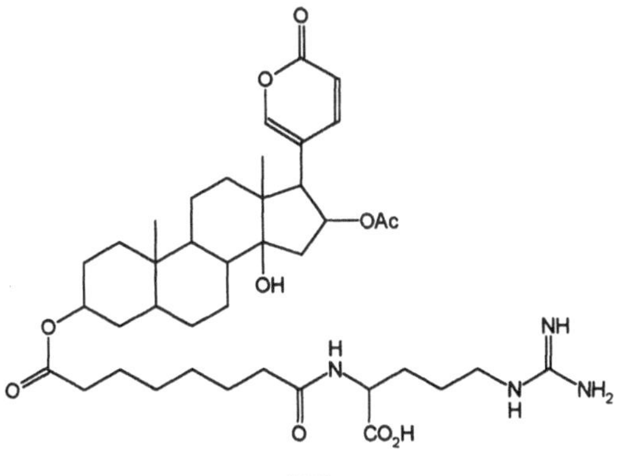

(673)

The general structure of the bufotoxins consists of a steroid aglycone with an α-pyrone or an α,β-unsaturated-γ-lactone side chain, the steroid being linked at the position 3 to an amino acid ester. The aglycone moiety

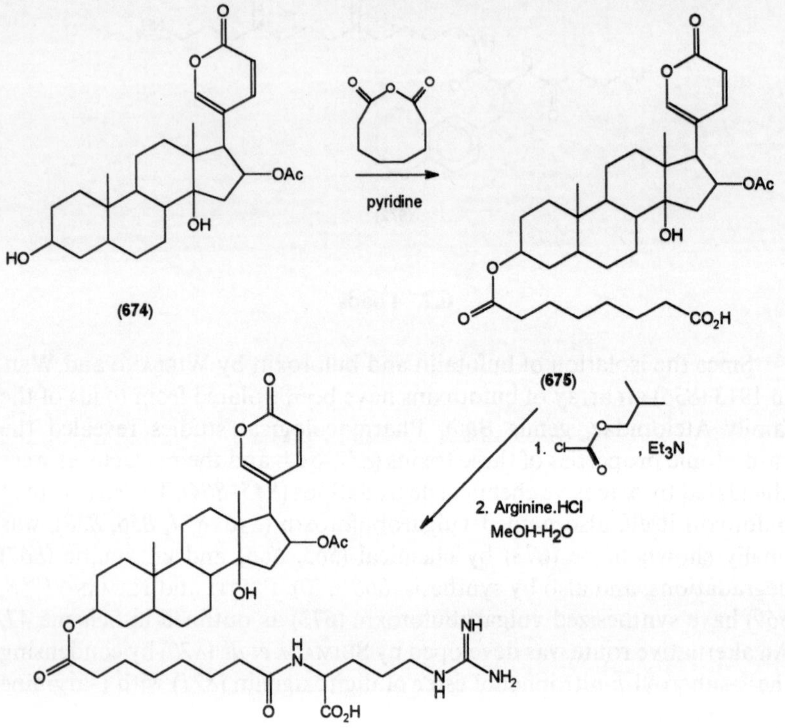

Scheme 47. Synthesis of vulgarobufotoxin

is also called bufadienolide or bufagin. Early work in this field has been reviewed by CHEN and KOVARIKOVA *(872)* and also by MARSHALL *(873)*.

A number of bufadienolides *(874–880)* and bufotoxins have been isolated from *Bufo* toads. Some examples of bufotoxins from *Bufo vulgaris formosus (881)* are the 3-succinoylarginine ester of resibufogenin (676) *(882)*; 3-suberoylarginine ester of gamabufotalin (677) *(883)* and three of its homologues, (678) *(884)*, (679) *(884)* and (680) *(885)*, the cardenobufotoxins (681) and (682) *(886)*; three bufalitoxin homologues (683) (684) and (685) *(887)*; four cinobufotalitoxin homologues (686), (687), (688) and (689) *(887)*; arenobufotoxin (690) *(887)* and cinobufotalitoxin (691) *(887)*. SHIMADA and NAMBARA *(888)* have isolated marinobufagin 3-pimeloylarginine ester (692), telocinobufagin 3-suberoylarginine ester (693) and marinobufagin 3-suberoylarginine ester (694) from *Bufo marinus* (L.) Schneider. Cinobufagin 3-glutaroyl-L-arginine ester (695) was isolated from *Bufo*

bufo gargarizans Cantor (*889*), while hellebritoxin (**696**) was isolated from *Bufo viridis* Laur. (*890*). A series of marinobufotoxin homologues (**697–699**) as well as telocinobufotoxin (**700**) were isolated from *Bufo bakorensis* Borbour (*891*).

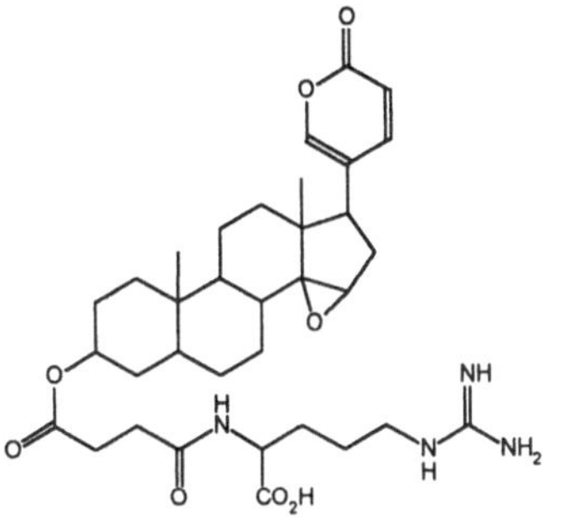

(**676**)

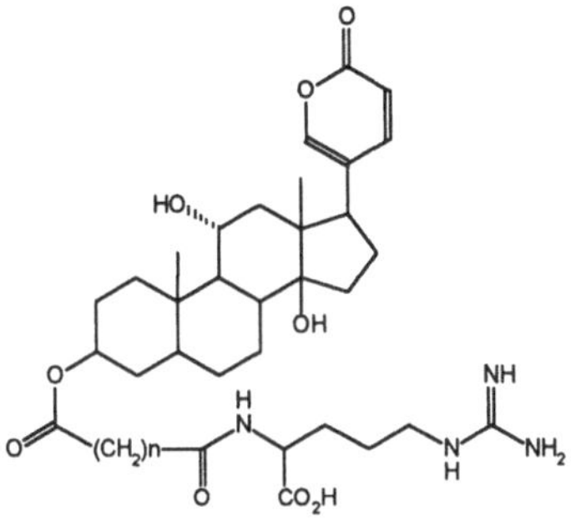

(**677**): n=6
(**678**): n=4
(**679**): n=2
(**680**): n=5

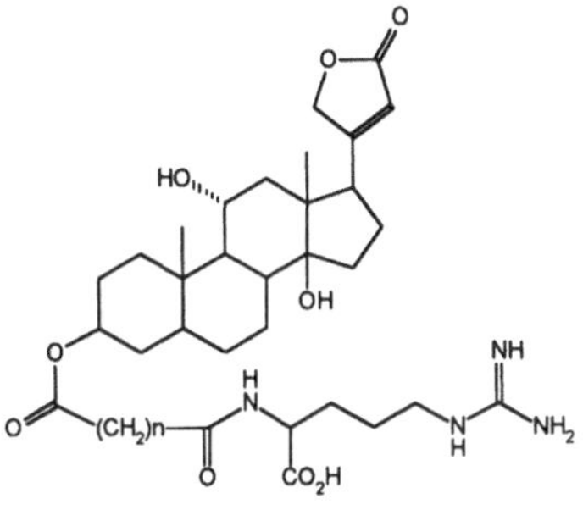

(**681**): n=6
(**682**): n=5

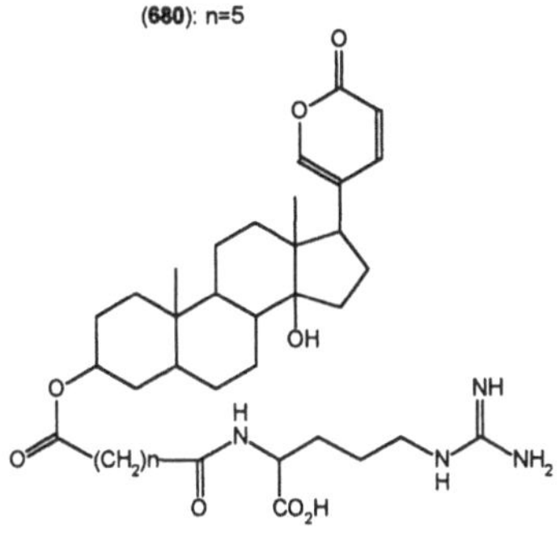

(**683**): n=6
(**684**): n=5
(**685**): n=4

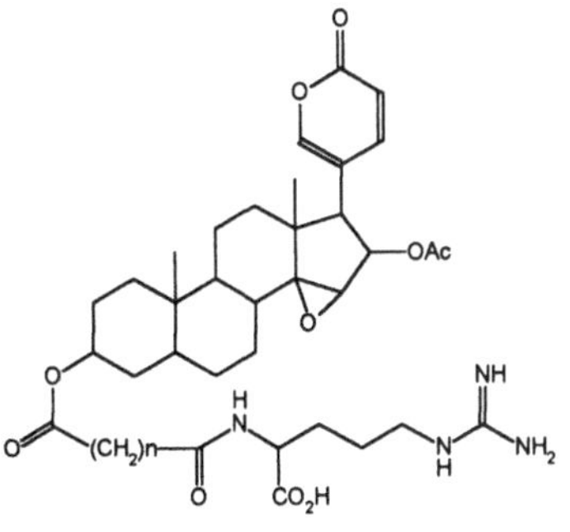

(**686**): n=6
(**687**): n=5
(**688**): n=4
(**689**): n=2

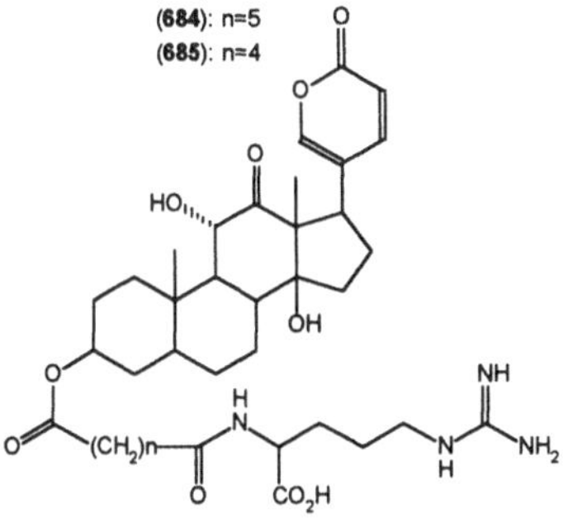

(**690**)· n=6

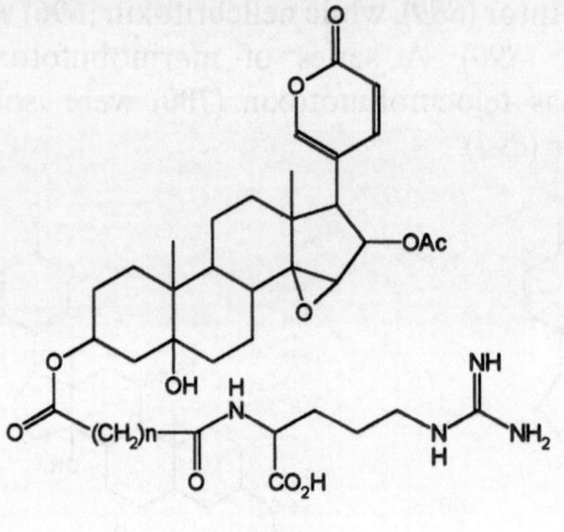

(691): n=6

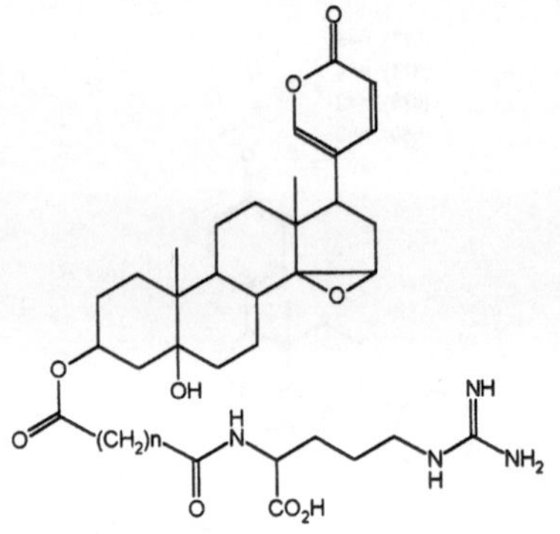

(692): n=5
(694): n=6

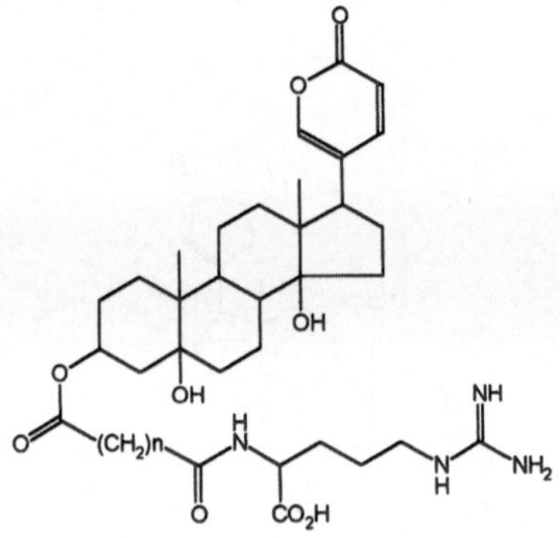

(693): n=6

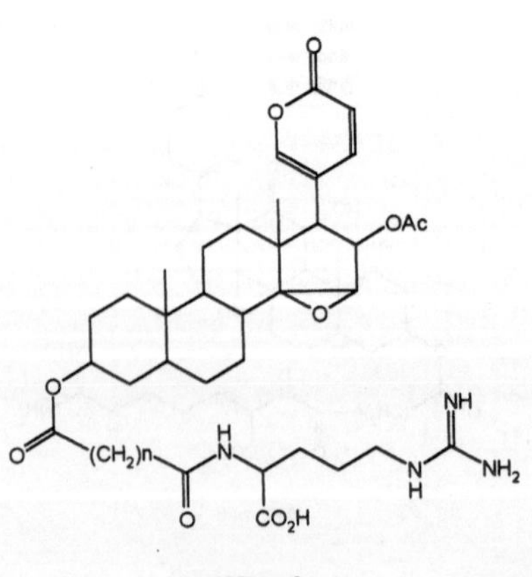

(695): n=3

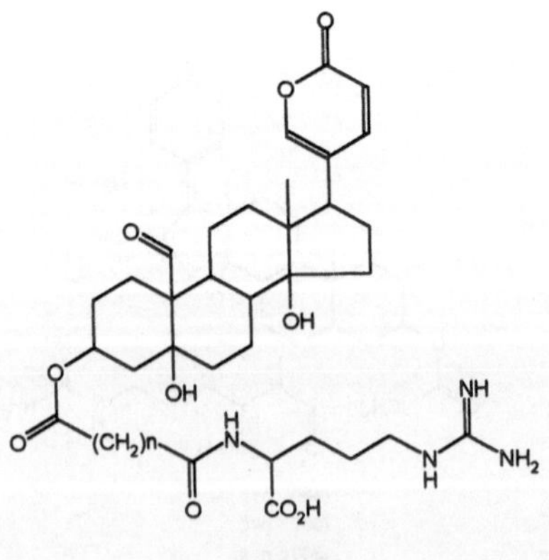

(696): n=6

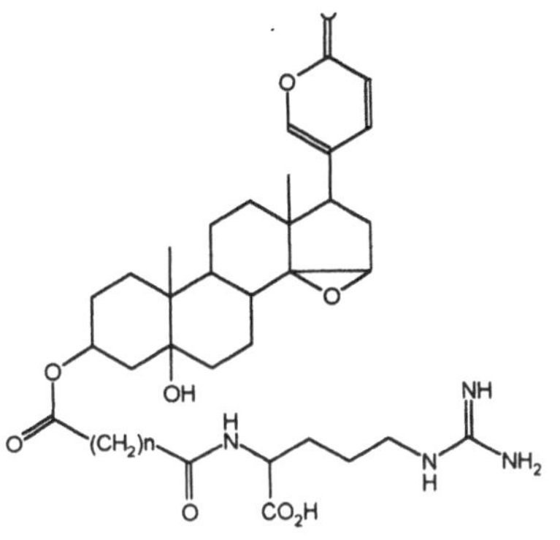

(697): n=2
(698): n=3
(699): n=4

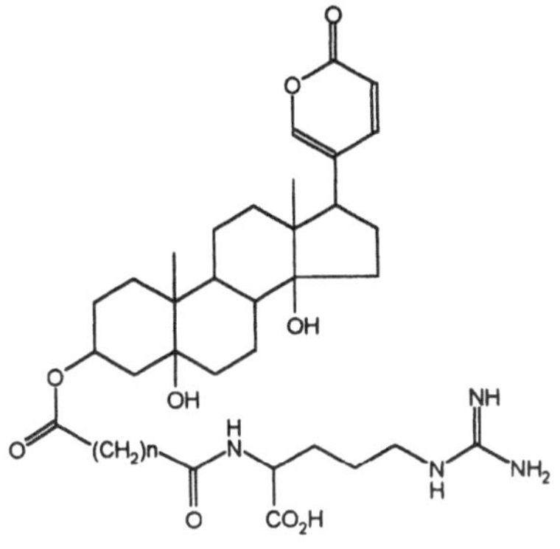

(700): n=3

Bufotoxins having amino acids other than arginine have been isolated from other *Bufo* species (*892–895*). The syntheses of some bufotoxins and bufadienolides have appeared (*896, 897*). Biosynthetic experiments with *Bufo marinus* showed that marinobufagin and marinobufotoxin incorporated [4-^{14}C]cholesterol (*866*). The separation methods employed for bufotoxin isolation have been reviewed by Japanese authors (*898, 899*), and many of these compounds display potent antiviral activity (*900*).

7. ^{13}C-NMR of Guanidine Natural Products

In reviewing the literature of guanidine natural products, it is evident that the chemical shift of the guanidine carbon depends on the guanidine substitution pattern (*901*). Nevertheless, in many papers dealing with the structure determination of naturally occurring guanidines it is stated that the presence of a guanidine function may be inferred by a "typical" ^{13}C guanidine carbon chemical shift. Moreover, the articles which presumably provided this information are rarely cited. As a consequence, some contradictory assignments are found in the literature. For example, the guanidine carbon chemical shift of blasticidin S (**126**) is reported (*297*) as 168.3 ppm, but in other articles (*e.g.*, *284, 290, 290, 292*) the guanidine carbon chemical shift of related compounds reported lies between 158.3 ppm and 158.6 ppm. The guanidine carbon chemical shift of A37812 (= *N*-methylstreptothricin F) and of albothricin (**17**) is reported (*94, 95*) as 163.3 ppm and 163.8 ppm, respectively; however, in the other members of streptothricin family of antibiotics, the chemical shift of the guanidine carbon lies between 158.7 ppm and 160.5 ppm. In aplysinopsin (**373**) the guanidine carbon is assigned two very different chemical shifts, *i.e.* 150.8 ppm (*699*) and 162.1 ppm (*698*). As the two assignments were based as selective decoupling experiments, it is difficult to decide which is the correct one. Finally, dibromoagelaspongin (**365**) is the only "phakellin" derivative with a strained 2-amino-4,5-dihydroimidazolyl ring whose guanidine carbon chemical shift at 161.3 ppm is further downfield than 157.2 ppm (*634*).

8. Conclusion

The chemistry of natural products incorporating a guanidine function continues to provide many surprises. Since guanidine derivatives are very polar compounds, research on their isolation has contributed to the development of new chromatographic techniques for the isolation of polar molecules. Synthesis of these highly polar derivatives is not simple, but several such accomplishments have been reported with the aim of obtaining models for the investigation of structure-activity relationships. Research in this field has impressive results which may well become of practical value as, for example, in the case of neurotoxins. Certainly, the chemistry and applications of guanidine natural products will continue to fascinate researchers interested in multidisciplinary aspects of contemporary science.

Acknowledgments

I am indebted to Prof. DÉSIRÉ DALOZE (Université Libre de Bruxelles) who encouraged me to write this review, to Prof. WERNER HERZ (The Florida State University) and Dr. BENJAMIN GILBERT (Companhia de Desenvolvimento Tecnológico) who patiently revised the text and made useful criticisms. I am also indebted to Prof. K.L. RINEHART (University of Illinois at Urbana-Champaign), Prof. B. B. SNIDER (Brandeis University), Prof. BART MARESCAU (Universitaire Instelling Antwerpen), Prof. PETER N. R. USHERWOOD (Nottingham University), Prof. WAINE W. CARMICHAEL (Wright State University) and Dr. T. TOKI (Daicel Chemical Industries) who provided reprints of various articles.

References

1. GUGGENHEIM, H: Die biogenen Amine. Basel: S. Karger. 1951.
2. THOAI, N., and J. ROCHE: Dérivés guanidiques biologiques. Fortschr. Chem. org. Naturstoffe **18**, 83 (1960).

3. REINBOTHE, H., and K. MOTHES: Urea, Ureides and Guanidines in Plants. Ann. Rev. Plant Physiol. **13**, 129 (1962).
4. WOLSTENHOLME, G.E.W., and M.P. CANORON: Comparative Biochemistry of Arginine and Derivatives. London: Churchill. 1965.
5. THOAI, N.V.: Nitrogenous Bases. In: Comprehensive Biochemistry, Vol. 6 (M. FLORKIN and E.H. STOTZ, eds.). Amsterdam: Elsevier. 1965.
6. THOAI, N.V., and Y. ROBIN: Guanidine Compounds and Phosphagens. In: Chemical Zoology, Vol. IV (M. FLORKIN and B.T. SCHEER, eds.). New York: Academic Press. 1969.
7. NEEDHAM, A.E.: Nitrogen Metabolism in Annelida. In: Comparative Biochemistry of Nitrogen Metabolism, Vol. I. (J.W. CAMPBELL, ed.). London: Academic Press. 1970.
8. MORI, A.: Natural Occurrence and Analyses of Guanidino Compounds. Jpn. J. Clin. Chem. **9**, 232 (1980).
9. MORI, A.: Guanidino Compounds and Neurological Disorders. Neurosci. **9**, 149 (1983).
10. CHEVOLOT, L.: Guanidine Derivatives. In: Marine Natural Products: Chemical and Biological Perspectives, Vol. 4 (P.J. SCHEUER, ed.), p. 53. London: Academic Press. 1981.
11. MARESCAU, B.: Guanidinederivaten: Metabool-biochemisch, diagnostisch, therapeutisch en pathophysiologisch Belang. Proefschrift. Antwerpen: Universitaire Instelling. 1990.
12. ROBIN, Y., and B. MARESCAU: Natural Guanidino Compounds. In: Guanidines (A. MORI, B.D. COHEN, and A. LOWENTHAL, eds.), p. 383. New York: Plenum Press. 1985.
13. COHEN, B.D.: Guanidines as Drugs. In: Guanidines, Vol. 2 (A. MORI, B.D. COHEN, and H. KOIDE, eds.), p.109. London: Plenum Press. 1987.
13a. GREENHILL, J.V.: Guanidines in Medicinal Chemistry. In: Development of Drugs and Modern Medicines (J.W. GORROD, G.G. GIBSON, and M. MITCHARD, eds.), pp. 49–62. Weinheim: VCH Publishers/Chichester: Ellis Horwood. 1986.
14. DURANT, G.J.: Guanidine Derivatives Acting at Histaminergic Receptors. Chem. Soc. Rev. **14**, 375 (1985).
15. BAYS, D.E., and H. FINCH: Inhibitors of Gastric Acid Secretion. Nat. Prod. Rep. **7**, 409 (1990).
16. Merck Index, 11th Ed. Merck and Co. 1989: Entries numbers 1208, 2837, 4469–4474, 4479, 4481 and 4482.
17. MULLER, G.W., D.E. WALTERS, and G.E. DUBOIS: N,N'-Disubstituted Guanidine High-Potency Sweeteners. J. Med. Chem. **35**, 740 (1992).
18. HUANG, S.-M., and Y.-C. HUANG: Chromatography and Electrophoresis of Creatinine and Other Guanidino Compounds. J. Chromatography **429**, 235 (1988).
19. PAULING, L.: The Nature of the Chemical Bond, p. 286. Ithaca: Cornell University Press. 1960.
20. ANGYAL, S.J., and W.K. WARBURTON: The Basic Strengths of Methylated Guanidines. J. Chem. Soc. (London) **1951**, 2492.
21. NEIVELT, B., E.C. MAYO, J.H. TIERS, D.H. SMITH, and G.W. WHELAND: The Base Strengths of N,N'-Dialkylguanidines. J. Amer. Chem. Soc. **73**, 3475 (1951).
22. DAVIS, T.L., and R.C. ELDERFIELD: The Determination of the Ionization Constants of Guanidine and Some of Its Alkylated Derivatives. J. Amer. Chem. Soc. **54**, 1499 (1932).
23. GOTO, T., N. KATO, Y. HIRATA, and Y. HAYASHI: The Structure of Leonurine. Tetrahedron Letters **1962**, 545.
24. GOTO, T., Y. KISHI, S. TAKAHASHI, Y. HIRATA: Tetrodotoxin. Tetrahedron **21**, 2059 (1965).

25. SHARMA, G., and B. MAGDOFF-FAIRCHILD: Natural Products of Marine Sponges, 7: The Constitution of Weakly Basic Guanidine Compounds, Dibromophakellin and Monobromophakellin. J. Organ. Chem. (USA) 42, 4118 (1977).
26. ROGERS, R.S., and H. RAPPOPORT: pKas of Saxitoxin. J. Amer. Chem. Soc. 102, 7335 (1980).
27. SZILAGYI, I., T. VALYI-NAGY, I. SZABO, and T. KERESZTES: Infrared and Ultraviolet Spectra of Primycin. Nature 193, 243 (1962).
28. ANGELL, C.L., N. SHEPPARD, A. YAMAGUCHI, T. SHIMANOUCHI, T. MIYAZAWA, and S. MIZUSHIMA: The Infrared Spectrum, Structure, and Normal Vibrations of the Guanidinium Ion. Trans. Faraday Soc. 53, 589 (1957).
29. GOTO, T., K. NAKANISHI, and M. OHASHI: An Account of the Infrared Absorption of Guanidiniums. Bull. Chem. Soc. Japan 30, 723 (1957).
30. GREENHILL, J.V., M.J. ISMAIL, P.N. EDWARDS, and P.J. TAYLOR: Conformational and Tautomeric Studies of Acylguanidines, Part 2: Vibrational and Carbon-13 Nuclear Magnetic Resonance Spectroscopy. J. Chem. Soc. Perkin. Trans. II 1985, 1265.
31. NAKANISHI, K.: Infrared Absorption Spectroscopy, p. 39. San Francisco: Holden-Day. 1962.
32. PRETSCH, E., J. SEIBL, W. SIMON, and T. CLERC: Tables for Structure Determination of Organic Compounds, 2nd Ed. Berlin-Heidelberg-New York: Springer. 1989.
33. SZILAGYI, I., T. VALYI-NAGY, and T. KERESZTES: Ultraviolet Absorption Spectra of Simple Aliphatic Guanidines. Nature 196, 376 (1962).
34. SZILAGYI, I., T. VALYI-NAGY, I. SZABO, and T. KERESZTES: Identification of Primycin and Other Guanidino Antibiotics. Nature 201, 81 (1964).
35. MATSUMOTO, K., and H. RAPOPORT: The Preparation and Properties of Some Acylguanidines. J. Organ. Chem. (USA) 33, 552.
36. GREENHALGH, R., and R.A.B. BANNARD: Guanidine Compounds, IV: Acetylation of Some Alkyl-Substituted Guanidines with Acetic Anhydride and Ethyl Acetate. Canad. J. Chem. 39, 1017 (1961).
37. CONN, R.B., Jr., and R.B. DAVIS: Green Fluorescence of Guanidinium Compounds with Ninhydrin. Nature 183, 1053 (1959).
38. FAURE, F., and B. BLANQUET: Substances à Réaction de Strecker et Fluorescence des Protides. Bull. soc. chim. biol. (Paris) 43, 953 (1961).
39. BENYON, J.H., J.A. HOPKINSON, and A.E. WILLIAMS: The Mass Spectra of Some Guanidines. Org. Mass Spectrom. 1, 169 (1968).
40. HART, N.K., S.R. JOHNS, J.A. LAMBERTON, and R.I. WILLING: Alkaloids of Alchornea javanensis (Euphorbiaceae). The Isolation of Hexahydroimidazo[1,2-a]pyrimidines and Guanidines. Austral. J. Chem. 23, 1679 (1970).
41. DELL, A., H.R. MORRIS, S.M. HECHT, and M.K. LEVIN: Characterisation of Guanidino-containing Antibiotics: Field Desorption Mass Spectrometry of Bleomycin B_2 and Phleomycins D_1 and E. Biochem. Biophys. Res. Comm. 97, 987 (1980).
42. DELL, A., H.R. MORRIS, M.D. LEVIN, and S.M. HECHT: Field Desorption and Fast Atom Bombardment Mass Spectrometry of Bleomycins and Their Derivatives. Biochem. Biophys. Res. Comm. 102 730 (1981).
43. MORRIS, H.R., A. DELL, and R.A. McDOWELL: Extended Performance Using a High Field Magnet Mass Spectrometer. Biomed. Mass Spectrom. 8, 463 (1981).
44. HEDGE, V.R., M.G. PATEL, H. WITTREICH, V.P. GULLO, M.S. PUAR, and P. BARTNER: Isolation and Structure of an Antifungal, Sch 40873. J. Organ. Chem. (USA) 54, 2402 (1989).
45. HUDSON, H.R., A. LAVORENTI, M. PIANKA, and C. REID: Fast-Atom Bombardment Mass Spectrometry of Guanidine Fungicides and Related Compounds: The Forma-

tion of [MH + 12]$^+$ Ions from Secondary Amines and from 1,1'-Iminodioctamethy-lenediguanidine, and of [MH + 24]$^+$ Ions from 9-Aza-1,17-diaminoheptadecane. Chem. and Ind. **1991**, 131.

46. OHUCHI, S., H. SUDA, H. NAGANAWA, T. TAKITA, T. AOYAGI, H. UMEZAWA, H. NAKAMURA, and Y. IITAKA: Arphamenines A and B, New Inhibitors of Aminopeptidase B, Produced by Bacteria. J. Antibiotics **36**, 1576 (1983).

47. OGAWA, K., H. NAGANAWA, H. IIMURA, T. AOYAGI, and H. UMEZAWA: The Structure of Histargin. J. Antibiotics **37**, 984 (1984).

48. CHILLEMI, R., R. MORRONE, A. PATTI, M. PIATTELLI, and S. SCIUTO: Nicaeesin, a New Amidinoureido Compound from the Red Alga *Schottera nicaeensis*. J. Nat. Prod. (Lloydia) **53**, 1220 (1990).

49. CARTER, J.D., and K.L. RINEHART, Jr.: Acarnidines, Novel Antiviral and Antimicrobial Compounds from the Sponge *Acarnus erithacus* (de Laubenfels). J. Amer. Chem. Soc. **100**, 4302 (1978).

50. SAMAIN, D., J.C. COOK, Jr., and K.L. RINEHART, Jr.: Structure of Scopafungin, a Potent Nonpolyene Antifungal Antibiotic. J. Amer. Chem. Soc. **104**, 4129 (1982).

51. ROWAN, D.D., M.B. HUNT, and D.L. GAYNOR: Peramine, a Novel Insect Feeding Deterrent from Ryegrass Infected with the Endophyte *Acremonium loliae*. Chem. Commun. **1986**, 935.

52. BAUER, V.J., W. FULMOR, G.O. MORTON, and S.R. SAFIR: Restricted Rotation in Guanidines. J. Amer. Chem. Soc. **90**, 6846 (1968).

53. KESSLER, H.: Detection of Hindered Rotation and Inversion by NMR Spectroscopy. Angew. Chem. Int. Ed. Engl. **9**, 219 (1970).

54. KENYON, G.L., and G.L. ROWLEY: Tautomeric Preferences Among Glycocyamidines. J. Amer. Chem. Soc. **93**, 5552 (1971).

55. KESSLER, H.: Thermal Isomerization About Double Bonds. Rotation and Inversion. Tetrahedron **30**, 1861 (1974).

56. CORRAL, R.A., O.O. ORAZI, and M.E. GONZALES: Detection and Substitution – Pattern Determination of Guanidine Compounds by ^{1}H Nuclear Magnetic Resonance. Rev. Latinoamer. Quim. **9**, 184 (1978).

57. BREITMEIER, E., and W. VOELTER: Carbon-13 NMR Spectroscopy. Weinheim: VCH. 1990.

58. KALINOWSKI, H.-O., and H. KESSLER: Mesomere Kationen, IV: ^{13}C-NMR-Spektren von Uronium-, Thiouronium- und Guanidiniumsalzen sowie einigen Guanidinen. Org. Magnetic Reson. **7**, 128 (1975).

59. SMITH, R.L., D.W. COCHRAN, P. GUND, and E.J. CRAGOE, Jr.: Proton, Carbon-13 and Nitrogen-15 Nuclear Magnetic Resonance and CNDO/2 Studies on the Tautomerism and Conformation of Amiloride, a Novel Acylguanidine. J. Amer. Chem. Soc. **101**, 191 (1979).

60. GREENHILL, J.V.: Guanidines in Medicinal Chemistry. In: Development of Drugs and Modern Medicines (J.W. GORROD, G.G. GIBSON, and M. MITCHARD, eds.), p. 1. Weinheim: VCH. 1986.

61. JACKMAN, L.M., and T. JEN: ^{1}H and ^{13}C Nuclear Magnetic Resonance Studies on the Tautomerism, Geometrical Isomerism, and Conformation of Some Cyclic Amidines, Guanidines and Related Systems. J. Amer. Chem. Soc. **97**, 2811 (1975).

62. MOLD, J.D., J.M. LADINO, and E.J. SCHANTZ: The Sakaguchi and Biacetyl Reactions for the Identification of Alkyl Guanidines. J. Amer. Chem. Soc. **75**, 6321 (1953).

63. HESSING, A., and K. HOPPE: Sakaguchi- und Fearon-Reaktion: Die Struktur der Farbstoffe, ihr Bildungsmechanismus und die Spezifität der Reaktion. Chem. Ber. **100**, 3469 (1967).

64. WENDISH, D.A.W.: Acronyms and Abbreviations in Molecular Spectroscopy, p. 49. Berlin-Heidelberg-New York: Springer. 1990.
65. Ref. (*64*), p. 129.
66. Ref. (*64*), p. 164.
67. WAKSMAN, S.A., and H.B. WOODRUFF: Streptothricin, a New Selective Bacteriostatic and Bactericidal Agent, Particularly Active Against Gram-Negative Bacteria. Proc. Soc. Exp. Biol. Med. **49**, 207 (1942).
68. PECK, R.L., A. WALTI, R.P. GRABER, E. FLYNN, C.E. HOFFHINE, Jr., V. ALLFREY, and K. FOLKERS: Streptomyces Antibiotics, VI: Isolation of Streptothricin. J. Amer. Chem. Soc. **68**, 772 (1946).
69. RIVETT, R.W., and W.H. PETERSON: Streptolin, a New Antibiotic from a Species of Streptomyces. J. Amer. Chem. Soc. **69**, 3006 (1947).
70. LARSON, L.M., H. STERNBERG, and W.H. PETERSON: Production, Isolation and Components of the Antibiotic Streptolin. J. Amer. Chem. Soc. **75**, 2036 (1953).
71. KUEHL, F.A., Jr., R.L. PECK, A. WALTI, and K. FOLKERS: Streptomyces Antibiotics, I: Crystalline Salts of Streptomycin and Streptothricin. Science **102**, 34 (1945).
72. FRIED, J., and O. WINTERSTEINER: Crystalline Reineckates of Streptothricin and Streptomycin. Science **102**, 613 (1945).
73. CARTER, H.E., W.R. HEARN, E.M. LANSFORD, Jr., A.C. PAGE, Jr., N.P. SALZMAN, D. SHAPIRO, and W.R. TAYLOR: Structure of the Diaminohexanoic Acid from Streptothricin. J. Amer. Chem. Soc. **74**, 3704 (1952).
74. VAN TAMELEN, E.E., and E.E. SMISSMAN: Streptolin. The Structure and Synthesis of Isolysine. J. Amer. Chem. Soc. **74**, 3713 (1952).
75. VAN TAMELEN, E.E., and E.E. SMISSMAN: Streptolin. The Structure and Synthesis of Isolysine. J. Amer. Chem. Soc. **75**, 2031 (1953).
76. SMISSMAN, E.E., R.W. SHARPE, B.F. AYCOCK, E.E. VAN TAMELEN, and W.H. PETERSON: Streptolin. Preliminary Investigation and Separation of Acid Hydrolysis Products. J. Amer. Chem. Soc. **75**, 2029 (1953).
77. CARTER, H.E., R.K. CLARK, Jr., P. KOHN, J.W. ROTHROCK, W.R. TAYLOR, C.A. WEST, G.B. WHITFIELD, and W.G. JACKSON: Streptothricin, I: Preparation, Properties and Hydrolysis Products. J. Amer. Chem. Soc. **76**, 566 (1954).
78. NAKANISHI, K., T. ITO, and Y. HIRATA: Structure of a New Amino Acid Obtained from Roseothricin. J. Amer. Chem. Soc. **76**, 2845 (1954).
79. NAKANISHI, K., T. ITO, M. OHASHI, I. M. ORIMOTO, and Y. HIRATA: Two Uncommon Amino Acids Obtained from Roseothricin. Bull. Chem. Soc. Japan **27**, 539 (1954).
80. NAKANISHI, K., and M. OHASHI: Confirmation of Structure of Roseonine. Bull. Chem. Soc. Japan **30**, 725 (1957).
81. VAN TAMELEN, E.E., J.R. DYER, H.E. CARTER, J.V. PIERCE, and E.E. DANIELS: Structure of the Aminosugar Derived from Streptothricin and Streptolin B. J. Amer. Chem. Soc. **78**, 4817 (1956).
82. GOTO, T., Y. HIRATA, S. HOSOYA, and N. KOMATSU: Structure of Roseothricin A. Bull. Chem. Soc. Japan **30**, 729 (1957).
83. CARTER, H.E., C.C. SWEELEY, E.E. DANIELS, J.E. McNARY, C.P. SCHAFFNER, C.A. WEST, E.E. VAN TAMELEN, J.R. DYER, and H.A. WHALEY: Streptothricin and Streptolin: The Structure of Streptolidine (Roseonine). J. Amer. Chem. Soc. **83**, 4296 (1961).
84. CARTER, H.E., J.V. PIERCE, G.B. WHITFIELD, Jr., J.E. McNARY, E.E. VAN TAMELEN, J.R. DYER, and H.A. WHALEY: N-*guan*-Streptolidyl Gulosamidine, a Degradation Product of the Streptothricin Antibiotic Group. J. Amer. Chem. Soc. **83**, 4287 (1961).
85. VAN TAMELEN, E.E., J.R. DYER, H.A. WHALEY, H.E. CARTER, and G.B. WHITFIELD, Jr.: Constitution of the Streptolin-Streptothricin Group of *Streptomyces* Antibiotics. J. Amer. Chem. Soc. **83**, 4295 (1961).

86. JOHNSON, A.W., and J.W. WESTLEY: The Streptothricin Group of Antibiotics, Part I: The General Structural Pattern. J. Chem. Soc. (London) **1962**, 1642.

87. BOWIE, J.H., E. BULLOCK, and A.W. JOHNSON: The Structure of Streptolidine. J. Chem. Soc. (London) **1963**, 4260.

88. TANIYAMA, H., and F. MIYOSHI: The Structure of Racemomycin A. Chem. Pharm. Bull. (Japan) **10**, 156 (1962).

89. BYCROFT, B.W., and T.J. KING: Crystal Structure of Streptolidine, a Guanidine-containing Amino-acid. Chem. Commun. **1972**, 652.

90. KAWAKAMI, Y., K. YAMASAKI, and S. NAKAMURA: The Structures of Component A_1 (= LL-AB664) and Component A_2 (= LL-AC541), Streptothricin-like Antibiotics. J. Antibiotics **34**, 921 (1981).

91. KUSUMOTO, S., Y. KAMBAYASHI, S. IMAOKA, K. SHIMA, and T. SHIBA: Total Chemical Structure of Streptothricin. J. Antibiotics **35**, 925 (1982).

92. BORDERS, D.B., K.J. SAX, J.E. LANCASTER, W.K. HAUSMANN, L.A. MITSCHER, E.R. WETZEL, and E.L. PATTERSON: Structures of LL-AC541 and LL-AB664 New Streptothricin-type Antibiotics. Tetrahedron **26**, 3123 (1970).

93. SAWADA, Y., S. KAWAKAMI, H. TANIYAMA, K. HAMANO, R. ENOKITA, S. IWADO, and M. ARAI: Glycinothricin, a New Streptothricin-class Antibiotic from *Streptomyces griseus*. J. Antibiotics **30**, 460 (1977).

94. HUNT, A.H., R.L. HAMILL, J.R. DEBOER, and E.A. PRESTI: A37812: *N*-methylstreptothricin F. J. Antibiotics **38**, 987 (1985).

95. OHBA, K., H. NAKAYAMA, K. FURIHATA, K. FURIHATA, A. SHIMAZU, H. SETO, N. OTAKE, Z.-Z. YANG, L.-S. XU, and W.-S. XU: Albothricin, a New Streptothricin Antibiotic. J. Antibiotics **39**, 872 (1986).

96. KIDO, Y., T. FURUIE, K. SUZUKI, K. SAKAMOTO, Y. YOKOYAMA, M. UYEDA, J. KINJYO, S. YAHARA, T. NOHARA, and M. SHIBATA: A Streptothricin-like Antibiotic Mixture, A-269A (and A-269A'). J. Antibiotics **40**, 1698 (1987).

97. ANDO, T., S. MIYASHIRO, K. HIRAYAMA, T. KIDA, H. SHIBAI, A. MURAI, and S. UDAKA: New Streptothricin-group Antibiotics AN-201 I, II and III, II: Chemical Studies. J. Antibiotics **40**, 1140 (1987).

98. GRÄFE, U., G. REINHARDT, H. BOCKER, and H. THRUM: Biosynthesis of Streptolidine Moiety of Streptothricins by *Streptomyces noursei* JA 3890b. J. Antibiotics **30**, 106 (1977).

99. BYCROFT, B.W.: Structural Relationship in Microbial Peptides. Nature **224**, 595 (1969).

100. CARTER, J.H., II, R.H. DU BUS, J.R. DYER, J.C. FLOYD, K.C. RICE, and P.D. SHAW: Biosynthesis of Viomycin, II: Origin of β-Lysine and Viomycidine. Biochemistry **13**, 1227 (1974).

101. SAWADA, Y., S. NAKASHIMA, H. TANIYAMA, Y. INAMORI, S. SUNAGAWA, and M. TSURUGA: Biosynthesis of Streptothricin Antibiotics, IV: On the Incorporation of L-Arginine into Streptolidine Moiety by *Streptomyces lavendulae* OP-2. Chem. Pharm. Bull. (Japan) **25**, 1161 (1977).

102. SAWADA, Y., S. KAWAKAMI, H. TANIYAMA, and Y. INAMORI: Incorporation of Carboxyl and Methyl Carbon-13 Labeled Acetates into Racemomycin A by *Streptomyces lavendulae* ISP 5069. J. Antibiotics **30**, 630 (1977).

103. GOULD, S.J., K.J. MARTINKUS, and C.-H. TANN: Biosynthesis of Streptothricin F, I: Observing the Interaction of Primary and Secondary Metabolism with $[1,2\text{-}^{13}C_2]$ Acetate. J. Amer. Chem. Soc. **103**, 2871 (1981).

104. GOULD, S.J., K.J. MARTINKUS, and C.-H. TANN: Studies of Nitrogen Metabolism Using ^{13}C NMR Spectroscopy, 2: Incorporation of L-[*guanido*-^{13}C, $^{15}N_2$]Arginine and DL-[*guanido*-^{13}C, 2-^{15}N]Arginine into Streptothricin F. J. Amer. Chem. Soc. **103**, 4639 (1981).

105. GOULD, S.J., and T.K. THIRUVENGADAM: Studies of Nitrogen Metabolism Using
    ^{13}C NMR Spectroscopy, 3: Synthesis of DL-[3-^{13}C, 2-^{15}N]Lysine and Its Incorpora-
    tion into Streptothricin F. J. Amer. Chem. Soc. **103**, 6752 (1981).
106. THIRUVENGADAM, T.K., S.J. GOULD, D.J. ABERHART, and H.-J. LIU: Biosynthesis of
    Streptothricin F, 5: Formation of β-Lysine by Streptomyces L-1689-3. J. Amer. Chem.
    Soc. **105**, 5470 (1983).
107. MARTINKUS, K.J., C.-H. TANN, and S.J. GOULD: The Biosynthesis of the Streptolidine
    Moiety in Streptothricin F. Tetrahedron **39**, 3493 (1983).
108. PALANISWAMY, V.A., and S.J. GOULD: Biosynthesis of Streptothricin F, Part 6:
    Formation and Intermediacy of D-Glucosamine in Streptomyces L-1689-23. J. Chem.
    Soc. Perkin Trans. I **1988**, 2283.
109. GOULD, S.J., J. LEE, and J. WITYAK: Biosynthesis of Streptothricin F, 7: The Fate of the
    Arginine Hydrogens. Bioorg. Chem. **19**, 333 (1991).
110. GOULD, S.J., and D.A. MINOTT: Biosynthesis of Capreomycin, 1: Incorporation of
    Arginine. J. Organ. Chem. (USA) **57**, 5214 (1992).
111. HERBERT, R.B.: The Biosynthesis of Plant Alkaloids and Nitrogenous Microbial
    Metabolites. Nat. Prod. Rep. **10**, 575 (1993).
112. SAWADA, Y., T. KUBO, and H. TANIYAMA: Biosynthesis of Streptothricin Antibiotics, I:
    Incorporation of ^{14}C-Labeled Compound into Racemomycin-A and Distribution of
    Radioactivity. Chem. Pharm. Bull. (Japan) **24**, 2163 (1976).
113. SAWADA, Y., S. NAKASHIMA, H. TANIYAMA, and Y. INAMORI: Biosynthesis of Strepto-
    thricin Antibiotics, III: Incorporation of D-Glucosamine into D-Gulosamine Moiety of
    Racemomycin-A. Chem. Pharm. Bull. (Japan) **25**, 1478 (1977).
114. PETERSON, D.H., and L.M. REINEKE: A Paper Chromatographic Technique and Its
    Application to the Study of New Antibiotics. J. Amer. Chem. Soc. **72**, 3598 (1950).
115. HOROWITZ, M.I., and C.P. SCHAFFNER: Paper Chromatography of Streptothricin
    Antibiotics. Analyt. Chem. **30**, 1616 (1958).
116. TANIYAMA, H., Y. SAWADA, and T. KITAGAWA: Chromatography of Racemomycins on
    Dextran Gel. J. Chromatography **56**, 360 (1971).
117. KHOKHLOV, A.S., and P.D. RESHETOV: Chromatography of Streptothricins on Car-
    boxymethylcellulose. J. Chromatography **14**, 495 (1964).
118. RESHETOV, P.D., and A.S. KHOKHLOV: Studies on Streptothricins, VI: Preparation and
    Properties of Individual Streptothricins. Khim. Prir. Soedin (English Ed.) **1**, 31 (1965).
119. RESHETOV, P.D., TS. A. EGOROV, and A.S. KHOKHLOV: Studies on Streptothricins, VII:
    Determination of the Ninhydrin-positive Fragments of Streptothricin Antibiotics.
    Khim. Prir. Soedin. (English Ed.) **1**, 91 (1965).
120. KUSUMOTO, S., S. TSUJI, and T. SHIBA: Synthesis of Streptolidine (Roseonine,
    Geamine). Tetrahedron Letters **1974**, 1417.
121. KUSUMOTO, S., S. IMAOKA, Y. KAMBAYASHI, K. YOSHIZAWA, and T. SHIBA: Synthesis of
    N^g-Streptolidyl Gulosamisidine – A New Evidence for the Proposed Structure of
    Antibiotic Streptothricin. Chem. Lett. **1981**, 1317.
122. KUSUMOTO, S., S. IMAOKA, Y. KAMBAYASHI, and T. SHIBA: Total Synthesis of
    Antibiotic Streptothricin F. Tetrahedron Letters **23**, 2961 (1982).
123. SAWADA, Y., and H. TANIYAMA: Studies on the β-Lysine Peptide, IV: Preparation of
    Semi-synthetic Racemomycins and Their Antimicrobial Activities. Chem. Pharm.
    Bull. (Japan) **25**, 1302 (1977).
124. WAKSMAN, S.A.: Nomenclature of Streptomycin Preparations. Science **107**, 233 (1948).
125. SCHATZ, A., E. BUGIE, and S.A. WAKSMAN: Streptomycin, a Substance Exhibiting
    Antibiotic Activity Against Gram-Positive and Gram-Negative Bacteria. Proc. Soc.
    Exp. Biol. Med. **55**, 66 (1944).

126. SCHATZ, A., and S.A. WAKSMAN: Effect of Streptomycin and Other Antibiotics Substances Upon *Mycobacterium tuberculosis* and Related Organisms. Proc. Soc. Exp. Biol. Med. **57**, 244 (1944).

127. JONES, D., H.J. METZGER, A. SCHATZ, and S.A. WAKSMAN: Control of Gram-Negative Bacteria in Experimental Animals by Streptomycin. Science **100**, 103 (1944).

128. ROBINSON, H.J., D.G. SMITH, and O.E. GRAESSELE: Chemotherapeutic Properties of Streptomycin. Proc. Soc. Exp. Biol. Med. **57**, 226 (1944).

129. BRINK, N.G., F.A. KUEHL, Jr., and K. FOLKERS: Streptomyces Antibiotics, III: Degradation of Streptomycin to Streptobiosamine Derivatives. Science **102**, 506 (1945).

130. PECK, R.L., N.G. BRINK, F.A. KUEHL, Jr., E.H. FLYNN, A. WALTI, and K. FOLKERS: Streptomycin Antibiotics, II: Crystalline Streptomycin Trihydrochloride-Calcium Chloride Double Salt. J. Amer. Chem. Soc. **67**, 1866 (1945).

131. FRIED, J., G.A. BOYACK, and O. WINTERSTEINER: Streptomycin: The Chemical Nature of Streptidine. J. Biol. Chem. **162**, 391 (1945).

132. SCHENCK, J.R., and M.A. SPIELMAN: The Formation of Maltol by the Degradation of Streptomycin. J. Amer. Chem. Soc. **67**, 2276 (1945).

133. CARTER, H.E., R.K. CLARK, S.R. DICKMAN, Y.H. LOO, J.S. MEEK, P.S. SKELL, W.A. STRONG, J.T. ALBERI, Q.R. BARTZ, S.B. BINKLEY, H.M. CROOKS, Jr., I.R. HOOPER, and M.C. REBSTOCK: Degradation of Streptomycin and the Structure of Streptidine and Streptamine. Science **103**, 53 (1946).

134. CARTER, H.E., R.K. CLARK, Jr., S.R. DICKMAN, Y.H. LOO, P.S. SKELL, and W.A. STRONG: Degradation of Streptomycin and the Structure of Streptidine and Streptamine. Science **103**, 540 (1946).

135. PECK, R.L., C.E. HOFFHINE, Jr., E.W. PEEL, R.P. GRABER, F.W. HOLLY, R. MOZINGO, and K. FOLKERS: Streptomyces Antibiotics, VII: The Structure of Streptidine. J. Amer. Chem. Soc. **68**, 776 (1946).

136. WINTERSTEINER, O., and A. KLINSBERG: Streptomycin, VII: Degradation of O-Tetramethylstreptamine to D,L-Dimethoxysuccinic Acid. J. Amer. Chem. Soc. **70**, 885 (1948).

137. WINTERSTEINER, O., and A. KLINSBERG: Streptomycin, XIII: New Derivatives of Streptamine, the Oxidative Degradation of O-Tetramethylstreptamine. J. Amer. Chem. Soc. **73**, 2917 (1951).

138. PECK, R.L., R.P. GRABER, A. WALTI, E.W. PEEL, C.E. HOFFHINE, Jr., and K. FOLKERS: Streptomyces Antibiotics, IV: Hydrolytic Cleavage of Streptomycin to Streptidine. J. Amer. Chem. Soc. **68**, 29 (1946).

139. KUEHL, F.A., Jr., E.H. FLYNN, F.W. HOLLY, R. MOZINGO, and K. FOLKERS: Streptomyces Antibiotics, V: N-Methyl-L-glucosamine from Streptomycin. J. Amer. Chem. Soc. **68**, 536 (1946).

140. KUEHL, F.A., Jr., E.H. FLYNN, F.W. HOLLY, R. MOZINGO, and K. FOLKERS: Streptomyces Antibiotics, XV: N-Methyl-L-glucosamine. J. Amer. Chem. Soc. **68**, 3032 (1946).

141. KUEHL, F.A., Jr., R.L. PECK, C.E. HOFFHINE, Jr., R.P. GRABER, and K. FOLKERS: Streptomyces Antibiotics, VIII: Isolation of Streptomycin. J. Amer. Chem. Soc. **68**, 1460 (1946).

142. R.L. PECK, C.E. HOFFHINE, Jr., and K. FOLKERS: Streptomyces Antibiotics, IX: Dihydrostreptomycin. J. Amer. Chem. Soc. **68**, 1390 (1946).

143. KUEHL, F.A., Jr., E.H. FLYNN, N.G. BRINK, and K. FOLKERS: Streptomyces Antibiotics, X: The Degradation of Streptomycin and Dihydrostreptomycin with Ethyl Mercaptan. J. Amer. Chem. Soc. **68**, 2096 (1946).

144. HOOPER, I.R., L.H. KLEMM, W.J. POLGLASE, and M.L. WOLFROM: Degradative Studies on Streptomycin. J. Amer. Chem. Soc. **68**, 2120 (1946).

145. BARTZ, Q.R., J. CONTROULIS, H.M. CROOKS, Jr., and M.C. REBSTOCK: Dihydrostreptomycin. J. Amer. Chem. Soc. **68**, 2163 (1946).

146. BRINK, N.G., F.A. KUEHL, Jr., E.H. FLYNN, and K. FOLKERS: Streptomyces Antibiotics, XII: The Degradation of Streptomycin and Dihydrostreptomycin with Methanol. J. Amer. Chem. Soc. **68**, 2557 (1946).

147. BRINK, N.G., F.A. KUEHL, Jr., E.H. FLYNN, and K. FOLKERS: Streptomycin Antibiotics, XI: The Structure of Tetraacetylbisdesoxystreptobiosamine. J. Amer. Chem. Soc. **68**, 2405 (1946).

148. FRIED, J., D.E. WALZ, and O. WINTERSTEINER: Streptomycin, III: 4-Desoxy-L-erythrose (Threose) Phenylosazone from Streptobiosamine. J. Amer. Chem. Soc. **68**, 2746 (1946).

149. KUEHL, F.A., Jr., E.H. FLYNN, N.G. BRINK, and K. FOLKERS: Streptomyces Antibiotics, XIII: The Structure of Streptobiosamine. J. Amer. Chem. Soc. **68**, 2679 (1946).

150. LEMIEUX, R.V., W.J. POLGLASE, C.W. DEWALT, and M.L. WOLFROM: Degradative Studies on Streptomycin. J. Amer. Chem. Soc. **68**, 2747 (1946).

151. HOOPER, I.R., L.H. KLEMM, W.J. POLGLASE, and M.L. WOLFROM: Degradative Studies on Streptomycin, I. J. Amer. Chem. Soc. **69**, 1052 (1947).

152. FRIED, J., and O. WINTERSTEINER: Streptomycin, II: Reduction and Oxidation Products of Streptomycin and of Streptobiosamine. J. Amer. Chem. Soc. **69**, 79 (1947).

153. LEMIEUX, R.V., C.W. DEWALT, and M.L. WOLFROM: Degradative Studies on Streptomycin. J. Amer. Chem. Soc. **69**, 1838 (1947).

154. HEDING, H.: Methyl-Steptomycin, a New Hydrogenation Product of Streptomycin. Tetrahedron Letters **1969**, 2831.

155. CARTER, H.E., Y.H. LOO, and P.S. SKELL: Streptomycin. The Linkage Between Streptidine and Streptobiosamine. J. Biol. Chem. **168**, 401 (1947).

156. KUEHL, F.A., Jr., R.L. PECK, C.E. HOFFHINE, Jr., E.W. PEEL, and K. FOLKERS: Streptomyces Antibiotics, XIV: The Position of the Linkage of Streptobiosamine to Streptidine in Streptomycin. J. Amer. Chem. Soc. **69**, 1234 (1947).

157. WOLFROM, M.L., and W. POLGLASE: A Synthesis of Streptidine. J. Amer. Chem. Soc. **70**, 1672 (1948).

158. WOLFROM, M.L., S.M. OLIN, and W.J. POLGLASE: A Synthesis of Streptidine. J. Amer. Chem. Soc. **72**, 1724 (1950).

159. TATSUOKA, S., S. HORII, K.L. RINEHART, Jr., and T. NAKABAYASHI: The Absolute Configurations of Streptidine in Dihydrostreptomycin and of Deoxystreptamine in Kanamycin A. J. Antibiotics **17A**, 88 (1964).

160. HOLLY, F.W., R. MOZINGO, and K. FOLKERS: Streptomyces Antibiotics, XX: Conversion of Streptamine Into Streptidine. J. Amer. Chem. Soc. **70**, 3944 (1948).

161. KUEHL, F.A., Jr., M.N. BISHOP, E.H. FLYNN, and K. FOLKERS: Streptomyces Antibiotics, XIX: Dihydrostreptosonic Acid Lactone and Configuration of Streptose and Streptobiosamine. J. Amer. Chem. Soc. **70**, 2613 (1948).

162. KUEHL, F.A., Jr., R.L. CLARK, M.N. BISHOP, E.H. FLYNN, and K. FOLKERS: Streptomyces Antibiotics, XXII: Configuration of Streptose and Streptobiosamine. J. Amer. Chem. Soc. **71**, 1445 (1949).

163. PECK, R.L., F.A. KUEHL, Jr., C.E. HOFFHINE, Jr., E.W. PEEL, and K. FOLKERS: Streptomyces Antibiotics, XVII: Heptabenzoylstreptidine from Streptomycin. J. Amer. Chem. Soc. **70**, 2321 (1948).

164. KUEHL, F.A., Jr., R.L. PECK, C.E. HOFFHINE, Jr., and K. FOLKERS: Streptomyces Antibiotics, XVIII: Structure of Streptomycin. J. Amer. Chem. Soc. **70**, 2325 (1948).

165. WOLFROM, M.L., and C.W. DEWALT: The Configuration of Streptose. J. Amer. Chem. Soc. **70**, 3148 (1948).
166. FRIED, J., and D.E. WALZ: Streptomycin, XI: Synthesis of 3,6-Dimethyl-N-methyl-D-glucosamine. J. Amer. Chem. Soc. **74**, 5468 (1952).
167. WOLFROM, H.L., M.J. CRON, C.W. DEWALT, and R.M. HUSBAND: Configuration of the Glycosidic Unions in Streptomycin. J. Amer. Chem. Soc. **76**, 3675 (1954).
168. DYER, J.R., and A.W. TODD: The Absolute Configuration of Streptidine in Streptomycin. J. Amer. Chem. Soc. **85**, 3896 (1963).
169. DYER, J.R., W.E. MCGONIGAL, and K.C. RICE: Streptomycin, II: Streptose. J. Amer. Chem. Soc. **87**, 654 (1965).
170. MCGILVERAY, I.J., and K.L. RINEHART, Jr.: The Anomeric Linkage of Streptose in Streptomycin and Bluensomycin. J. Amer. Chem. Soc. **87**, 4003 (1965).
171. FRIED, J., and E. TITUS: Streptomycin B, an Antibiotically Active Constituent of Streptomycin Concentrates. J. Biol. Chem. **168**, 391 (1947).
172. TITUS, E., and J. FRIED: The Use of "Counter-Current Distribution" for the Analysis of Streptomycin Preparations. J. Biol. Chem. **168**, 393 (1947).
173. FRIED, J., and E. TITUS: Streptomycin, VIII: Isolation of Mannosidostreptomycin (Streptomycin B). J. Amer. Chem. Soc. **70**, 3615 (1948).
174. FRIED, J., and H.E. STAVELY: Streptomycin, V: Degradation of Streptomycin B to Streptidine, Streptobiosamine and D-Mannose. J. Amer. Chem. Soc. **70**, 1549 (1948).
175. PECK, R.L., C.E. HOFFHINE, Jr., P. GALE, and K. FOLKERS: Streptomyces Antibiotics, XXI: Linkage of Mannosidostreptobiosamine to Streptidine in Mannosidostreptomycin. J. Amer. Chem. Soc. **70**, 3968 (1948).
176. STAVELY, H.E., and J. FRIED: Streptomycin, IX: The Stepwise Degradation of Mannosidostreptomycin. J. Amer. Chem. Soc. **71**, 135 (1949).
177. O'KEEFFE, A.E., M.A. DOLLIVER, and E.T. STILLER: Separation of Streptomycins. J. Amer. Chem. Soc. **71**, 2452 (1949).
178. FRIED, J., and H.E. STAVELY: Streptomycin, X: The Structure of Mannosidostreptomycin. J. Amer. Chem. Soc. **74**, 5461 (1952).
179. GRUNDY, W.E., J.R. SCHENCK, R.K. CLARK, Jr., M.P. HARGIE, R.K. RICHARDS, and J.C. SYLVESTER: A Note on a New Antibiotic. Arch. Biochem. **28**, 150 (1950).
180. BENEDICT, R.G., F.H. STODOLA, O.L. SHOTWELL, A.M. BORUD, and L.A. LINDENFELSER: A New Streptomycin. Science **112**, 77 (1950).
181. STODOLA, F.H., O.L. SHOTWELL, A.M. BORUD, R.G. BENEDICT, and A.C. RILEY, Jr.: Hydroxystreptomycin, a New Antibiotic from *Streptomyces griseocarneus*. J. Amer. Chem. Soc. **73**, 2290 (1951).
182. BANNISTER, B., and A.D. ARGOUDELIS: The Chemistry of Bluensomycin, I: The Structure of Bluensidine. J. Amer. Chem. Soc. **85**, 119 (1963).
183. BANNISTER, B., and A.D. ARGOUDELIS: The Chemistry of Bluensomycin, II: The Structure of Bluensomycin. J. Amer. Chem. Soc. **85**, 234 (1963).
184. BARLOW, C.D., and L. ANDERSON: A Study of the Structure of Bluensomycin with the Tetraminocopper Reagent. J. Antibiotics **25**, 281 (1972).
185. UMEZAWA, S., Y. TAKAHASHI, T. USUI, and T. TSUCHIYA: Total Synthesis of Streptomycin. J. Antibiotics **27**, 997 (1974).
186. UMEZAWA, S., T. TSUCHIYA, T. YAMASAKI, H. SANO, and Y. TAKAHASHI: Total Synthesis of Dihydrostreptomycin. J. Amer. Chem. Soc. **96**, 920 (1974).
187. SILVERMAN, M., and S.V. RIEDER: The Formation of N-Methyl-L-glucosamine from D-Glucose by *Streptomyces griseus*. J. Biol. Chem. **235**, 1251 (1960).
188. CANDY, D.J., N.L. BLUMSOM, and J. BADDILEY: The Biosynthesis of Streptomycin – Incorporation of ^{14}C-Labelled Compounds into Streptose and N-Methyl-L-glucosamine. Biochem. J. **91**, 31 (1964).

189. HORNER, W.H.: Biosynthesis of Streptomycin, I: Origin of the Guanidine Group. J. Biol. Chem. **239**, 578 (1964).

190. WALKER, M.S., and J.B. WALKER: Enzymic Studies on the Biosynthesis of Streptomycin – Transamidination of Inosamine and Streptamine Derivatives. J. Biol. Chem. **241**, 1262 (1966).

191. BRUTON, J., W.H. HORNER, and G.A. RUSS: Biosynthesis of Streptomycin, IV: Further Studies on the Biosynthesis of Streptidine and N-Methyl-L-glucosamine. J. Biol. Chem. **242**, 813 (1967).

192. WALKER, M.S., and J.B. WALKER: Streptomycin Biosynthesis. Conversion of Myo-Inositol to O-Phosphorylstreptidine. Biochim. Biophys. Acta **136**, 272 (1967).

193. WALKER, J.B., and M.S. WALKER: Enzymatic Synthesis of Streptidine from $Scyllo$-Inosamine. Biochemistry **6**, 3821 (1967).

194. BRUCE, R.M., H.S. RAGHEB, and H. WEINER: Biosynthesis of Streptomycin: Origin of Streptidine from D-Glucose. Biochim. Biophys. Acta **158**, 499 (1968).

195. WALKER, J.B., and M.S. WALKER: Streptomycin Biosynthesis: Enzymatic Synthesis of $Scyllo$-Inosamine from $Scyllo$-Inosose and L-Glutamine. Biochim. Biophys. Acta **170**, 219 (1968).

196. WALKER, J.B., and M.S. WALKER: Streptomycin Biosynthesis. Transamination Reactions Involving Inosamines and Inosadiamines. Biochemistry **8**, 763 (1969).

197. BRUTON, J., and W.H. HORNER: Biosynthesis of Streptomycin, V: Origin of the Formyl Carbon Atom of Streptose. Biochim. Biophys. Acta **184**, 641 (1969).

198. HORNER, W.H., and G.A. RUSS: Biosynthesis of Streptomycin, VI: Myo-Inosose-2, an Intermediate in Streptidine Biosynthesis. Biochim. Biophys. Acta **192**, 352 (1969).

199. WALKER, M.S., and J.B. WALKER: Streptomycin Biosynthesis. Separation and Substrate Specificities of Phosphatases Acting on Guanidino Deoxy-$scyllo$-inositol Phosphate and Streptomycin-($Streptidino$)phosphate. J. Biol. Chem. **246**, 7034 (1971).

200. WALKER, J.B.: Biosynthesis of the Monoguanidinated Inositol Moiety of Bluensomycin, a Possible Evolutionary Precursor of Streptomycin. J. Biol. Chem. **249**, 2397 (1974).

201. WALKER, J.B.: Pathways of Biosynthesis of the Guanidinated Inositol Moieties of Streptomycin and Bluensomycin. Methods in Enzymology **43**, 429 (1975).

202. MAIER, S., and H. GRISEBACH: Biosynthesis of Streptomycin. Enzymic Oxidation of Dihydrostreptomycin (6-Phosphate) to Streptomycin (6-Phosphate) with a Particulate Fraction of $Streptomyces\ griseus$. Biochim. Biophys. Acta **586**, 231 (1979).

203. KNIEP, B., and H. GRISEBACH: Biosynthesis of Streptomycin. Enzymatic Formation of Dihydrostreptomycin 6-Phosphate from Dihydrostreptosyl Streptidine 6-Phosphate. J. Antibiotics **33**, 416 (1980).

204. KNIEP, B., and H. GRISEBACH: Biosynthesis of Streptomycin. Purification and Properties of a dTDP-L-Dihydrostreptose: Streptidine-6-phosphate Dihydrostreptosyltransferase from $Streptomyces\ griseus$. Eur. J. Biochem. **105**, 139 (1980).

205. OHNUKI, T., T. IMANAKA, and S. AIBA: Isolation of Streptomycin-Nonproducing Mutants Deficient in Biosynthesis of the Streptidine Moiety or Linkage Between Steptidine-6-Phosphate and Dihydrostreptose. Antimicrob. Agents Chemot. **27**, 367 (1985).

206. CARTER, H.E., R.K. CLARK, Jr., S.R. DICKMAN, Y.H. LOO, P.S. SKELL, and W.A. STRONG: Isolation and Purification of Streptomycin. J. Biol. Chem. **160**, 337 (1945).

207. TITUS, E., and J. FRIED: Counter-Current Distribution Studies on Streptomycin: The Tautomerism of Streptomycin. J. Biol. Chem. **174**, 57 (1948).

208. PLAUT, G.W.E., and R.B. MCCORMACK: A Counter-Current Distribution System for the Separation and Determination of Streptomycin Types. J. Amer. Chem. Soc. **71**, 2264 (1949).

209. LEMIEUX, R.V., and M.L. WOLFROM: The Chemistry of Streptomycin. Adv. Carbohydrate Chem. **3**, 337 (1948).
210. DUTCHER, J.D.: Chemistry of the Amino Sugars Derived from Antibiotic Substances. Adv. Carbohydrate Chem. **18**, 259 (1963).
211. UMEZAWA, H.: Recent Advances in Chemistry and Biochemistry of Antibiotics. Microbial Chemistry Research Foundation, p. 67. Tokyo. 1964.
212. HORNER, W.H.: Streptomycin. In: Antibiotics, Vol. II. (D. GOTTLIEB and P.D. SHAW, eds.). Berlin-Heidelberg-New York: Springer. 1967.
213. JACOBY, G.A., and L. GORINI: The Effect of Streptomycin and Other Aminoglycoside Antibiotics on Protein Synthesis. In: Antibiotics, Vol. I (D. GOTTLIEB and P.D. SHAW, eds.). Berlin-Heidelberg-New York: Springer. 1967.
214. SCHLESSINGER, D., and G. MEDOFF: Streptomycin, Dihydrostreptomycin and the Gentamycins. In: Antibiotics, Vol. III (J.W. CORCORAN and F.E. HAHN, eds.). Berlin-Heidelberg-New York: Springer. 1975.
215. TANAKA, N.: Aminoglycoside Antibiotics. In: Antibiotics, Vol. III (J.W. CORCORAN and F.H. HAHN, eds.). Berlin-Heidelberg-New York: Springer. 1975.
216. PEARCE, C.L., and K.L. RINEHART, Jr.: Biosynthesis of Aminocyclitol Antibiotics. In: Antibiotics, Vol. IV (J.W. CORCORAN, ed.). Berlin-Heidelberg-New York: Springer. 1981.
217. ASSELINAU, J., and J.-P. ZALTA: Les Antibiotiques—Structure et Exemples de Mode d'Action. Paris: Hermann. 1976.
218. WAKSMAN, S.A.: Streptomycin. Baltimore: Williams and Wilkins. 1949.
219. HASKELL, T.H., S.A. FUSARI, R.P. FROHARDT, and Q.R. BARTZ: The Chemistry of Viomycin. J. Amer. Chem. Soc. **74**, 599 (1952).
220. MAYER, R.L., P.C. EISMAN, and E.A. KONOPKA: Antituberculosis Activity of Vinactame. Experientia **10**, 335 (1954).
221. DYER, J.R., H.B. HAYES, E.G. MILLER, Jr., and R.F. NASSAR: Viomycin, I: The Structure of Viomycidine. J. Amer. Chem. Soc. **86**, 5363 (1964).
222. BOWIE, J.H., A.W. JOHNSON, and G. THOMAS: The Chromophore of Viomycin. Tetrahedron Letters **1964**, 863.
223. BOWIE, J.H., D.A. COX, A.W. JOHNSON, and G. THOMAS: Viomycin. Tetrahedron Letters **1964**, 3305.
224. DYER, J.R., C.K. KELLOGG, R.F. NASSAR, and W.E. STREETMAN: Viomycin, II: The Structure of Viomycin. Tetrahedron Letters **1965**, 585.
225. KITAGAWA, T., Y. SAWADA, T. MIURA, T. OZASA, and H. TANIYAMA: Viomycin, I: The Amino Acid Sequence of Viomycin. Tetrahedron Letters **1968**, 109.
226. BYCROFT, B.W., D. CAMERON, L.R. CROFT, A.W. JOHNSON, T. WEBB, and P. COGGON: Viomycin. Further Degradative Studies. Tetrahedron Letters **1968**, 2925.
227. BYCROFT, B.W., L.R. CROFT, A.W. JOHNSON, and T. WEBB: The Structure, Stereochemistry and Reactions of the Guanidine Moiety of Viomycin. J. Antibiotics **22**, 133 (1969).
228. FLOYD, J.C., J.A. BERTRAND, and J.R. DYER: The Crystal Structure of Viomycidine. Chem. Commun. **1968**, 998.
229. TAKITA, T., and K. MAEDA: The Structure of the Guanido Amino Acid Moiety of Viomycin. J. Antibiotics **21**, 512 (1968).
230. KOYAMA, G., H. NAKAMURA, S. OMOTO, T. TAKITA, K. MAEDA, and Y. IITAKA: The Molecular and Crystal Structure of Viomycidine. J. Antibiotics **22**, 34 (1969).
231. BYCROFT, B.W., D. CAMERON, L.R. CROFT, A. HASSANALI-WALJI, A.W. JOHNSON, and T. WEBB: The Chromophore and Partial Structure of Viomycin. Tetrahedron Letters **1968**, 5901.
232. BYCROFT, B.W., D. CAMERON, A. HASSANALI-WALJI, and A.W. JOHNSON: Synthesis of

a Model Relating to the Chromophores of Capreomycin and Viomycin. Tetrahedron Letters **1969**, 2539.

233. COGGON, P.: Structure and Absolute Configuration of Viocidic Acid: X-Ray Analysis of Viocidic Acid Dihydrobromide. J. Chem. Soc. (London) (B) **1970**, 838.

234. BÜCHI, G., and J.A. RALEIGH: The Structure of Viomycidine. J. Organ. Chem. (USA) **36**, 873 (1971).

235. BYCROFT, B.W., D. CAMERON, L.R. CROFT, A. HASSANALI-WALJI, A.W. JOHNSON, and T. WEBB: The Total Structure of Viomycin, a Tuberculostatic Peptide Antibiotic. Experientia **27**, 501 (1971).

236. YOSHIOKA, H., T. AOKI, H. GOKO, K. NAKATSU, T. NODA, H. SAKAKIBARA, T. TAKE, A. NAGATA, J. ABE, T. WAKAMIYA, T. SHIBA, and T. KANEKO: Chemical Studies on Tuberactinomycin, II: The Structure of Tuberactinomycin O. Tetrahedron Letters **1971**, 2043.

237. BYCROFT, B.W., L.R. CROFT, A.W. JOHNSON, and T. WEBB: Viomycin, Part I: The Structure of the Guanidine-Containing Unit. J. Chem. Soc. Perkin I **1972**, 820.

238. BYCROFT, B.W., D. CAMERON, L.R. CROFT, A. HASSANALI-WALJI, A.W. JOHNSON, and T. WEBB: Viomycin, Part II: The Structure of the Chromophore. J. Chem. Soc. Perkin I **1972**, 827.

239. BYCROFT, B.W.: The Crystal Structure of Viomycin, a Tuberculostatic Antibiotic. Chem. Commun. **1972**, 660.

240. KITAGAWA, T., T. MIURA, K. FUJIWARA, and H. TANIYAMA: The Total Structure of Viomycin by Sequential Analysis. Chem. Pharm. Bull. (Japan) **20**, 2215 (1972).

241. KITAGAWA, T., T. MIURA, and H. TANIYAMA: Characterization of Viomycin and Its Acyl Derivatives. Chem. Pharm. Bull. (Japan) **20**, 2176 (1972).

242. KITAGAWA, T., T. MIURA, Y. SAWADA, K. FUJIWARA, R. ITO, and H. TANIYAMA: Studies on Viomycin, VII: Oxidative Modifications of Viomycin. Chem. Pharm. Bull. (Japan) **22**, 1827 (1974).

243. KITAGAWA, T., T. MIURA, T. TAKAISHI, and H. TANIYAMA: Studies on Viomycin, VIII: Selective Modifications of the Terminal Amino Groups of Viomycin. Chem. Pharm. Bull. (Japan) **23**, 2123 (1975).

244. KITAGAWA, T., T. MIURA, C. TAKAISHI, and H. TANIYAMA: Studies on Viomycin, IX: Amino Acid Derivatives of Viomycin. Chem. Pharm. Bull. (Japan) **24**, 1324 (1976).

245. KITAGAWA, T., T. MIURA, K. MORI, H. TANIYAMA, K. KAWANO, and Y. KYOGOKU: ^{13}C-Nuclear Magnetic Resonance Studies on Viomycin and Its Related Compounds. Chem. Pharm. Bull. (Japan) **25**, 280 (1977).

246. KITAGAWA, T., T. MIURA, S. TANAKA, and H. TANIYAMA: Relationships Between Antimicrobial Activities and Chemically Reactive Functions in Viomycin. J. Antibiotics **25**, 429 (1972).

247. KITAGAWA, T., T. MIURA, S. TANAKA, and H. TANIYAMA: Relationships Between Antimicrobial Activities and Chemical Structures of Reduced Products of Viomycin. J. Antibiotics **26**, 528 (1973).

248. VIGLINO, P., C. FRANCONI, A. LAI, E. BROSIO, and F. CONTI: NMR Studies of Viomycin. Org. Magnetic Reson. **4**, 237 (1972).

249. HERR, E.B., Jr., M.E. HAREY, G.E. PITTENGER, and C.E. HIGGENS: Isolation and Characterization of a New Peptide Antibiotic. Proc. Indian Acad. Sci. **69**, 134 (1959).

250. HERR, E.B., Jr., and M.O. REDSTONE: Chemical and Physical Characterization of Capreomycin. Ann. New York Acad. Sci. **135**, 940 (1966).

251. HERR, E.B., Jr.: Chemical and Biological Properties of Capreomycin and Other Peptide Antibiotics. Antimic. Agents Chemoth. **1962**, 201.

252. BYCROFT, B.W., D. CAMERON, L.R. CROFT, and A.W. JOHNSON: Synthesis and Stereochemistry of Capreomycidine [α-(2-Iminohexahydro-4-pyrimidyl)glycine]. Chem. Commun. **1968**, 1301.

253. BYCROFT, B.W., D. CAMERON, and A.W. JOHNSON: Synthesis of Capreomycidine and Epicapreomycidine, the Epimers of α-(2-Iminohexahydropyrimid-4-yl)glycine. J. Chem. Soc. (London) (C) **1971**, 3040.

254. SHIBA, T., T. UKITA, K. MIZUNO, T. TESHIMA, and T. WAKAMIYA: Total Synthesis of L-Capreomycidine. Tetrahedron Letters **1977**, 2681.

255. BYCROFT, B.W., D. CAMERON, L.R. CROFT, A. HASSANALI-WALJI, A.W. JOHNSON, and T. WEBB: Total Structure of Capreomycin IB, a Tuberculostatic Peptide Antibiotic. Nature **132**, 301 (1971).

256. SHIBA, T., S. NOMOTO, T. TESHIMA, and T. WAKAMIYA: Revised Structure and Total Synthesis of Capreomycin. Tetrahedron Letters **1976**, 3907.

257. SHIBA, T., S. NOMOTO, and T. WAKAMIYA: The Chemical Structure of Capreomycin. Experientia **32**, 1109 (1976).

258. NOMOTO, S., T. TESHIMA, T. WAKAMIYA, and T. SHIBA: The Revised Structure of Capreomycin. J. Antibiotics **30**, 955 (1977).

259. NOMOTO, S., T. TESHIMA, T. WAKAMIYA, and T. SHIBA: Total Synthesis of Capreomycin. Tetrahedron **34**, 921 (1978).

260. WANG, M., and S.J. GOULD: Biosynthesis of Capreomycin, 2: Incorporation of L-Serine, L-Alanine and L-2,3-Diaminopropionic Acid. J. Organ. Chem. (USA) **58**, 5176 (1993).

261. NAGATA, A., T. ANDO, R. IZUMI, H. SAKAKIBARA, T. TAKE, K. HAYANO, and J.N. ABE: Studies on Tuberactinomycin (Tuberactin), a New Antibiotic, I: Taxonomy of Producing Strain, Isolation and Characterization. J. Antibiotics **21**, 681 (1968).

262. ANDO, T., K. MATSUURA, R. IZUMI, T. NODA, T. TAKE, A. NAGATA, and J. ABE: Studies on Tuberactinomycin, II: Isolation and Properties of Tuberactinomycin-N, a New Tuberactinomycin Group Antibiotic. J. Antibiotics **24**, 680 (1971).

263. IZUMI, R., T. NODA, T. ANDO, T. TAKE, and A. NAGATA: Studies on Tuberactinomycin, III: Isolation and Characterization of Two Minor Components, Tuberactinomycin B and Tuberactinomycin O. J. Antibiotics **25**, 201 (1972).

264. WAKAMIYA, T., T. SHIBA, T. KANEKO, H. SAKAKIBARA, T. TAKE, and J. ABE: Chemical Studies on Tuberactinomycin, I: The Structure of Tuberactidine, Guanidino Amino Acid Component. Tetrahedron Letters **1970**, 2497.

265. WAKAMIYA, T., T. SHIBA, T. KANEKO, H. SAKAKIBARA, T. NODA, and T. TAKE: Chemical Studies on Tuberactinomycin, V: Structures of Guanidino Amino Acids in Tuberactinomycins. Bull. Chem. Soc. Japan **46**, 949 (1973).

266. NODA, T., T. TAKE, A. NAGATA, T. WAKAMIYA, and T. SHIBA: Chemical Studies on Tuberactinomycin, III: The Chemical Structure of Viomycin (Tuberactinomycin B). J. Antibiotics **25**, 427 (1972).

267. TESHIMA, T., S. NOMOTO, T. WAKAMIYA, and T. SHIBA: Chemical Studies on Tuberactinomycins, X: Total Synthesis of Tuberactinomycin O. Tetrahedron Letters **1976**, 2343.

268. WAKAMIYA, T., and T. SHIBA: Chemical Studies on Tuberactinomycins, IX: Nuclear Magnetic Resonance Studies on Tuberactinomycins and Tuberactinamine N. Bull. Chem. Soc. Japan **48**, 2502 (1975).

269. NOMOTO, S., and T. SHIBA: Chemical Studies on Tuberactinomycin, XIII: Modification of β-Ureidodehydroalanine Residue in Tuberactinomycin N. J. Antibiotics **30**, 1008 (1977).

270. TESHIMA, T., S. NOMOTO, T. WAKAMIYA, and T. SHIBA: Chemical Studies on Tuberactinomycin, XII: Syntheses and Antimicrobial Activities of [Ala³, Ala⁴]-, [Ala³]- and [Ala⁴]-Tuberactinomycin O. Bull. Chem. Soc. Japan **50**, 3372 (1977).

271. MCGAHREN, W. J., G.O. MORTON, M.P.KUNSTMANN, and G.A. ELLESTAD: Carbon-13 Nuclear Magnetic Resonance Studies on a New Antitubercular Peptide Antibiotic LLBM547β. J. Organ. Chem. (USA) **42**, 1282 (1977).

272. PAUNCZ, J.K.: Thin-Layer Chromatography of Basic Water-Soluble Antibiotics on Resin-Coated Chromatoplates. J. Antibiotics **25**, 677 (1972).

273. TAKEUCHI, S., K. HIRAYAMA, K. UEDA, H. SAKAI, and H. YONEHARA: Blasticidin S, a New Antibiotic. J. Antibiotics **11A**, 1 (1958).

274. YONEHARA, H., S. TAKEUCHI, N. OTAKE, T. ENDO, Y. SAKAGAMI, and Y. SUMIKI: Chemical Studies on Blasticidin S, I: Hydrolysis of Blasticidin S. J. Antibiotics **16A**, 195 (1963).

275. ENDO, T., N. OTAKE, S. TAKEUCHI, and H. YONEHARA: The Structure of Blastidone, a Degradative Component of Blasticidin S. J. Antibiotics **17A**, 172 (1964).

276. OTAKE, N., S. TAKEUCHI, T. ENDO, and H. YONEHARA: Structure of Cytosinine, A Structural Component of Blasticidin S. Tetrahedron Letters **1965**, 1405.

277. OTAKE, N., S. TAKEUCHI, T. ENDO, and H. YONEHARA: Structure of Blasticidin S. Tetrahedron Letters **1965**, 1411.

278. ONUMA, S., Y. NAWATA, and Y. SAITO: An X-Ray Analysis of Blasticidin S Mono-hydrobromide. Bull. Chem. Soc. Japan **39**, 1091 (1966).

279. SWAMINATHAN, V., J.L. SMITH, M. SUNDARALINGAM, C. COUTSOGEORGOPOULOS, and G. KARTHA: Crystal and Molecular Structure of the Antibiotic Blasticidin S Hydro-chloride Pentahydrate. Biochim. Biophys. Acta **655**, 335 (1981).

280. FOX, J.J., and K.A. WATANABE: Nucleosides, XXXII: On the Structure of Blasticidin S, a Nucleoside Antibiotic. Tetrahedron Letters **1966**, 897.

281. YONEHARA, H., and N. OTAKE: Absolute Configuration of Blasticidin S. Tetrahedron Letters **1966**, 3785.

282. IWASA, T., K. SUETOMI, and T. KUSAKA: Taxonomic Study and Fermentation of Producing Organism and Antimicrobial Activity of Mildiomycin. J. Antibiotics **31**, 511 (1978).

283. HARADA, S., and T. KISHI: Isolation and Characterization of Mildiomycin, a New Nucleoside Antibiotic. J. Antibiotics **31**, 519 (1978).

284. HARADA, S., E. MIZUTA, and T. KISHI: Structure of Mildiomycin, a New Antifungal Nucleoside Antibiotic. J. Amer. Chem. Soc. **100**, 4895 (1978).

285. HARADA, S., E. MIZUTA, and T. KISHI: Structure of Mildiomycin, a New Antifungal Nucleoside Antibiotic. Tetrahedron **37**, 1317 (1981).

286. TASHIRO, S., N. SUGITA, T. IWASA, and H. SAWADA: Structure of Mildiomycin D. Agric. Biol. Chem. **48**, 881 (1984).

287. SETO, H., M. KOYAMA, H. OGINO, T. TSURUOKA, S. INOUYE, and N. OTAKE: The Structure of Novel Nucleoside Antibiotics Miharamycin A and Miharamycin B. Tetrahedron Letters **24**, 1805 (1983).

288. TAKAHASHI, A., N. SAITO, K. HOTTA, Y. OKAMI, and H. UMEZAWA: Bagougeramines A and B, New Nucleoside Antibiotics Produced by a Strain of *Bacillus circulans*, I: Taxonomy of the Producing Organism and Isolation and Biological Properties of the Antibiotics. J. Antibiotics **39**, 1033 (1986).

289. TAKAHASHI, A., D. IKEDA, H. NAGANAWA, Y. OKAMI, and H.UMEZAWA: Bagouger-amines A and B, New Nucleoside Antibiotics Produced by a Strain of *Bacillus circulans*, II: Physico-Chemical Properties and Structure Determination. J. Antibiotics **39**, 1041 (1986).

290. ARGOUDELIS, A.D., L. BACZYNSKYJ, M.T. KUO, A.L. LABORDE, O.K. SEBEK, S.E. TRUESDELL, and F.B. SHILLIDAY: Arginomycin: Production, Isolation, Characterization and Structure. J. Antibiotics **40**, 750 (1987).

291. GULLO, V., M. CONOVER, R. COOPER, C. FEDERBUSH, A.C. HORAN, T. KUNG, J. MARQUEZ, M. PATEL, and A. WATNIK: Sch 36605, a Novel Anti-Inflammatory Compound. Taxonomy, Fermentation, Isolation and Biological Properties. J. Antibiotics **41**, 20 (1988).

292. COOPER, R., M. CONOVER, and M. PATEL: Sch 36605, Structure of a Novel Nucleoside. J. Antibiotics **41**, 123 (1988).

293. SETO, H., and H. YONEHARA: Studies on the Biosynthesis of Blasticidin S, VI: The Isolation and Structure of Blasticidin H. J. Antibiotics **30**, 1019 (1977).

294. SETO, H., and H. YONEHARA: Studies on the Biosynthesis of Blasticidin S, VII: Isolation of Demethylblasticidin S. J. Antibiotics **3**, 1022 (1977).

295. SETO, H., I. YAMAGUCHI, N. OTAKE, and H. YONEHARA: Biogenesis of Blasticidin S. Tertrahedron Letters **1966**, 3793.

296. SETO, H., I. YAMAGUCHI, N. OTAKE, and H. YONEHARA: Studies on the Biosynthesis of Blasticidin S, Part I: Precursors of Blasticidin S Biosynthesis. Agr. Biol. Chem. **32**, 1292 (1968).

297. PRABHAKARAN, P.C., N.T. WOO, P. YORGEY, and S.J. GOULD: Studies on Nitrogen Metabolism Using ^{13}C NMR Spectroscopy, 5: Metabolism of L-α-Arginine in the Biosynthesis of Blasticidin S. Tetrahedron Letters **27**, 3815 (1986).

298. GUO, J., and S.J. GOULD: Biosynthesis of Blasticidin S from Cytosylglucuronic Acid (CGA). Isolation of Cytosine/UDPglucuronosyltransferase and Incorporation of CGA by *Streptomyces griseochromogenes*. J. Amer. Chem. Soc. **113**, 5898 (1991).

299. GUO, J., and S.J. GOULD: Biosynthesis of Blasticidin S. Cell-Free Demonstration of N-Methylation as the Last Step. Bioorg. Med. Chem. Lett. **1**, 497 (1991).

300. GOODY, R.S., K.A. WATANABE, and J.J. FOX: Nucleosides, LXVI: Synthetic Studies on Nucleoside Antibiotics, 4: Synthesis of Methyl 4-Amino-2,3,4-trideoxy-α-D-*erythro*-hex-2-enopyranosiduronic Acid, the Carbohydrate Moiety of Blasticidin S. Tetrahedron Letters **1970**, 2594.

301. WATANABE, K.A., I. WEMPEN, and J.J. FOX: Nucleosides, LXIX: Synthetic Studies on Nucleoside Antibiotics, 6: Syntheses of 1-(4-Amino-2,3,4-trideoxy-β-D-*erythro*-hex-2-enopyranosyl)cytosine and Derivatives Related to the Nucleoside Moiety of Blasticidin S. Chem. Pharm. Bull. (Japan) **18**, 2368 (1970).

302. FOX, J.J., and K.A. WATANABE: Studies Directed Towards the Total Synthesis of the Nucleoside Antibiotics Gougerotin and Blasticidin S. Pure Appl. Chem. **28**, 475 (1971).

303. KONDO, T., H. NAKAI, and T. GOTO: Synthesis of Cytosinine, the Nucleoside Component of Antibiotic Blasticidin S. Tetrahedron Letters **1972**, 1881.

304. FINLAY, A.C., F.A. HOCHSTEIN, B.A. SOBIN, and F.X. MURPHY: Netropsin, a New Antibiotic Produced by a Streptomyces. J. Amer. Chem. Soc. **73**, 341 (1951).

305. COSAR, C., L. NINET, S. PINNERT-SINDICO, and J. PREUD'HOMME: Activité Trypanocide d'un Antibiotique Produit par un Streptomyces. C.R. séances hebd. Acad. Sci. **234**, 1498 (1952).

306. JULIA, M., and N. PRÉAU-JOSEPH: Premières Études sur la Structure Chimique d'un Nouvel Antibiotique, la Congocidine. C.R. séances hebd. Acad. Sci. **243**, 961 (1956).

307. VAN TAMELEN, E.E., D.M. WHITE, I.C. KOGON, and A.D.G. POWELL: Structural Studies on the Antibiotic Netropsin. J. Amer. Chem. Soc. **78**, 2157 (1956).

308. WALLER, C.W., C.F. WOLF, W.J. STEIN, and B.L. HUTCHINGS: The Structure of Antibiotic T-1384. J. Amer. Chem. Soc. **79**, 1265 (1957).

309. WEISS, M.J., J.S. WEBB, and J.M. SMITH, Jr.: The Structure of Antibiotic T-1384. Synthesis of the Degradation Fragments. J. Amer. Chem. Soc. **79**, 1266 (1957).

310. VAN TAMELEN, E.E., and A.D.G. POWELL: The Structure of Netropsin. Chem. and Ind. **1957**, 365.

311. JULIA, M., and N. PRÉAU-JOSEPH: Structure et Synthèse de la Congocidine. C. R. séances hebd. Acad. Sci. **257**, 1115 (1963).

312. JULIA, M., and N. PRÉAU-JOSEPH: Amidines et Guanidines Apparentées à la Congocidine, I: Structure de la Congocidine. Bull. soc. chim. France **1967**, 4348.

313. NAKAMURA, S., H. YONEHARA, and H. UMEZAWA: On the Structure of Netropsin. J. Antibiotics **17A**, 220 (1964).

314. RENTZEPERIS, D., and L.A. MARKY: Netropsin Binding as a Thermodynamic Probe of the Grooves of Parallel DNA. J. Amer. Chem. Soc. **115**, 1645 (1993).

315. PARRICK, J., and M. PORSSA: Synthesis of a Nitro Oligo-N-Methylimidazole Carboxamide Derivative: a Radiosensitiser Targeted to DNA. Tetrahedron Letters **34**, 5011 (1993).

316. RHEE, Y., C. WANG, B.L. GAFFNEY, and R.A. JONES: Nitrogen-15-Labeled Oligodeoxynucleotides, 6: Use of ^{15}N NMR to Probe Binding of Netropsin and Distamycin to {d[CGCGAATTCGCG]}$_2$. J. Amer. Chem. Soc. **115**, 8742 (1993).

317. VALYI-NAGY, T., J. URI, and I. SZILAGYI: Primycin, a New Antibiotic. Nature **174**, 1105 (1964).

318. ABERHART, J., T. FEHR, R.C. JAIN, P. DE MAYO, O. MOTL, L. BACZYNSKYJ, D.E.F. GRACEY, D.B. MACLEAN, and I. SZILAGYI: Primycin. J. Amer. Chem. Soc. **92**, 5816 (1970).

319. ABERHART, J., R.C. JAIN, T. FEHR, P. DE MAYO, and I. SZILAGYI: The Constitution of Primycin, I: Characterization, Functional Groups, and Degradation to the Secoprimycins. J. Chem. Soc. Perkin I **1974**, 816.

320. GRACEY, D.E.F., L. BACZYNSKYJ, T.I. MARTIN, and D.B. MACLEAN: The Constitution of Primycin, Part II: The Mass Spectra of the Secoprimycins. J. Chem. Soc. Perkin I **1974**, 827.

321. FEHR, T., R.C. JAIN, P. DE MAYO, O. MOTL, I. SZILAGYI, L. BACZYNSKYJ, D.E.F. GRACEY, H.L. HOLLAND, and D.B. MACLEAN: The Constitution of Primycin, Part III: Degradation of Methylated Primycin, and the Structure of Primycin. J. Chem. Soc. Perkin I **1974**, 836.

322. URI, J.V., and P. ACTOR: Crystallization and Antifungal Activity of Primycin. J. Antibiotics **32**, 1207 (1979).

323. ARAI, M.: Azalomycins B and F, Two New Antibiotics, I: Production and Isolation. J. Antibiotics **13A**, 46 (1960).

324. ARAI, M.: Azalomycins B and F, Two New Antibiotics, II: Properties of Azalomycins B and F. J. Antiobiotics **13A**, 51 (1960).

325. ARAI, M., and K. HAMANO: Isolation of Three Main Components, F_3, F_4 and F_5, from Azalomycin F-Complex. J. Antibiotics **23**, 107 (1970).

326. NAMIKOSHI, M., K. SASAKI, Y. KOISO, K. FUKUSHIMA, S. IWASAKI, S. NOZOE, and S. OKUDA: Studies on Macrocyclic Lactone Antibiotics, I: Physicochemical Properties of Azalomycin F_{4a}. Chem. Pharm. Bull. (Japan) **30**, 1653 (1982).

327. NAMIKOSHI, M., S. IWASAKI, K. SASAKI, M. YANO, K. FUKUSHIMA, S. NOZOE, and S. OKUDA: Studies on Macrocyclic Lactone Antibiotics, II: Partial Structures of Azalomycin F_{4a}. Chem. Pharm. Bull. (Japan) **30**, 1658 (1982).

328. IWASAKI, S., M. NAMIKOSHI, K. SASAKI, M. YANO, K. FUKUSHIMA, S. NOZOE, and S. OKUDA: Studies on Macrocyclic Lactone Antibiotics, III: Skeletal Structure of Azalomycin F_{4a}. Chem. Pharm. Bull. (Japan) **30**, 1669 (1982).

329. IWASAKI, S., M. NAMIKOSHI, K. SASAKI, K. FUKUSHIMA, and S. OKUDA: Studies on

Macrocyclic Lactone Antibiotics, V: The Structures of Azalomycins F_{3a} and F_{5a}. Chem. Pharm. Bull. (Japan) **30**, 4006 (1982).

330. JOHNSON, L.E., and A. DIETZ: Scopafungin, a Crystalline Antibiotic Produced by *Streptomyces hygroscopicus* var. *enhygrus* var. nova. Appl. Microbiol. **22**, 303 (1971).

331. BERGY, M.E., and H. HOEKSMA: Scopafungin, a Crystalline Endomycin Component. J. Antibiotics **25**, 39 (1972).

332. ARAI, T., S. KURODA, H. OHARA, Y. KATO, and H. KAJI: Copiamycin, a New Antifungal Antibiotic Derived from *S. hygroscopicus* var. *crystallogenes*. J. Antibiotics **18A**, 63 (1965).

333. FUKUSHIMA, K., T. ARAI, S. IWASAKI, M. NAMIKOSHI, and S. OKUDA: Studies on Macrocyclic Lactone Antibiotics, VI: Skeletal Structure of Copiamycin. J. Antibiotics **35**, 1480 (1982).

334. ARAI, T., J. UNO, I. HORIMI, and K. FUKUSHIMA: Isolation of Neocopiamycin A from *Streptomyces hygroscopicus* var. *crystallogenes*, the Copiamycin Source. J. Antibiotics **37**, 103 (1984).

335. FUKAI, T., T. NOMURA, J. UNO, and T. ARAI: Demalonylcopiamycin, a New Antibiotic Produced by *Streptomyces hygroscopicus* var. *crystallogenes*, the Copiamycin Source. Heterocycles **24**, 3351 (1986).

336. BASSI, L., B. JOOS, P. GASSMANN, H.-P. KAISER, H. LEUENBERGER, and W. KELLER-SCHIERLEIN: Stoffwechselprodukte von Mikroorganismen, 218. Mitteilung: Versuche zur Strukturaufklärung von Niphimycin, 1. Teil: Reinigung und Charakterisierung der Niphimycine Iα und Iβ sowie Abbau mit Salpetersäure. Helv. Chim. Acta **66**, 92 (1983).

337. KELLER-SCHIERLEIN, W., B. JOOS, H.-P. KAISER, and P. GASSMANN: Stoffwechselprodukte von Mikroorganismen, 219. Mitteilung: Versuche zur Strukturaufklärung von Niphimycin, 2. Teil: Die Konstitution von Desmalonylniphimycin, I. Helv. Chim. Acta **66**, 226 (1983).

338. GASSMANN, P., L. HAGMANN, W. KELLER-SCHIERLEIN, and D. SAMAIN: Stoffwechselprodukte von Mikroorganismen, 226. Mitteilung: Versuche zur Strukturaufklärung von Niphimycin, 3. Teil: Identität von Scopafungin mit Niphimycin I und Lage des Malonylrestes in Niphimycin und Copiamycin. Helv. Chim. Acta **67**, 696 (1984).

339. TAKESAKO, K., and T. BEPPU: Studies on New Antifungal Antibiotics Guanidylfungins A and B, I: Taxonomy, Fermentation, Isolation and Characterization. J. Antibiotics **37**, 1161 (1984).

340. TAKESAKO, K., and T. BEPPU: Studies on New Antifungal Antibiotics Guanidylfungins A and B, II: Structure Elucidation and Biosynthesis. J. Antibiotics **37**, 1170 (1984).

341. FUKAI, T., C. TAKAHASHI, T. NOMURA, J. UNO, and T. ARAI: Guanidolide A, a Novel Antibiotic Produced by *Streptomyces hygroscopicus* var. *crystallogenes*, the Copiamycin Source. Heterocycles **27**, 2333 (1988).

342. FRÉCHET, D., M. DANZER, F. DEBU, B. MONEGIER DU SORBIER, D. REISDORF, C. SNOZZI, and M. VUILHORGNE: Structure Elucidation of RP 63834, a New Macrocyclic Lactone Antibiotic. Tetrahedron **47**, 61 (1991).

343. KUMASAWA, S., Y. ASAMI, K. AWANA, H. OHTANI, and C. FUKUCHI: Structure Elucidation of Macrocyclic Lactone Antibiotics MBA 028–24A and MBA 028–24B. Anal. Sci. Suppl. **7**, 809 (1991).

344. KOSHINO, H., K. KOBINATA, J. UZAWA, M. URAMOTO, K. ISONO, and H. OSADA: Structure of Malolactomycins A and B, Novel 40-Membered Macrolide Antibiotics. Tetrahedron **49**, 8827 (1993).

345. IWASAKI, S., K. SASAKI, M. NAMIKOSHI, and S. OKUDA: Studies on Macrocyclic Lactone Antibiotics, Part IV: Biosynthetic Studies on Azalomycin F_{4a} Using ^{13}C-Labelled Acetate and Propionate. Heterocycles **17**, 331 (1982).

346. TAKESAKO, K., T. BEPPU, T. NAKAMURA, and A. OBAYASHI: Demalonyl Derivatives of

Guanidylfungin A and Copiamycin: Their Synthesis and Antifungal Activity. J. Antibiotics **38**, 1363 (1985).

347. TAKESAKO, K., T. NAKAMURA, A. OBAYASHI, S. IWASAKI, M. NAMIKOSHI, S. OKUDA, and T. BEPPU: Demalonyl Derivatives of Azalomycin F_4 and Scopafungin. J. Antibiotics **39**, 713 (1986).

348. UNO, J., M.L. SHIGEMATSU, and T. ARAI: Novel Synergism of Two Antifungal Agents, Copiamycin and Imidazole. Antimicrob. Agents Chemot. **24**, 552 (1983).

349. KURODA, S., J. UNO, and T. ARAI: Target Substances of Some Antifungal Agents in the Cell Membrane. Antimicrob. Agents Chemot. **13**, 454 (1978).

350. THOMPSON, R.Q., and M.S. HUGHES: Stendomycin. A New Antibiotic. J. Antibiotics **16A**, 187 (1963).

351. BODANSKY, M., I. MURAMATSU, and A. BODANSKY: Fatty Acid Constituents of the Antifungal Antibiotic Stendomycin. J. Antibiotics **20A**, 384 (1967).

352. MURAMATSU, I., and M. BODANSKY: The Occurrence of Dehydrobutyrine in Stendomycin. J. Antibiotics **21**, 68 (1968).

353. BODANSKY, M., I. MURAMATSU, A. BODANSKY, M. LUKIN, and M.R. DOUBLER: Amino Acid Constituents of Stendomycin. J. Antibiotics **21**, 77 (1968).

354. BODANSKY, M., G.G. MARCONI, and G.C. COLMAN: On the N-Methyl-L-threonine Residue in Stendomycin. J. Antibiotics **21**, 668 (1968).

355. THOMAS, D.W., E. LEDERER, M. BODANSKY, J. IZDEBSKI, and I. MURAMATSU: Partial Structure of the Peptide Antibiotics Stendomycin as Determined by Mass Spectrometry. Nature **220**, 580 (1968).

356. BODANSKY, M., J. IZDEBSKI, and I. MURAMATSU: The Structure of the Peptide Antibiotic Stendomycin. J. Amer. Chem. Soc. **91**, 2351 (1969).

357. BODANSKY, M., G.G. MARCONI, and A. BODANSKY: The Structure of Stendomycidine. J. Antibiotics **22**, 40 (1969).

358. PITNER, T.P., and D.W. URRY: Conformational Studies of Polypeptide Antibiotics. Proton Magnetic Resonance of Stendomycin. Biochemistry **11**, 4132 (1972).

359. UMEZAWA, H.: Natural and Artificial Bleomycins: Chemistry and Antitumour Activities. Pure Appl. Chem. **28**, 665 (1971).

360. UMEZAWA, H.: Studies on Bleomycin: Chemistry and the Biological Action. Biomedicine **18**, 459 (1973).

361. UMEZAWA, H.: Chemistry and Mechanism of Action of Bleomycin. Federat. Proc. (Amer. Soc. Exp. Biol.) **33**, 2296 (1974).

362. UMEZAWA, H.: Recent Studies on Bleomycin. J. Nat. Prod. (Lloydia) **40**, 67 (1977).

363. MULLER, W.E.G., and R.K. ZAHN: Bleomycin, an Antibiotic That Removes Thymine from Double-Stranded DNA. Prog. Nucl. Acid Res. Mol. Biol. **20**, 21 (1977).

364. CARTER, S.K., S.T. CROOKE, and H. UMEZAWA: Bleomycin: Current Status and New Developments. New York: Academic Press. 1978.

365. HECHT, S.M.: Bleomycin: Chemical, Biochemical and Biological Aspects. New York: Springer. 1979.

366. HECHT, S.M.: The Chemistry of Activated Bleomycin. Accounts Chem. Res. **19**, 383 (1986).

367. STUBBE, J., and J.W. KOZARICH: Mechanisms of Bleomycin-Induced DNA Degradation. Chem. Rev. **87**, 1107 (1987).

368. TAKITA, T., Y. UMEZAWA, S. SAITO, H. MORISHIMA, H. NAGANAWA, H. UMEZAWA, T. TSUCHIYA, T. MIYAKE, S. KAGEYAMA, S. UMEZAWA, Y. MURAOKA, M. SUZUKI, M. OTSUKA, M. NARITA, S. KOBAYASHI, and M. OHNO: Total Synthesis of Bleomycin A2. Tetrahedron Letters **23**, 521 (1982).

369. AOYAGI, Y., K. KATANO, H. SUGUNA, J. PRIMEAU, L.-H. CHANG, L. CHANG, and S.M. HECHT: Total Synthesis of Bleomycin. J. Amer. Chem. Soc. **104**, 5537 (1982).

370. TAN, J.D., S.E. HUDSON, S.J. BROWN, M.M. OLMSTEAD, and P.K. MASCHARAK: Syntheses, Structures and Reactivities of Synthetic Analogues of the Three Forms of Co(III)-Bleomycin: Proposed Mode of Light-Induced DNA Damage by the Co(III) Chelate of the Drug. J. Amer. Chem. Soc. **114**, 3841 (1992).

371. URATA, H., Y. UEDA, Y. USAMI, and M. AKAGI: Enantiospecific Recognition of DNA by Bleomycin. J. Amer. Chem. Soc. **115**, 7135 (1993).

372. HAMAMICHI, N., and S.M. HECHT: Determination of the Absolute Configuration of the Thiazolinylthiazole Moiety of Phleomycin. J. Amer. Chem. Soc. **115**, 12605 (1993).

373. KURYLO-BOROWSKA, Z., and E.L. TATUM: Biosynthesis of Edeine by *Bacillus brevis* Vm4 *in vivo* and *in vitro*. Biochim. Biophys. Acta **113**, 206 (1966).

374. BOROWSKI, E., H. CHMARA, and E. JARECZEK-MORAWSKA, The Antibiotic Edeine, VI: Paper and Thin-layer Chromatography of Components of the Edeine Complex. Biochim. Biophys. Acta **130**, 560 (1966).

375. RONCARI, G., Z. KURYLO-BOROWSKA, and L.C. CRAIG: On the Chemical Nature of the Antibiotic Edeine. Biochemistry **5**, 2153 (1966).

376. HETTINGER, T.P., and L.C. CRAIG: Edeine, II: The Composition of the Antibiotic Peptide Edeine A. Biochemistry **7**, 4147 (1968).

377. HETTINGER, T.P., Z. KURYLO-BOROWSKA, and L.C. CRAIG: Edeine, III: The Composition of the Antibiotic Peptide Edeine B. Biochemistry **7**, 4153 (1968).

378. HETTINGER, T.P., and L.C. CRAIG: Edeine, IV: Structures of the Antibiotic Peptides Edeines A_1 and B_1. Biochemistry **9**, 1224 (1970).

379. ROBIN, Y., N.V. THOAI, L.A. PRADEL, and J. ROCHE: Sur les Constituants Guanidiques des Oeufs d'*Audouinia tentaculata* Mtg. C. R. séances Soc. Biol. **150**, 1892 (1956).

380. ROCHE, J., N.V. THOAI, Y. ROBIN, and L.A. PRADEL: Sur la Présence d'un Dérivé Guanidique Nouveau dans le Muscle de la Sangsue, *Hirudo medicinalis* L. C. R. séances Soc. Biol. **150**, 1684 (1956).

381. ROBIN, Y., N.V. THOAI, and J. ROCHE: Sur la Présence d'Arcaïne chez la Sangsue, *Hirudo medicinalis* L. C. R. séances Soc. Biol. **151**, 2015 (1957).

382. ROBIN, Y., and N.V. THOAI: Structure et Synthèse de l'Hirudonine [Diamidinospermidine ou *N*-(3-Guanidopropyl)-4-aminobutyl-guanidine]. C. R. séances hebd. Acad. Sci. **252**, 1224 (1961).

383. ROBIN, Y., C. AUDIT, S. ZAPPACOSTA, and N.V. THOAI: Biogénèse de l'Hirudonine-4. Comp. Biochem. Physiol. **7**, 221 (1962).

384. ROBIN, Y., C. AUDIT, and N.V. THOAI: Biogénèse de l'Arcaïne chez la Sangsue, *Hirudo medicinalis* L. C. R. séances Soc. Biol. **156**, 1232 (1962).

385. ROBIN, Y., and J. ROCHE: Répartition Biologique des Guanidines Substituées chez les Vers Terrestres et d'Eau Douce (Oligochètes, Hirudinées, Turbellariées) Récoltés en Hongrie. Comp. Biochem. Physiol. **14**, 453 (1965).

386. ROCHE, J., C. AUDIT, and Y. ROBIN: Isolement et Identification d'un Nouveau Dérivé Diguanidique Biologique, l'Adouine (1.5-Diamidinocadavérine), et de l'Arcaïne (1,4-Diamidinoputrescine) chez une Annélide Polychète Marine, *Andouinia tentaculata* Montagu. C. R. séances hebd. Acad. Sci. **260**, 7023 (1965).

387. ROBIN, Y., and C. AUDIT: Biogénèse des Derivés Guanidiques chez *Andouinia tentaculata* Mtg. C. R. séances Soc. Biol. **160**, 1410 (1966).

388. AUDIT, C., B. VIALA, and Y. ROBIN: Biogénèse des Derivés Diguanidiques chez la Sangsue, *Hirudo medicinalis* L., I: Origine des Groupements Guanidiques et de la Chaîne Carbonée. Comp. Biochem. Physiol. **22**, 775 (1967).

389. ROBIN, Y., C. AUDIT, and M. LANDON: Biogénèse des Derivés Diguanidiques chez la Sangsue, *Hirudo medicinalis* L, II: Mécanisme de la Double Transamidination. Comp. Biochem. Physiol. **22**, 787 (1967).

390. HIGASHIDE, E., K. HATANO, and M. SHIBATA: Enduracidin, a New Antibiotic, I: *Streptomyces fungicidicus* No B5477, an Enduracidin-Producing Organism. J. Antibiotics **21**, 126 (1968).

391. ASAI, M., M. MUROI, N. SUGITA, H. KAWASHIMA, K. MIZUNO, and A. MIYAKE: Enduracidin, a New Antibiotic, II: Isolation and Characterization. J. Antibiotic **21**, 138 (1968).

392. HORII, S., and Y. KAMEDA: Enduracidin, a New Antibiotic, V: Structures of New Basic Amino Acid Enduracididine and Alloenduracididine. J. Antibiotics **21**, 665 (1968).

393. HORI, M., N. SUGITA, and M. MIYAZAKI: Enduracidin, a New Antibiotic, VI: Separation and Determination of Enduracidins A and B by Column Chromatography. Chem. Pharm. Bull (Japan) **21**, 1171 (1973).

394. HORI, M., H. IWASAKI, S. HORII, I. YOSHIDA, and T. HONGO: Enduracidin, a New Antibiotic, VII: Primary Structure of the Peptide Moiety. Chem. Pharm. Bull. (Japan) **21**, 1175 (1973).

395. IWASAKI, H., S. HORII, M. ASAI, K. MIZUNO, J. UEYANAGI, and A. MIYAKE: Enduracidin, a New Antibiotic, VIII: Structures of Enduracidins A and B. Chem. Pharm. Bull. (Japan) **21**, 1184 (1973).

396. KAMIYA, K., M. NISHIKAWA, H. MATSUMARU, M. ASAI, and K. MIZUNO: X-Ray Analysis of an Unusual Amino Acid Isolated from the Hydrolysate of a New Antibiotic, Enduracidin. Chem. Pharm. Bull. (Japan) **16**, 2303 (1968).

397. OHUCHI, S., H. SUDA, H. NAGANAWA, T. TAKITA, T. AOYAGI, H. UMEZAWA, H. NAKAMURA, and Y. IITAKA: The Structures of Arphamenines A and B. J. Antibiotics **36**, 1572 (1983).

398. OHUCHI, S., A. OKUYAMA, H. NAGANAWA, T. AOYAGI, and H. UMEZAWA: Biosynthetic Studies of Arphamenines A and B. J. Antibiotics **37**, 518 (1984).

399. OKUYAMA, A., S. OHUCHI, T. TANAKA, H. NAGANAWA, T. AOYAGI, and H. UMEZAWA: Isolation of Intermediates of Arphamenine A Biosynthesis. Biochem. Intern. **12**, 361 (1986).

400. OKUYAMA, A., S. OHUCHI, T. TANAKA, T. AOYAGI, H. UMEZAWA, and H. NAGANAWA: Cell-Free Biosynthesis of Arphamenine A. Biochem. Intern. **12**, 485 (1986).

401. OGITA, T., S. GUNJI, Y. FUKAZAWA, A. TERAHARA, T.T. KINOSHITA, H. NAGAKI, and T. BEPPU: The Structures of Fosfazinomycins A and B. Tetrahedron Letters **24**, 2283 (1983).

402. UMEZAWA, H., T. AOYAGI, K. OGAWA, H. IINUMA, H. NAGANAWA, M. HAMADA, and T. TAKEUCHI: Histargin, a New Inhibitor of Carboxypeptidase B, Produced by Actinomycetes. J. Antibiotics **37**, 1088 (1984).

403. MORIGUCHI, M., Y. UMEDA, K. MIYAZAKI, T. NAKAMURA, K. OGAWA, F. KOJIMA, H. IINUMA, and T. AOYAGI: Synthesis of Histargin and Related Compounds and Their Inhibition of Enzymes. J. Antibiotics **41**, 1823 (1988).

404. RAPP, C., G. JUNG, M. KUNGLER, and W. LOEFFLER: Rhizocticins — New-Phosphono-Oligopeptides with Antifungal Activity. Liebigs Ann. Chem. **1988**, 655.

405. KONDO, E., T. KATAYAMA, Y. KAWAMURA, Y. YASUDA, K. MATSUMOTO, K. ISHII, T. TANIMOTO, H. HINOO, T. KATO, H. KYOTANI, and J. SHOJI: Isolation and Characterization of New Antibiotics Resorcinomycins A and B. J. Antibiotics **42**, 1 (1989).

406. MASAKI, S., T. KONISHI, N. TSUJI, and J. SHOJI: New Antibiotics Resorcinomycins A and B: Antibacterial Activity of Resorcinomycin A Against Mycobacteria *in vitro*. J. Antibiotics **42**, 463 (1989).

407. KATAYAMA, N., S. FUKUSUMI, Y. FUNABASHI, T. IWAHI, and H. ONO: TAN-1057 A–D, New Antibiotics with Potent Antibacterial Activity Against Methicillin-Resistant *Staphylococcus aureus*. Taxonomy, Fermentation and Biological Activity. J. Antibiotics **46**, 606 (1993).

408. FUNABASHI, Y., S. TSUBOTANI, K. KOYAMA, N. KATAYAMA, and S. HARADA: A New Anti-MRSA Dipeptide, TAN-1057 A. Tetrahedron **49**, 13 (1993).

409. AOYAGI, T., M. HATSU, F. KOJIMA, C. HAYASHI, M. HAMADA, and T. TAKEUCHI: Benarthin: A New Inhibitor of Pyroglutamyl Peptidase, I: Taxonomy, Fermentation, Isolation and Biological Activities. J. Antibiotics **45**, 1079 (1992).

410. HATSU, M., H. NAGANAWA, T. AOYAGI, and T. TAKEUCHI: Benarthin: A New Inhibitor of Pyroglutamyl Peptidase, II: Physico-Chemical Properties and Structure Determination. J. Antibiotics **45**, 1084 (1992).

411. HATSU, M., M. TUDA, Y. MURAOKA, T. AOYAGI, and T. TAKEUCHI: Benarthin: A New Inhibitor of Pyroglutamyl Peptidase, III: Synthesis and Structure-Activity Relationships. J. Antibiotics **45**, 1088 (1992).

412. KAMIYAMA, T., T. UMINO, N. NAKAYAMA, Y. ITEZONO, T. SATOH, Y. YAMASHITA, A. YAMAGUCHI, and K. YOKOSE: Ro 09-1679, a Novel Thrombin Inhibitor. J. Antibiotics **45**, 424 (1992).

413. KOMORI, T., M. EZAKI, E. KINO, M. KOHSAKA, H. AOKI, and H. IMANAKA: Lavendomycin, a New Antibiotic, I: Taxonomy, Isolation and Characterization. J. Antibiotics **38**, 691 (1985).

414. USHIDA, I., N. SHIGEMATSU, M. EZAKI, and M. HASHIMOTO: Structure of Lavendomycin, a New Peptide Antibiotic. Chem. Pharm. Bull. (Japan) **33**, 3053 (1985).

415. SCHMIDT, U., K. MUNDIGER, R. MANGOLD, and A. LIEBERKNECHT: Lavendomycin: Total Synthesis and Assignment of Configuration. Chem. Commun. **1990**, 1216.

416. EZAKI, M., N. SHIGEMATSU, M. YAMASHITA, T. KOMORI, K. UMEHARA, and H. IMANAKA: Biphenomycin C, a Precursor of Biphenomycin A in Mixed Culture. J. Antibiotics **46**, 135 (1993).

417. EZAKI, M., M. IWAMI, M. YAMASHITA, S. HASHIMOTO, T. KOMORI, K. UMEHARA, Y. MINE, M. KOHSAKA, H. AOKI, and H. IMANAKA: Biphenomycins A and B, Novel Peptide Antibiotics, I: Taxomomy, Fermentation, Isolation and Characterization. J. Antibiotics **38**, 1453 (1985).

418. USHIDA, I., N. SHIGEMATSU, M. EZAKI, M. HASHIMOTO, H. AOKI, and H. IMANAKA: Biphenomycins A and B, Novel Peptide Antibiotics, II: Structural Elucidation of Biphenomycins A and B. J. Antibiotics **38**, 1462 (1985).

419. USHIDA, I., M. EZAKI, N. SHIGEMATSU, and M. HASHIMOTO: Structure of WS-43708A, a Novel Cyclic Peptide Antibiotic. J. Organ. Chem. (USA) **50**, 1341 (1985).

420. SCHMIDT, U., R. MEYER, U. LEITENBERGER, A. LIEBERKNECHT, and H. GRIESSER: The Synthesis of Biphenomycin B. Chem. Commun. **1991**, 275.

421. KANNAN, R., and D.W. WILLIAMS: Stereochemistry of the Cyclic Tripeptide Antibiotic WS-43708A. J. Organ. Chem. (USA) **52**, 5435 (1987).

422. HEMPEL, J.C., and F.K. BROWN: NMR Template Analysis of Biphenomycin: The Prediction of Conformational Domains Defined by Clustered Distance Constraints. J. Amer. Chem. Soc. **111**, 7323 (1989).

423. BROWN, F.K., J.C. HEMPEL, J.S. DIXON, S. AMATO, L. MUELLER, and P.W. JEFFS: Structure of Biphenomycin A Derived from Two-Dimensional NMR Spectroscopy and Molecular Modeling. J. Amer. Chem. Soc. **111**, 7328 (1989).

424. TAKITA, T., N. SHIMADA, N. YAGISAWA, K. KATO, H. NAGANAWA, K. MAEDA, and H. UMEZAWA: Chemistry of Pheganomycins, New Peptide Antibiotics. Proceedings of the 15th Symposium of Peptide Chemistry, Osaka, Japan (1977).

425. UMEZAWA, H., T. AOYAGI, H. MORISHIMA, S. KUNIMOTO, M. MATSUZAKI, M. HAMADA, and T. TAKEUCHI: Chymostatin, a New Chymotrypsin Inhibitor Produced by Actinomycetes. J. Antibiotics **23**, 425 (1970).

426. TATSUTA, K., N. MIKAMI, K. FUJIMOTO, S. UMEZAWA, H. UMEZAWA, and T. AOYAGI: The Structure of Chymostatin, a Chymotrypsin Inhibitor. J. Antibiotics **26**, 625 (1973).

427. UMEZAWA, H., T. AOYAGI, A. OKURA, H. MORISHIMA, T. TAKEUCHI, and Y. OKAMI: Elastatinal, a New Elastase Inhibitor Produced by Actinomycetes. J. Antibiotics **26**, 787 (1973).

428. OKURA, A., H. MORISHIMA, T. TAKITA, T. AOYAGI, T. TAKEUCHI, and H. UMEZAWA: Structure of Elastatinal, an Elastase Inhibitor of Microbial Origin. J. Antibiotics **28**, 337 (1975).

429. AOYAGI, T., T. TAKEUCHI, A. MATSUZAKI, K. KAWAMURA, S. KONDO, M. HAMADA, K. MAEDA, and H. UMEZAWA: Leupetins, New Protease Inhibitors from Actinomycetes. J. Antibiotics **22**, 283 (1969).

430. KONDO, S., K. KAWAMURA, J. IWANAGA, M. HAMADA, T. AOYAGI, K. MAEDA, T. TAKEUCHI, and H. UMEZAWA: Isolation and Characterization of Leupeptins Produced by Actinomycetes. Chem. Pharm. Bull. (Japan) **17**, 1896 (1969).

431. KAWAMURA, K., S. KONDO, K. MAEDA, and H. UMEZAWA: Structures and Synthesis of Leupeptins Pr-LL and Ac-LL. Chem. Pharm. Bull. (Japan) **17**, 1902 (1969).

432. YAMASHITA, K., K. WATANABE, H. TAKAYAMA, S. MIZUGUCHI, M. ISHIBASHI, H. MIYAZAKI, W. TANAKA, and H. UMEZAWA: Assay of Plasma Leupeptin Using the Reversible Binding of Leupeptin to Bovine Pancreatic Trypsin. Anal. Biochem. **156**, 503 (1986).

433. SUZUKI, K., S. TSUJI, and S. ISHIURA: Effect of Ca^{2+} on the Inhibition of Calcium-Actived Neutral Protease by Leupeptin, Antipain and Epoxisuccinate Derivatives. FEBS Letters **136**, 119 (1981).

434. SUDA, H., T. AOYAGI, M. HAMADA, T. TAKEUCHI, and H. UMEZAWA: Antipain, a New Protease Inhibitor Isolated from Actinomycetes. J. Antibiotics **25**, 263 (1972).

435. UMEZAWA, S., K. TATSUTA, K. FUJIMOTO, T. TSUCHIYA, H. UMEZAWA, and H. NAGANAWA: Structure of Antipain, a New Sakaguchi-Positive Product of Streptomyces. J. Antibiotics **25**, 267 (1972).

436. HANADA, K., M. TAMAI, M. YAMAGISHI, S. OHMURA, J. SAWADA, and I. TANAKA: Isolation and Characterization of E-64, a New Thiol Protease Inhibitor. Agric. Biol. Chem. **42**, 523 (1978).

437. HANADA, K., M. TAMAI, S. OHMURA, J. SAWADA, T. SEKI, and I. TANAKA: Structure and Synthesis of E-64, a New Thiol Protease Inhibitor. Agric. Biol. Chem. **42**, 529 (1978).

438. HANADA, K., M. TAMAI, S. MORIMOTO, T. ADACHI, S. OHMURA, J. SAWADA, and I. TANAKA: Inhibitory Activities of E-64 Derivatives on Papain. Agric. Biol. Chem. **42**, 537 (1978).

439. MURAO, S., and T. WATANABE: Novel Microbial Alkaline Protease Inhibitor, MAPI, Produced by *Streptomyces* sp. No WT-27. Agric. Biol. Chem. **41**, 1313 (1977).

440. WATANABE, T., and S. MURAO: Purification and Characterization of Crystalline Microbial Alkaline Proteinase Inhibitors (MAPI), Produced by *Streptomyces nigrescens* WT-27. Agric. Biol. Chem. **43**, 243 (1979).

441. SHIN-WATANABE, T., K. FUKUHARA, and S. MURAO: The Structure of β-MAPI, a Novel Proteinase Inhibitor. Tetrahedron **38**, 1775 (1982).

442. HUANG, L., G. ROWIN, J. DUNN, R. SYKES, R. DOBNA, B.A. MAYLES, D.M. GROSS, and R.W. BURG: Discovery, Purification and Characterization of the Angiotensin Converting Enzyme Inhibitor, L-681,176, Produced by *Streptomyces* sp. MA 5143a. J. Antibiotics **37**, 462 (1984).

443. HENSENS, O.D., and J.M. LIESCH: Structure Elucidation of Angiotensin Converting Enzyme Inhibitor L-681,176 from *Streptomyces* sp. MA 5143a. J. Antibiotics **37**, 466 (1984).

444. OGURA, K., M. MAEDA, M. NAGAI, T. TANAKA, K. NOMOTO, and T. MURACHI: Purification and Structure of a Novel Cysteine Proteinase Inhibitor, Strepin P-1. Agric. Biol. Chem. **49**, 799 (1985).

445. MURAO, S., T. SHIN, Y. KATSU, S. NAKATANI, and K. MIRAYAMA: Novel Thiol Proteinase Inhibitor, Thiolstatin, Produced by a Strain of *Bacillus cereus*. Agric. Biol. Chem. **49**, 895 (1985).

446. UMEZAWA, H.: Low-Molecular-Weight Enzyme Inhibitors of Microbial Origin. Ann. Rev. Microbiol. **36**, 75 (1982).

447. UMEZAWA, H.: Enzyme Inhibitors Produced by Microorganisms. In: Natural Products Isolation (Journal of Chromatography Library, Vol. 43) (G.H. WAGMAN and R. COOPER, eds.), p. 481. Amsterdam: Elsevier. 1989.

448. UMEZAWA, S., K. TATSUTA, T. TSUCHIYA, H. UMEZAWA, and H. NAGANAWA: The Structure of Arglecin, a New Metabolite of *Streptomyces*. Tetrahedron Letters **1971**, 259.

449. TATSUTA, K., T. TSUCHIYA, T. SOMENO, S. UMEZAWA, H. UMEZAWA, and H. NAGANAWA: Arglecin, a New Microbial Metabolite. Isolation and Chemical Structure. J. Antibiotics **24**, 735 (1971).

450. TATSUTA, K., T. TSUCHIYA, S. UMEZAWA, H. NAGANAWA, and H. UMEZAWA: Revised Structure for Arglecin. J. Antibiotics **25**, 674 (1972).

451. TATSUTA, K., K. FUJIMOTO, M. YAMASHITA, T. TSUCHIYA, S. UMEZAWA, and H. UMEZAWA: Argvalin, a New Microbial Metabolite: Isolation and Structure. J. Antibiotics **26**, 606 (1973).

452. MACDONALD, J.C., and G.G. BISHOP: Biosynthesis of Arglecin and Related Compounds. Canad. J. Biochem. **55**, 165 (1977).

453. OHTA, A., Y. AOYAGI, Y. KURIHARA, K. YUASA, M. SHIMAZAKI, T. KURIHARA, and H. MIYAMAE: Synthesis of Arglecin. Heterocycles **26**, 3181 (1987).

454. OHTA, A., Y. AOYAGI, Y. KURIHARA, A. KOJIMA, K. YUASA, and M. SHIMAZAKI: Synthesis of Argvalin and Its Related Compounds. Heterocycles **27**, 437 (1988).

455. ELLESTAD, G.A., J.H. MARTIN, G.O. MORTON, M.L. SASSIVER, and J.E. LANCASTER: Structure of LL-BM123α, a New *Myo*-Inosamine-2 Containing Antibiotic. J. Antibiotics **30**, 678 (1977).

456. TRESNER, H.D., J.H. KORSHALLA, A.A. FANTINI, J.D. KORSHALLA, J.P. KIRBY, J.J. GOODMAN, R.A. KELE, A.J. SHAY, and D.B. BORDERS: Glycocinnamoylspermidines, a New Class of Antibiotics, I: Description and Fermentation of the Organism Producing the LL-BM123 Antibiotics. J. Antibiotics **31**, 394 (1978).

457. MARTIN, J.H., M.P. KUNSTMANN,' F. BARBATSCHI, M. HERTZ, G.A. ELLESTAD, M. DANN, G.S. REDIN, A.C. DORNBUSH, and N.A. KUCK: Glycocinnamoylspermidines, a New Class of Antibiotics, II: Isolation, Physicochemical and Biological Properties of LL-BM123β, γ_1 and γ_2. J. Antibiotics **31**, 398 (1978).

458. ELLESTAD, G.A., D.B. COSULICH, R.W. BROSCHARD, J.H. MARTIN, M.P. KUNSTMANN, G.O. MORTON, J.E. LANCASTER, W. FULMOR, and F.M. LOVELL: Glycocinnamoylspermidines, a New Class of Antibiotics, 3: The Structures of LL-BM123β, γ_1 and γ_2. J. Amer. Chem. Soc. **100**, 2515 (1978).

459. HLAVKA, J.J., P. BITHA, J. BOOTHE, and T. FIELDS: Glycocinnamoylspermidines, a New Class of Antibiotics, IV: Chemical Modification of LL-BM123γ. J. Antibiotics **31**, 477 (1978).

460. KUCK, N.A., and G.S. REDIN: Glycocinnamoylspermidines, a New Class of Antibiotics, V: Antibacterial Evaluation of the Isopropyl Derivative of LL-BM123γ. J. Antibiotics **31**, 405 (1978).

461. CHIU, S.H.L., R. FIALA, R. KENNETT, L. WOZNIAK, and M.W. BULLOCK: Biosynthesis of the Glycocinnamoylspermidine Antibiotic, Cinodine. J. Antibiotics **37**, 1000 (1984).

462. CHIU, S.H.L., R. FIALA, and M.W. BULLOCK: Biosynthesis of the Spermidine Moiety of Glycocinnamoylspermidine Antibiotic Cinodine. J. Antibiotics **37**, 1079 (1984).

463. DOBASHI, K., N. MATSUDA, M. HAMADA, H. NAGANAWA, T. TAKITA, and T. TAKEUCHI: Novel Antifungal Antibiotics Octacosamicins A and B, I: Taxonomy, Fermentation and Isolation, Physico-Chemical Properties and Biological Activities. J. Antibiotics **41**, 1525 (1988).

464. DOBASHI, K., H. NAGANAWA, Y. TAKAHASHI, T. TAKITA, and T. TAKEUCHI: Novel Antifungal Antibiotics Octacosamicins A and B, II: The Structure Elucidation Using Various NMR Spectroscopic Methods. J. Antibiotics **41**, 1533 (1988).

465. ARGOUDELIS, A.D., F. REUSSER, H.A. WHALLEY, L. BACZYNSKYJ, S.A. MIZSAK, and R.J. WNUK: Antibiotics Produced by *Streptomyces ficellus*, I: Ficellomycin. J. Antibiotics **29**, 1001 (1976).

466. KUO, M.S., D.A. YUREK, and S.A. MIZSAK: Structure Elucidation of Ficellomycin. J. Antibiotics **42**, 357 (1989).

467. HAYAKAWA, Y., N. KANAMARU, N. MORISAKI, H. SETO, and K. FURIHATA: Structure of Lydicamycin, a New Antibiotic of a Novel Skeletal Type. Tetrahedron Letters **32**, 213 (1991).

468. MCINERNEY, B.V., W.C. TAYLOR, M.J. LACEY, R.J. AKHURST, and R.P. GREGSON: Biologically Active Metabolites from *Xenorhabdus spp.*, Part 2: Benzopyran-1-one Derivatives with Gastroprotective Activity. J. Nat. Prod. (Lloydia) **54**, 785 (1991).

469. CHRISTOPHERSEN, C: Marine Alkaloids. In: The Alkaloids, Vol. 24 (A. BROSSI, ed.), p. 25. Orlando: Academic Press. 1985.

470. PIETRA, F.: Total Synthesis of Marine Natural Products: A Powerful Contribution to the Understanding and Development of Marine Organic Chemistry. Gazz. Chim. Ital. **115**, 443 (1985).

471. FAULKNER, D.J.: Marine Natural Products: Metabolites of Marine Algae and Herbivorous Marine Molluscs. Nat. Prod. Rep. **1**, 251 (1984).

472. FAULKNER, D.J.: Marine Natural Products: Metabolites of Marine Invertebrates. Nat. Prod. Rep. **1**, 551 (1984).

473. FAULKNER, D.J.: Marine Natural Products. Nat. Prod. Rep. **3**, 1 (1986).

474. FAULKNER, D.J.: Marine Natural Products. Nat. Prod. Rep. **4**, 539 (1987).

475. FAULKNER, D.J.: Marine Natural Products. Nat. Prod. Rep. **5**, 613 (1988).

476. FAULKNER, D.J.: Marine Natural Products. Nat. Prod. Rep. **7**, 269 (1990).

477. FAULKNER, D.J.: Marine Natural Products. Nat. Prod. Rep. **8**, 97 (1991).

478. FAULKNER, D.J.: Marine Natural Products. Nat. Prod. Rep. **9**, 323 (1992).

479. SCHEUER, P.J.: Chemistry of Marine Natural Products, p. 144. New York: Academic Press. 1973.

480. SHIMIZU, Y.: Dinoflagellate Toxins. In: Marine Natural Products: Chemical and Biological Perspectives, Vol. I (P.J. SCHEUER, ed.), p. 17. New York: Academic Press. 1981.

481. MOORE, R.E.: Constituents of Blue-Green Algae. In: Marine Natural Products: Chemical and Biological Perspectives, Vol. IV (P.J. SCHEUER, ed.), p. 10. New York: Academic Press. 1981.

482. NATORI, S., N. IKEKAWA, and M. SUZUKI, eds.: Red Tide Toxins: Assay and Isolation of the Toxic Components. In: Advances in Natural Products Chemistry: Extraction and Isolation of Biologically Active Compounds, p. 151. New York: Wiley. 1981.

483. NATORI, S., N. IKEKAWA, and M. SUZUKI, eds.: The Extraction and Isolation of

Tetrodotoxin from Fugu Ovaries. In: Advances in Natural Products Chemistry: Extraction and Isolation of Biologically Active Compounds, p. 511. New York: Wiley. 1981.

484. NAKANISHI, K., T. GOTO, S. ITO, S. NATORI, and S. NOZOE, eds.: Structure of Tetrodotoxin. In: Natural Products Chemistry, Vol. 2, p. 457. Tokyo: Kodansha, New York: Academic Press. 1975.

485. NAKANISHI, K., T. GOTO, S. ITO, S. NATORI, and S. NOZOE, eds.: Synthesis of Tetrodotoxin. In: Natural Products Chemistry, Vol. 2, p. 461. Tokyo: Kodansha, New York: Academic Press. 1975.

486. NAKANISHI, K., T. GOTO, S. ITO, S. NATORI, and S. NOZOE, eds.: Structure of Saxitoxin. In: Natural Products Chemistry, Vol. 2, p. 464. Tokyo: Kodansha, New York: Academic Press. 1975.

487. HALSTEAD, B.W.: Poisonous and Venomous Marine Animals of the World, 2nd Ed., p. 525. Princeton, NJ: Darwin Press. 1988.

488. KAO, C.Y., and S.R. LEVINSON, eds.: Tetrodotoxin, Saxitoxin, and the Molecular Biology of the Sodium Channel. Ann. New York Acad. Sci. 479 (1986).

489. FUHRMAN, F.A.: Tetrodotoxin, Tarinchatoxin, and Chiriquitoxin: Historical Perspectives. Ann. New York Acad. Sci. 479, 1 (1986).

490. SCHANTZ, E.J.: Chemistry and Biology of Saxitoxin and Related Toxins. Ann. New York Acad. Sci. 479, 15 (1986).

491. SHIMIZU, Y.: Chemistry and Biochemistry of Saxitoxin Analogues and Tetrodotoxin. Ann. New York Acad. Sci. 479, 24 (1986).

492. MOSHER, H.S.: The Chemistry of Tetrodotoxin. Ann. New York Acad. Sci. 479, 32 (1986).

493. YASUMOTO, T., H. NAGAI, D. YASUMURA, T. MICHISHITA, A. ENDO, M. YOTSU, and Y. KOTAKI: Interspecies Distribution and Possible Origin of Tetrodotoxin. Ann. New York Acad. Sci. 479, 44 (1986).

494. KAO, C.Y.: Structure-Activity Relations of Tetrodotoxin, Saxitoxin and Analogues. Ann. New York Acad. Sci. 479, 52 (1986).

495. ULBRICHT, W., H.-H. WAGNER, and J. SCHMIDTMAYER: Kinetics of TTX-STX Block Sodium Channels. Ann. New York Acad. Sci. 479, 68 (1986).

496. STRICHARTZ, G., T. RANDO, S. HALL, J. GITSCHIER, L. HALL, B. MAGNANI, and C.H. BAY: On the Mechanism by Which Saxitoxin Binds to and Blocks Sodium Channels. Ann. New York Acad. Sci. 479, 96 (1986).

497. ANGELIDES, K., S. TERAKAWA, and G.B. BROWN: Spatial Relations of the Neurotoxin Binding Sites on the Sodium Channel. Ann. New York Acad. Sci. 479, 221 (1986).

498. NARAHASHI, T.: Mechanism of Tetrodotoxin and Saxitoxin Action. In: Handbook of Natural Toxins, Vol. 3 (A.T. TU, ed.), p. 185. New York-Basel: Marcel Dekker. 1988.

499. TSUDA, K., R. TACHIKAWA, K. SAKAI, C. TAMURA, O. AMASAKU, M. KAWAMURA, and S. IKUMA: Über die Konstitution und Konfiguration des Anhydrotetrodotoxins. Chem. Pharm. Bull. (Japan) 12, 634 (1964).

500. TSUDA, K., R. TACHIKAWA, K. SAKAI, C. TAMURA, O. AMASAKU, M. KAWAMURA, and S. IKUMA: On the Structure of Tetrodotoxin. Chem. Pharm. Bull. (Japan) 12, 642 (1964).

501. TSUDA, K., S. IKUMA, M. KAWAMURA, R. TACHIKAWA, K. SAKAI, C. TAMURA, and O. AMASAKU: Tetrodotoxin, VII: On the Structures of Tetrodotoxin and Its Derivatives. Chem. Pharm. Bull. (Japan) 12, 1357 (1964).

502. TSUDA, K.: Über Tetrodotoxin, Giftstoff der Bowlfische. Naturwiss. 53, 171 (1966).

503. MOSHER, H.S., F.A. FUHRMAN, H.D. BUCHWALD, and H.G. FISCHER: Tarinchatoxin-Tetrodotoxin: A Potent Neurotoxin. Science 144, 1100 (1964).

504. WOODWARD, R.B.: The Structure of Tetrodotoxin. Pure Appl. Chem. **9**, 49 (1964).

505. WOODWARD, R.B., and J.Z. GOUGOUTAS: The Structure of Tetrodotoxin. J. Amer. Chem. Soc. **86**, 5030 (1964).

506. NAKAMURA, M., and T. YASUMOTO: Tetrodotoxin Derivatives in Puffer Fish. Toxicon **23**, 271 (1985).

507. YASUMOTO, T., M. YOTSU, M. MURATA, and H. NAOKI: New Tetrodotoxin Analogues from Newt *Cynops ensicauda*. J. Amer. Chem. Soc. **110**, 2344 (1988).

508. HIRATA, Y.: Bioactive Substances of Marine Animals: Polyoxygenated Substances. Pure Appl. Chem. **61**, 293 (1989).

509. ENDO, A., S.S. KHORA, M. MURATA, H. NAOKI, and T. YASUMOTO: Isolation of 11-Nortetrodotoxin-6(R)-ol and Other Tetrodotoxin Derivatives from the Puffer *Fugu niphobles*. Tetrahedron Letters **29**, 4127 (1988).

510. YASUMOTO, T., M. YOTSU, A. ENDO, M. MURATA, and H. NAOKI: Interspecies Distribution and Biogenetic Origin of Tetrodotoxin and Its Derivatives. Pure Appl. Chem. **61**, 505 (1989).

511. KHORA, S.S., and T. YASUMOTO: Isolation of 11-Oxotetrodotoxin from the Puffer *Arothron Nigropunctatus*. Tetrahedron Letters **30**, 4393 (1989).

512. NOGUCHI, T., J.-K. JEON, O. ARAKAWA, H. SUGITA, Y. DEGUCHI, Y. SHIDA, and K. HASHIMOTO: Occurrence of Tetrodotoxin and Anhydrotetrodotoxin in *Vibrio* sp. Isolated from the Intestines of a Xanthid Crab, *Atergatis floridus*. J. Biochem. **99**, 311 (1986).

513. YASUMOTO, T., D. YASUMURA, M. YOTSU, T. MICHISHITA, A. ENDO, and Y. KOTAKI: Bacterial Production of Tetrodotoxin and Anhydrotetrodotoxin. Agric. Biol. Chem. **50**, 793 (1986).

514. YOTSU, M., T. YAMAZAKI, Y. MEGURO, A. ENDO, M. MURATA, H. NAOKI, and T. YASUMOTO: Production of Tetrodotoxin and Its Derivatives by *Pseudomonas* sp. Isolated from the Skin of a Pufferfish. Toxicon **25**, 225 (1987).

515. NOGUCHI, T., D.F. HWANG, O. ARAKAWA, H. SUGITA, Y. DEGUCHI, Y. SHIDA, and K. HASHIMOTO: *Vibrio alginolyticus*, a Tetrodotoxin-Producing Bacterium, in the Intestines of the Fish *Fugu vermicularis vermicularis*. Mar. Biol. **94**, 625 (1987).

516. SHIMIDU, U., T. NOGUCHI, D.-F. HWANG, Y. SHIDA, and K. HASHIMOTO: Marine Bacteria Which Produce Tetrodotoxin. Appl. Environ. Microbiol. **53**, 1714 (1987).

517. HWANG, D.F., O. ARAKAWA, T. SAITO, T. NOGUCHI, U. SHIMIDU, K. TSUKAMOTO, Y. SHIDA, and K. HASHIMOTO: Tetrodotoxin-Producing Bacteria from the Blue-Ringed Octopus *Octopus maculosus*. Mar. Biol. **100**, 327 (1989).

518. FREITAS, J.C., T. OGATA, S. SATO, and M. KODAMA: The Occurrence of Tetrodotoxin and Paralytic Shellfish Toxins in Macroalgae from the Brazilian Coast. Proc. Japan. Assoc. Mycotoxicol. Suppl. **1**, 29 (1988).

519. FREITAS, J.C.: Biomedical Importance of Marine Natural Products. Ciência e Cultura **42**, 20 (1990).

520. FREITAS, J.C., T. OGATA, C.H. VEIT, and M. KODAMA: Occurrence of Guanidine Neurotoxins in *Ascidia nigra* (Tunicata, Ascidiacea). In: Recent Adv. Toxinol. Res., Vol. 2 (P. GOPALAKRISHNAKONE and C.K. TAN, eds.), p. 541. Singapore: University of Singapore. 1992.

521. FREITAS, J.C., M. OGATA, M. KODAMA, S.C.G. MARTINEZ, M.F. LIMA, and C.K. MONTEIRO: Possible Microbial Source of Guanidine Neurotoxins Found in the Mussel *Perna perna* (Mollusca, Bivalvia, Mytilidae). In: Recent Adv. Toxinol. Res., Vol. 2 (P. GOPALAKRISHNAKONE and C.K. TAN, eds.), p. 589. Singapore: University of Singapore. 1992.

522. FREITAS, J.C., S. SATO, T. OGATA, and M. KODAMA: Guanidine Neurotoxins in the Digestive Secretion of Crabs (Crustacea, Brachyura) (submitted).

523. KIM, Y.H., G.B. BROWN, H.S. MOSHER, and F.A. FUHRMAN: Tetrodotoxin: Occurrence in Atelopid Frogs of Costa Rica. Science 189, 151 (1975).

524. YOTSU, M., T. YASUMOTO, Y.H. KIM, H. NAOKI, and C.Y. KAO: The Structure of Chiriquitoxin from the Costa Rican Frog Atelopus chiriquensis. Tetrahedron Letters 31, 3187 (1990).

525. KOTAKI, Y., and Y. SHIMIZU: 1-Hydroxy-5,11-dideoxytetrodotoxin, the First N-Hydroxy and Ring Deoxy Derivative of Tetrodotoxin Found in the Newt Taricha granulosa. J. Amer. Chem. Soc. 115, 827 (1993).

526. SHIMIZU, Y., and M. KOBAYASHI: Apparent Lack of Tetrodotoxin in Captured Tarincha torosa and Tarincha granulosa. Chem. Pharm. Bull. (Japan) 31, 3625 (1983).

527. YASUMOTO, T., and T. MICHISHITA: Fluorometric Determination of Tetrodotoxin by High Performance Liquid Chromatography. Agric. Biol. Chem. 49, 3077 (1985).

528. YOTSU, M., A. ENDO, and T. YASUMOTO: An Improved Tetrodotoxin Analyser. Agric. Biol. Chem. 53, 893 (1989).

529. KISHI, Y., F. NAKATSUBO, M. ARATANI, T. GOTO, S. INOUE, and H. KAKOI: Synthetic Approach Towards Tetrodotoxin, II: A Stereospecific Synthesis of a Compound Having the Same Six Chiral Centers on the Cyclohexane Ring as Those of Tetrodotoxin. Tetrahedron Letters 1970, 5129.

530. KISHI, Y., M. ARATANI, T. FUKUYAMA, F. NAKATSUBO, T. GOTO, S. INOUE, H. TANINO, S. SUGIURA, and H. KAKOI: Synthetic Studies on Tetrodotoxin and Related Compounds, III: A Stereospecific Synthesis of an Equivalent of Acetylated Tetrodamine. J. Amer. Chem. Soc. 94, 9217 (1972).

531. KISHI, Y., T. FUKUYAMA, M. ARATANI, F. NAKATSUBO, T. GOTO, S. INOUE, H. TANINO, S. SUGIURA, and H. KAKOI: Synthetic Studies on Tetrodotoxin and Related Compounds, IV: Stereospecific Total Syntheses of DL-Tetrodotoxin. J. Amer. Chem. Soc. 94, 9219 (1972).

532. KEANA, J.F.W., F.P. MASON, and J.S. BLAND: Synthetic Intermediates Potentially Useful for the Synthesis of Tetrodotoxin and Derivatives. J. Organ. Chem. (USA) 34, 3705 (1969).

533. KEANA, J.F.W., and C.U. KIM: Synthetic Intermediates Potentially Useful for the Synthesis of Tetrodotoxin and Derivatives, II: Reaction of Diazomethane with Some Shikimic Acid Derivatives. J. Organ. Chem. (USA) 35, 1093 (1970).

534. KEANA, J.F.W., and C.U. KIM: Synthetic Intermediates Potentially Useful for the Synthesis of Tetrodotoxin and Derivatives, III: Synthesis of a Key Lactone Intermediate from Shikimic Acid. J. Organ. Chem. (USA) 36, 118 (1971).

535. KEANA, J.F.W., J.S. BLAND, P.E. ECKLER, and V. NELSON: Diels-Alder Reactions Involving Heterocyclic Dienophiles. Synthesis of Substituted Hydroquinazolines and 1,3-Diazaspiro[4.5]decadienes. J. Organ. Chem. (USA) 41, 2124 (1976).

536. KEANA, J.F.W., and P.E. ECKLER: A New Furan and Dihydro-4-pyrone Synthesis via Diels-Alder Reactions Between Methyl 2-[2'-Acetamido-4'(1'H)-pyrimidon-6'-yl]glyoxylate and Diethyl Oxomalonate and Oxygenated 1,3-Dienes. J. Organ. Chem. (USA) 41, 2850 (1976).

537. KEANA, J.F.W., P.J. BOYLE, M. ERION, R. HARTLING, J.R. THUSMAN, J.E. RICHMAN, R.B. ROMAN, and R.M. WAH: Synthetic Intermediates Potentially Useful for the Synthesis of Tetrodotoxin and Derivatives, 8: A Series of Highly Functionalized Pyrimidones. J. Organ. Chem. (USA) 48, 3621 (1983).

538. KEANA, J.F.W., J.S. BLAND, P.J. BOYLE, M. ERION, R. HARTLING, J.R. HUSMAN, and

R.B. ROMAN: Synthetic Intermediates Potentially Useful for the Synthesis of Tetrodo-toxin and Derivatives, 9: Hydroquinazolines Possessing the Carbon Skeleton of Tetrodotoxin. J. Organ. Chem. (USA) **48**, 3627 (1983).

539. ISOBE, M., T. NISHIKAWA, S. PIKUE, and T. GOTO: Synthetic Studies on Tetrodotoxin, 1: Stereocontrolled Synthesis of the Cyclohexane Moiety. Tetrahedron Letters **28**, 6485 (1987).

540. ISOBE, M., T. NISHIKAWA, M. FUKAMI, and T. GOTO: Synthetic Studies on a Stereochemically Complex Natural Product: Designs for the Total Synthesis of (−)-Tetrodotoxin. Pure Appl. Chem. **59**, 399 (1987).

541. ISOBE, M., Y. FUKUDA, T. NISHIKAWA, P. CHABERT, T. KAWAI, and T. GOTO: Synthetic Studies on (−)-Tetrodotoxin, 3: Nitrogenation Through Overman Rearrangement and Guanidine Ring Formation. Tetrahedron Letters **31**, 3327 (1990).

542. TANINO, H., S. INOUE, M. ARATANI, and Y. KISHI: Synthetic Studies on Tetrodotoxin and Related Compounds, V: The Protecting Group of the C_9-Hydroxy Group. Tetrahedron Letters **1974**, 335.

543. FUNABASHI, M., H. WAKAI, K. SATO, and J. YOSHIMURA: Branched-Chain Sugars, Part 15: Synthesis of 1L-(1,2,3',4,5/3,6)-3-Hydroxymethyl-4,5-O-isopropylidene-3,3'-O-methylene-6-nitro-2,3,4,5-tetrahydroxycyclohexenecarbaldehyde Dimethyl Acetal, A Potential Key Compound for the Total Synthesis of Optically Active Tetrodotoxin. J. Chem. Soc. Perkin Trans. I **1980**, 14.

544. NIMITZ, J.S.: Part 1: Synthetic and Mechanistic Studies of Macrolide Ring Closure. Part 2: Studies on the Synthesis of Tetrodotoxin. Diss. Abs. Int. **42-B**, 637 (1981).

545. NACHMAN, R.J.: Synthetic Approaches to Tetrodotoxin and Analogs. Diss. Abs. Int. **42-B**, 1895 (1981).

546. FRASER-REID, B., and R.C. ANDERSON: Carbohydrate Derivatives in the Asymmetric Synthesis of Natural Products. Fortschr. Chem. org. Naturstoffe **39**, 1 (1980).

547. SATO, K., Y. KAJIHARA, Y. NAKAMURA, and J. YOSHIMURA: Synthesis of the Function-alized Cyclohexanecarbaldehyde Derivative. A Potential Key Compound for the Total Synthesis of Optically Active Tetrodotoxin. Chem. Lett. **1991**, 1559.

548. ALONSO, R.A., C.S. BURGEY, B. VENKATESWARA RAO, G.D. VITE, R. VOLLERTHUN, M.A. ZOTTOLA, and B. FRASER-REID: Carbohydrates to Carbocycles: Synthesis of the Densely Functionalized Carbocyclic Core of Tetrodotoxin by Radical Cyclization of an Anhydro Sugar Precursor. J. Am. Chem. Soc. **115**, 6666 (1993).

549. PAVELKA, L.A.: Analogues of Tetrodotoxin. Diss. Abs. Int. **41-B**, 4127 (1981).

550. CHICHEPORTICHE, R., M. BALERNA, A. LOMBET, G. ROMEY, and M. LAZDURSKI: Synthesis of New, Highly Radioactive Tetrodotoxin Derivatives and Their Binding Properties of the Sodium Channel. Eur. J. Biochem. **104**, 617 (1980).

551. ANGELIDES, K.J.: Fluorescent and Photoactivable Fluorescent Derivatives of Tetrodo-toxin to Probe the Sodium Channel of Excitable Membranes. Biochemistry **20**, 4107 (1981).

552. BALERNA, M., A. LOMBET, R. CHICHEPORTICHE, G. ROMEY, and M. LAZDURSKI: Synthesis and Properties of New Photoactivable Derivatives of Tetrodotoxin. Bio-chim. Biophys. Acta **664**, 219 (1981).

553. PAVELKA, L.A., F.A. FUHRMAN, and H.S. MOSHER: 11-Nortetrodotoxin-6,6-diol and 11-Nortetrodotoxin-6-ol. Heterocycles **17**, 225 (1982).

554. NACHMAN, R.J., and H.S. MOSHER: 11-Nor-4,9-Anhydrotetrodotoxin-6,6-diol: A Synthon for Tetrodotoxin Analogs. Heterocycles **23**, 3055 (1985).

555. SHIMIZU, Y.: Recent Progress in Marine Toxin Research. Pure Appl. Chem. **54**, 1973 (1982).

556. SHIMIZU, Y., M. KOBAYASHI, A. GENENAH, and Y. OSHIMA: Isolation of the Side-Chain

Sulfated Saxitoxin Analogs. Their Significance in Interpretation of the Mechanism of Action. Tetrahedron **40**, 539 (1984).

557. SHIMIZU, Y.: Paralytic Shellfish Poisons. Fortschr. Chem. organ. Naturstoffe **45**, 235 (1984).

558. HORI, A., and Y. SHIMIZU: Biosynthetic ^{15}N-Enrichment and ^{15}N-NMR Spectra of Neosaxitoxin and Gonyautoxin, II: Application of Structure Determination. Chem. Commun. **1983**, 790.

559. SHIMIZU, Y., M. NORTE, A. HORI, A. GENENAH, and M. KOBAYASHI: Biosynthesis of Saxitoxin Analogues: The Unexpected Pathway. J. Amer. Chem. Soc. **106**, 6433 (1984).

560. SHIMIZU, Y.: Toxigenesis and Biosynthesis of Saxitoxin Analogues. Pure Appl. Chem. **58**, 257 (1986).

561. SHIMIZU, Y., S. GUPTA, K. MASUDA, L. MARANDA, C.K. WALKER, and R. WANG: Dinoflagellate and Other Microalgal Toxins: Chemistry and Biochemistry. Pure Appl. Chem. **61**, 513 (1989).

562. GUPTA, S., M. NORTE, and Y. SHIMIZU: Biosynthesis of Saxitoxin Analogues: The Origin and Introduction Mechanism of the Side-Chain Carbon. Chem. Commun. **1989**, 1421.

563. SHIMIZU, Y.: The Chemistry of Paralytic Shellfish Toxins. In: Handbook of Natural Toxins, Vol. 3 (A.T. TU, ed.), p. 63. New York-Basel: Marcel Dekker. 1988.

564. METTING, B., and J.W. PYNE: Biologically Active Compounds from Microalgae. Enzyme Microb. Technol. **8**, 386 (1986).

565. SCHANTZ, E.J., V.E. GHAZAROSSIAN, H.K. SCHNOES, F.M. STRONG, J.P. SPRINGER, J.O. PEZZANITE, and J. CLARDY: The Structure of Saxitoxin. J. Amer. Chem. Soc. **97**, 1238 (1975).

566. BORDNER, J., W.E. THIESSEN, H.A. BATES, and H. RAPOPORT: The Structure of a Crystalline Derivative of Saxitoxin. The Structure of Saxitoxin. J. Amer. Chem. Soc. **97**, 6008 (1975).

567. SHIMIZU, Y., M. ALAM, Y. OSHIMA, and W.E. FALLON: Presence of Four Toxins in Red Tide Infested Clams and Cultured *Gonyaulax tamarensis* Cells. Biochem. Biophys. Res. Commun. **66**, 731 (1975).

568. BUCKLEY, L.J., M. IKAWA, and J.J. SASNER, Jr.: Isolation of *Gonyaulax tamarensis* Toxins from Softshell Clams (*Mya arenaria*) and a Thin-Layer Chromatographic-Fluorometric Method for Their Detection. J. Agric. Food Chem. **24**, 107 (1976).

569. SHIMIZU, Y., L.J. BUCKLEY, M. ALAM, Y. OSHIMA, W.E. FALLON, H. KASAI, I. MIURA, V.P. GULLO, and K. NAKANISHI: Structure of Gonyautoxin-II from the East Coast Toxic Dinoflagellate, *Gonyaulax tamarensis*. J. Amer. Chem. Soc. **98**, 5414 (1976).

570. BOYER, G.L., E.J. SCHANTZ, and H.K. SCHNOES: Characterization of 11-Hydroxy-saxitoxin Sulphate, a Major Toxin in Scallops Exposed to Blooms of the Poisonous Dinoflagellate *Gonyaulax tamarensis*. Chem. Commun. **1978**, 889.

571. SHIMIZU, Y.: Red Tide Toxins (Symposium Lecture, No. 337). The American Chemical Society – Chemical Society of Japan Chemistry Congress, Honolulu, Hawaii (1979).

572. OSHIMA, Y., W.E. FALLON, Y. SHIMIZU, T. NOGUCHI, and Y. HASHIMOTO: Toxins of the *Gonyaulax* sp. and Infested Bivalves in Owase Bay. Bull. Jpn. Soc. Sci. Fish. **42**, 851 (1976).

573. SHIMIZU, Y., W.E. FALLON, J.C. WEKELL, D. GERBER, Jr., and E.J. GAUGLITZ, Jr.: Analysis of Toxic Mussels (*Mytilus* sp.) from the Alaskan Inside Passage. J. Agric. Food. Chem. **26**, 878 (1978).

574. HALL, S., P.B. REICHARDT, and R.A. NEVE: Toxins Extracted from an Alaskan Isolate of *Protogonyaulax* sp. Biochem. Biophys. Res. Commun. **97**, 649 (1980).

575. HARADA, T., Y. OSHIMA, H. KAMIYA, and T. YASUMOTO: Confirmation of Paralytic

Shellfish Toxins in the Dinoflagellate *Pyrodinium bahamense* var. *compressa* and Bivalves in Palau. Bull. Jpn. Soc. Sci. Fish. **48**, 821 (1982).

576. HARADA, H., Y. OSHIMA, and T. YASUMOTO: Structure of Two Paralytic Shellfish Toxins: Gonyautoxins-V and -VI, Isolated from a Tropical Dinoflagellate, *Pyrodinium bahamense* var. *compressa*. Agric. Biol. Chem. **46**, 1861 (1982).

577. KOEHN, F.E., S. HALL, C.F. WICHMANN, H.K. SCHNOES, and P.B. REICHARDT: Dinoflagellate Neurotoxins Related to Saxitoxin: Structure and Latent Activity of Toxins B1 and B2. Tetrahedron Letters **23**, 2247 (1982).

578. HSU, C.P., A. MARCHAND, Y. SHIMIZU, and G.G. SIMS: Paralytic Shellfish Toxins in Sea Scallops, *Placeopecten magellanicus* in the Bay of Fundy. J. Fish. Res. Board Can. **36**, 32 (1979).

579. SULLIVAN, J.J., W.T. IWAOKA, and J. LISTON: Enzymatic Transformation of PSP Toxins in the Little Neck (*Protothaca staminea*). Biochem. Biophys. Res. Commun. **114**, 465 (1983).

580. HARADA, H., Y. OSHIMA, and T. YASUMOTO: Natural Occurrence of Decarbamoylsaxitoxin in Tropical Dinoflagellate and Bivalves. Agric. Biol. Chem. **47**, 191 (1983).

581. KOBAYASHI, M., and Y. SHIMIZU: Gonyautoxin-VIII, A Cryptic Precursor of Paralytic Shellfish Poisons. Chem. Commun. **1981**, 827.

582. WICHMANN, C.F., W.P. NIEMCZURA, H.K. SCHNOES, S. HALL, P.B. REICHARDT, and S.D. DARLING: Structures of Two Novel Toxins from Protogonyaulax. J. Amer. Chem. Soc. **103**, 6977 (1981).

583. OSHIMA, Y., L.J. BUCKLEY, M. ALAM, and Y. SHIMIZU: Heterogeneity of Paralytic Shellfish Poisons. Three New Toxins from Cultured *Gonyaulax tamarensis* Cells, *Mya arenaria*, and *Saxidomus giganteus*. Comp. Biochem. Physiol. **57C**, 31 (1977).

583a. SHIMIZU, Y., and C.P. HSU: Confirmation of the Structures of Gonyautoxins I–IV by Correlation with Saxitoxin. Chem. Commun. **1981**, 314.

583b. WICHMANN, C.F., G.L. BOYER, C.L. DIVAN, E.J. SCHANTZ, and H.K. SCHNOES: Neurotoxins of *Gonyaulax excavata* and Bay of Fundy Scallops. Tetrahedron Letters **22**, 1941 (1981).

584. SHIMIZU, Y., C.P. HSU, W.E. FALLON, Y. OSHIMA, I. MIURA, and K. NAKANISHI: The Structure of Neosaxitoxin. J. Amer. Chem. Soc. **100**, 6791 (1978).

585. HALL, S., S.D. DARLING, G.L. BOYER, P.B. REICHARDT, and H.W. LIU: Dinoflagellate Neurotoxins Related to Saxitoxin: Structures of Toxins C3 and C4, and Confirmation of the Structure of Neosaxitoxin. Tetrahedron Letters **25**, 3537 (1984).

586. GARSON, M.J.: Biosynthetic Studies on Marine Natural Products. Nat. Prod. Rep. **6**, 143 (1989).

587. HERBERT, R.B.: The Biosynthesis of Plant Alkaloids and Nitrogenous Microbial Metabolites. Nat. Prod. Rep. **9**, 507 (1992).

588. SHIMIZU, Y.: Microalgal Metabolites. Chem. Rev. **93**, 1685 (1993).

589. GARSON, M.J.: The Biosynthesis of Marine Natural Products. Chem. Rev. **93**, 1699 (1993).

590. YASUMOTO, T., and M. MURATA: Marine Toxins. Chem. Rev. **93**, 1897 (1993).

591. MOORE, R.E., I. OHTANI, B.S. MOORE, C.B. DE KONING, W.Y. YOSHIDA, M.T.C. RUNNEGAR, and W.W. CARMICHAEL: Cyanobacterial Toxins. Gazz. Chim. Ital. **123**, 329 (1993).

592. NAKAMURA, M., Y. OSHIMA, and T. YASUMOTO: Occurrence of Saxitoxin in Puffer Fish. Toxicon **22**, 381 (1984).

593. MARUYAMA, J., T. NOGUCHI, S. MATSUNAGA, and K. HASHIMOTO: Fast Atom

Bombardment- and Secondary Ion-Mass Spectrometry of Paralytic Shellfish Poisons and Tetrodotoxin. Agric. Biol. Chem. **48**, 2783 (1984).

594. WHITE, K.D., J.A. SPHON, and S. HALL: Fast Atom Bombardment Mass Spectrometry of 12 Marine Toxins Isolated from *Protogonyaulax*. Analyt. Chem. **58**, 562 (1986).

595. TANINO, H., T. NAKATA, T. KANEKO, and Y. KISHI: A Stereospecific Total Synthesis of D,L-Saxitoxin. J. Amer. Chem. Soc. **99**, 2818 (1977).

596. KISHI, Y.: Total Synthesis of D,L-Saxitoxin. Heterocycles **14**, 1477 (1980).

597. JACOBI, P.A., A. BROWNSTEIN, M. MARTINELLI, and K. GROZINGER: A Mild Procedure for the Generation of Azomethine Imines. Stereochemical Factors in the Intramolecular 1,3-Dipolar Addition of Azomethine Imines and a Synthetic Approach to Saxitoxin. J. Amer. Chem. Soc. **103**, 239 (1981).

598. MARTINELLI, M., A. BROWNSTEIN, P. ALLEN, and P.A. JACOBI: The Azomethine Imine Route to Guanidines. Total Synthesis of (±)-Saxitoxin. Croat. Chem. Acta **59**, 267 (1986).

599. JACOBI, P.A., M.J. MARTINELLI, and (in part) S. POLANC: Total Synthesis of (±)-Saxitoxin. J. Amer. Chem. Soc. **106**, 5594 (1984).

600. JACOBI, P.A.: The Total Synthesis of Saxitoxin. In: Strategy and Tactics in Organic Synthesis (T. LINDBERG, ed.), p. 191. San Diego, CA: Academic Press. 1989.

601. HONG, C.Y., and Y. KISHI: Enantioselective Total Synthesis of (−)-Decarbamoyl-saxitoxin. J. Amer. Chem. Soc. **114**, 7001 (1992).

602. GHAZAROSSIAN, V.E., E.J. SCHANTZ, H.K. SCHNOES, and F.M. STRONG: A Biologically Active Acid Hydrolysis Product of Saxitoxin. Biochem. Biophys. Res. Commun. **68**, 776 (1976).

603. HANNICK, S.M., and Y. KISHI: Improved Procedure for the Blaise Reaction: A Short, Practical Route to the Key Intermediates of the Saxitoxin Synthesis. J. Organ. Chem. (USA) **48**, 3833 (1983).

604. SHIMOMURA, O., T. GOTO, and Y. HIRATA: Crystalline *Cypridina* Luciferin. Bull. Chem. Soc. Japan **30**, 929 (1957).

605. KISHI, Y., T. GOTO, Y. HIRATA, O. SHIMOMURA, and F.H. JOHNSON: *Cypridina* Luminescence, I: Structure of *Cypridina* Luciferin. Tetrahedron Letters **1966**, 3427.

606. KISHI, Y., T. GOTO, S. EGUCHI, Y. HIRATA, E. WATANABE, and T. AOYAMA: *Cypridina* Bioluminescence, II: Structural Studies of *Cypridina* Luciferin by Means of a High Resolution Mass Spectrometer and an Amino Acid Analyser. Tetrahedron Letters **1966**, 3437.

607. KISHI, Y., T. GOTO, S. INOUE, S. SUGIURA, and H. KISHIMOTO: *Cypridina* Bioluminescence, III: Total Synthesis of *Cypridina* Luciferin. Tetrahedron Letters **1966**, 3445.

607a. INOUE, S., S. SUGIURA, H. KAKOI, and T. GOTO: *Cypridina* Luminescence, VI: A New Route for the Synthesis of *Cypridina* Luciferin and Its Analogs. Tetrahedron Letters **1969**, 1609.

607b. GOTO, T., and Y. KISHI: Luciferins, Bioluminescent Substances. Angew. Chem., Int. Ed. Engl. **7**, 407 (1968).

608. GOTO, T.: Bioluminescence of Marine Organisms. In: Marine Natural Products: Chemical and Biological Perspectives, Vol. III (P.J. SCHEUER, ed.), p. 179. New York: Academic Press. 1980.

609. RINEHART, K.L., K. HARADA, M. NAMIKOSHI, C. CHEN, C.A. HARVIS, M.H.G. MUNRO, J.W. BLUNT, P.E. MULLIGAN, V.R. BEASLEY, A.M. DAHLEM, and W.W. CARMICHAEL: Nodularin, Microcystin and the Configuration of Adda. J. Amer. Chem. Soc. **110**, 8557 (1988).

610. MATSUNAGA, S., R.E. MOORE, W.P. NIEMCZURA, and W.W. CARMICHAEL: Anatoxin-a(s), a Potent Anticholinesterase from *Anabaena flos-aquae.* J. Amer. Chem. Soc. **111**, 8021 (1989).

611. MAHMOOD, N.A., and W.W. CARMICHAEL: Anatoxin-a(s), an Anticholinesterase from the Cyanobacterium *Anabaena flos-aquae* NRC-525-17. Toxicon **25**, 1221 (1987).

612. MAHMOOD, N.A., W.W. CARMICHAEL, and D. PHALER: Anticholinesterase Poisonings in Dogs from Cyanobacterial (Blue-Green Algae) Bloom Dominated by *Anabaena flos-aquae.* Am. J. Veterin. Res. **49**, 500 (1988).

613. COOK, W.O., V.R. BEASLEY, R.A. LOVELL, A.M. DAHLEM, S.B. HOOSER, N.A. MAHMOOD, and W.W. CARMICHAEL: Consistent Inhibition of Periferal Cholinesterases by Neurotoxins from the Freshwater Cyanobacterium *Anabaena flos-aquae:* Studies of Ducks, Swine, Mice and a Steer. Environ. Toxicol. Chem. **8**, 915 (1989).

614. MOORE, B.S., I. OHTANI, C.B. DE KONING, R.E. MOORE, and W.W. CARMICHAEL: Biosynthesis of Anatoxin-a(s). Origin of the Carbons. Tetrahedron Letters **33**, 6595 (1992).

615. CARMICHAEL, W.W.: Cyanobacteria Secondary Metabolites – The Cyanotoxins. J. Appl. Bacter. **72**, 445 (1992).

616. CARMICHAEL, W.W.: Toxins from Freshwater Algae. In: Handbook of Natural Toxins, Vol. 3 (A.T. TU, ed.), p. 121. New York-Basel: Marcel Dekker. 1988.

617. OHTANI, I., R.E. MOORE, and M.T.C. RUNNEGAR: Cylindrospermosin: A Potent Hepatotoxin from the Blue-Green Alga *Cylindrospermopsis raciborskii.* J. Amer. Chem. Soc. **114**, 7941 (1992).

618. WAKAMIYA, T., H. NAKAMOTO, and T. SHIBA: Structural Determination of Carnosadine, a New Cyclopropyl Amino Acid, from Red Alga *Grateloupia carnosa.* Tetrahedron Letters **25**, 4411 (1984); Tetrahedron Letters **26**, 2138 (1985).

619. WAKAMIYA, T., Y. ODA, H. FUJITA, and T. SHIBA: Synthesis and Stereochemistry of Carnosadine, a New Cyclopropyl Amino Acid from Red Alga *Grateloupia carnosa.* Tetrahedron Letters **27**, 2143 (1986).

620. AITKEN, D.J., D. GUILLAUME, and H.-P. HUSSON: First Asymmetric Synthesis of Carnosadine. Tetrahedron **49**, 6375 (1993).

621. AITKEN, D.J., J. ROYER, and H.-P. HUSSON: Asymmetric Synthesis of 2,3-Methanohomoserine: A General Approach to Chiral 2-Substituted Cyclopropane Amino Acids. J. Organ. Chem. (USA) **55**, 2814 (1990).

622. FATTORUSSO, E., and M. PIATTELLI: Amino Acids from Marine Algae. In: Marine Natural Products: Chemical and Biological Perspectives, Vol. III (P.J. SCHEUER, ed.), p. 95. New York: Academic Press. 1980.

623. IRELAND, C.M., T.F. MOLINSKI, D.M. ROLL, T.M. ZABRINSKIE, T.C. MCKEE, J.C. SWERSEY, and M.P. FOSTER: Natural Product Peptides from Marine Organisms. In: Bioorganic Marine Chemistry, Vol. 3 (P.J. SCHEUER, ed.), p. 1. Berlin-Heidelberg-New York: Springer. 1989.

624. BURKHOLDER, P.R., and G.M. SHARMA: Antimicrobial Agents from the Sea. J. Nat. Prod. (Lloydia) **32**, 466 (1969).

625. SHARMA, G.M., and P.R. BURKHOLDER: Structure of Dibromophakellin, a New Bromine-Containing Alkaloid from the Marine Sponge *Phakellia flabellata.* Chem. Commun. **1971**, 151.

626. FOLEY, L.H., and G. BÜCHI: Biomimetic Synthesis of Dibromophakellin. J. Amer. Chem. Soc. **104**, 1776 (1982).

627. SHARMA, G.M., J.S. BUYER, and M.W. POMERANTZ: Characterization of a Yellow Compound Isolated from the Marine Sponge *Phakellia flabellata.* Chem. Commun. **1980**, 435.

628. CIMINO, G., S. DE ROSA, S. DE STEFANO, L. MAZZARELLA, R. PULITI, and G. SODANO: Isolation and X-ray Crystal Structure of a Novel Bromo-Compound from Two Marine Sponges. Tetrahedron Letters **23**, 767 (1982).

629. KITAGAWA, I., M. KOBAYASHI, K. KITANAKA, M. KIDO, and Y. KYOGOKU: Marine Natural Products, XII: On the Chemical Constituents of the Okinawan Marine Sponge *Hymeniacidon aldis*. Chem. Pharm. Bull. (Japan) **31**, 2321 (1983).

630. PETTIT, G.R., C.L. HERALD, J.E. LEET, R. GUPTA, D.E. SCHAUFELBERGER, R.B. BATES, P.J. CLEWLOW, D.L. DOUBEK, K.P. MANFREDI, K. RUTZLER, J.M. SCHMIDT, L.P. TACKETT, F.B. WARD, M. BRUCK, and F. CAMOU: Antineoplastic Agents, 168: Isolation and Structure of Axinohydantoin. Canad. J. Chem. **68**, 1621 (1990).

631. SCHMITZ, F.J., S.P. GUNASEKERA, V. LAKSHMI, and L.M.V. TILLEKERATNE: Marine Natural Products: Pyrrololactams from Several Sponges. J. Nat. Prod. (Lloydia) **48**, 47 (1985).

632. DE NANTEUIL, G., A. AHOND, J. GUILHEM, C. POUPAT, E.T.H. DAU, P. POTIER, M. PUSSET, J. PUSSET, and J. LABOUTE: Invertebrés Marins du Lagon Neo-Calédonien, V: Isolement et Identification des Metabolites d'une Nouvelle Espèce de Spongiaire, *Pseudaxinyssa cantharella*. Tetrahedron **41**, 6019 (1985).

633. FEDOREYEV, S.A., N.K. UTKINA, S.G. ILYIN, M.V. RESHETNYAK, and O.B. MAXIMOV: The Structure of Dibromoisophakellin from the Marine Sponge *Acanthella carteri*. Tetrahedron Letters **27**, 3177 (1986).

634. FEDOREYEV, S.A., S.G. ILYIN, N.K. UTKINA, O.B. MAXIMOV, M.V. RESHETNYAK, M.Y. ANTIPIN, and Y.T. STRUCHKOV: The Structure of Dibromoagelaspongin – A Novel Bromine-Containing Guanidine Derivative from the Marine Sponge *Agelas* sp. Tetrahedron **45**, 3487 (1989).

635. BRAEKMAN, J.C., D. DALOZE, C. STOLLER, and R.W.M. VAN SOEST: Chemotaxonomy of Agelas (Porifera: Demospongiae). Biochem. Syst. Ecol. **20**, 417 (1992).

636. KOBAYASHI, J., M. TSUDA, and Y. OHIZUMI: A Potent Actomyosin ATPase Activator from the Okinawan Marine Sponge *Agelas* cf. *nemoechinata*. Experientia **47**, 301 (1991).

637. KEIFER, P.A., R.E. SCHWARTZ, M.E.S. KOKER, R.G. HUGHES, Jr., D. RITTSCHOF, and K.L. RINEHART: Bioactive Bromopyrrole Metabolites from the Caribbean Sponge *Agelas conifera*. J. Organ. Chem. (USA) **56**, 2965 (1991).

638. KINNEL, R.B., H.-P. GEHRKEN, and P.J. SCHEUER: Palau'amine: A Cytotoxic and Immunosupressive Hexacyclic Bisguanidine Antibiotic from the Sponge *Stylotella agminata*. J. Amer. Chem. Soc. **115**, 3376 (1993).

639. KAZLAUSKAS, R., P.T. MURPHY, R.J. QUINN, and R.J. WELLS: Aplysinopsin, a New Tryptophan Derivative from a Sponge. Tetrahedron Letters **1977**, 61.

640. HOLLENBEAK, K.H., and F.J. SCHMITZ: Aplysinopsin: Antineoplastic Tryptophan Derivative from the Marine Sponge *Verongia spengelli*. J. Nat. Prod. (Lloydia) **40**, 479 (1977).

641. DJURA, P., and D.J. FAULKNER: Metabolites of the Marine Sponge *Dercitus* sp. J. Organ. Chem. (USA) **45**, 735 (1980).

642. TYMIAK, A.A., K.L. RINEHART, Jr., and G.J. BAKUS: Constituents of Morphologically Similar Sponges: *Aplysina* and *Smenospongia* Species. Tetrahedron **41**, 1039 (1985).

643. DALKAFOUKI, A., J. ARDISSON, N. KUNESCH, L. LACOMBE, and J.E. POISSON: Synthesis of 2-Dimethylaminoimidazole Derivatives: A New Access to Indolylimidazole Alkaloids of Marine Origin. Tetrahedron Letters **32**, 5325 (1991).

644. GUELLA, G., I. MANCINI, H. ZIBROWIUS, and F. PIETRA: Novel Aplysinopsin-Type Alkaloids from the Scleratinian Corals of the Family Dendrophylliidae of the Mediterranean and the Philippines. Configurational-Assignment Criteria, Stereospecific Synthesis, and Photoisomerization. Helv. Chim. Acta **71**, 773 (1988).

645. BLUNT, J.W., M.H.G. MUNRO, and S.G. YORKE: A General Synthesis of the Acarni-
    dines. Tetrahedron Letters 23, 2793 (1982).
646. YORKE, S.C., J.W. BLUNT, M.H.G. MUNRO, J.C. COOK, and K.L. RINEHART, Jr.:
    Synthesis of Acarnidines: Guanidinated Spermidine Homologues Through Imine
    Intermediates. Austral. J. Chem. 39, 447 (1986).
647. BOUKOUVALAS, J., B.T. GOLDING, R.W. MCCABE, and P.K. SLAICH: Synthesis of
    Acylpolyamines: Acetylspermidines and $C_{12:0}$ Acarnidine. Angew. Chem. Suppl. 1983,
    860.
648. HARBOUR, G.C., A.A. TYMIAK, K.L. RINEHART, Jr., P.D. SHAW, R.G. HUGHES, Jr., S.A.
    MIZSAK, J.H. COATS, G.E. ZURENKO, L.H. LI, and S.L. KUENTZEL: Ptilocaulin and
    Isoptilocaulin, Antimicrobial and Cytotoxic Cyclic Guanidines from the Caribbean
    Sponge Ptilocaulis aff. P. spiculifer (Lamarck, 1814). J. Amer. Chem. Soc. 103, 5604
    (1981).
649. SNIDER, B.B., and W.C. FAITH: The Total Synthesis of (±)-Ptilocaulin. Tetrahedron
    Letters 24, 861 (1983).
650. SNIDER, B.B., and W.C. FAITH: Total Synthesis of (±) and (−)-Ptilocaulin. J. Amer.
    Chem. Soc. 106, 1443 (1984).
651. ROUSH, W.R., and A.E. WALTS: Total Synthesis of (−)-Ptilocaulin. J. Amer. Chem.
    Soc. 106, 721 (1984).
652. WALTS, A.E., and W.R. ROUSH: A Stereorational Synthesis of (−)-Ptilocaulin.
    Tetrahedron 41, 3463 (1985).
653. UYEHARA, T., T. FURUTA, Y. KABAWAWA, J. YAMADA, T. KATO, and Y. YAMAMOTO:
    Rearrangement Approaches to Cyclic Skeletons, 6: Total Synthesis of (±)-Ptilocaulin
    on the Basis of Formal Bridgehead Substitution and Photochemical [1,3] Acyl
    Migration of a Bicyclo[3.2.2]non-6-en-2-one System. J. Organ. Chem. (USA) 53, 3669
    (1988).
654. HASSNER, A., and K.S.K. MURTHY: Stereoselective Synthesis of Ptilocaulin and Its 7-
    Epimer. Tetrahedron Letters 27, 1407 (1986).
655. MURTHY, K.S.K., and A. HASSNER: Stereoselective Total Synthesis of (±)-Ptilocaulin
    and Its 7-Epimer. A Strategy Based on the Use of an Intramolecular Nitrile Oxide
    Olefin Cycloaddition (INOC) Reaction. Israel J. Chem. 31, 239 (1991).
656. ASAOKA, M., M. SAKURAI, and H. TAKEI: Total Synthesis of (+)-Ptilocaulin.
    Tetrahedron Letters 31, 4759 (1990).
657. SULLIVAN, B., D.J. FAULKNER, and L. WEBB: Siphonodictidine, a Metabolite of the
    Burrowing Sponge Siphonodictyon sp. That Inhibits Coral Growth. Science 221, 1175
    (1983).
658. JEFFORD, C.W., P.-Z. HUANG, J.-C. ROSSIER, A.W. SLEDESKI, and J. BOUKOUVALAS: A
    Practical, Versatile Approach to Marine Furanosesquiterpenes. Synthesis of Siphono-
    dictidine and Peraplysinin-2. Synlett 1990, 745.
659. JEFFORD, C.W.: Short, Novel Syntheses of Lactones and Furans of Marine Origin.
    Gazz. Chim. Ital. 123, 317 (1993).
660. NAKAMURA, H., H. WU, J. KOBAYASHI, Y. OHIZUMI, and Y. HIRATA: Agelasidine-A, a
    Novel Sesquiterpene Possessing Antispasmodic Activity from the Okinawan Sea
    Sponge Agelas sp. Tetrahedron Letters 24, 4105 (1983).
661. CAPON, R.J., and D.J. FAULKNER: Antimicrobial Metabolites from a Pacific Sponge,
    Agelas sp. J. Amer. Chem. Soc. 106, 1819 (1984).
662. NAKAMURA, H., H. WU, J. KOBAYASHI, M. KOBAYASHI, Y. OHIZUMI, and Y. HIRATA:
    Agelasidines, Novel Hypotaurocyamine Derivatives from the Okinawan Sea Sponge
    Agelas nakamurai Hoshino. J. Organ. Chem. (USA) 50, 2494 (1985).
663. MORALES, J.J., and A.D. RODRIGEZ: (−)-Agelasidine C and (−)-Agelasidine D, Two

New Hypotaurocyamine Diterpenoids from the Caribbean Sea Sponge *Agelas clathrodes*. J. Nat. Prod. (Lloydia) **55**, 389 (1992).

664. ICHIKAWA, Y.: First Synthesis of Agelasidine A. Tetrahedron Letters **29**, 4957 (1988).

665. MANES, L.V., S. NAYLOR, P. CREWS, and G.J. BAKUS: Suvanine, a Novel Sesterpene from an *Ircinia* Marine Sponge. J. Organ. Chem. (USA) **50**, 284 (1985).

666. MANES, L.V., P. CREWS, M.R. KERMAN, D.J. FAULKNER, F.R. FRONCZEK, and R.D. GANDOUR: Chemistry and Revised Structure of Suvanine. J. Organ. Chem. (USA) **53**, 570 (1988).

667. WRIGHT, A.E., P.J. MCCARTHY, and G.K. SCHULTE: Sulfircin: A New Sesterterpene Sulfate from a Deep-Water Sponge of the Genus *Ircinia*. J. Organ. Chem. (USA) **54**, 3472 (1989).

668. SAKEMI, S., T. ICHIBA, S. KOHMOTO, G. SANCY, and T. HIGA: Isolation and Structure Elucidation of Onnamide A, a New Bioactive Metabolite of a Marine Sponge, *Theonella* sp. J. Amer. Chem. Soc. **110**, 4851 (1988).

669. MATSUNAGA, S., N. FUSETANI, and Y. NAKAO: Eight New Cytotoxic Metabolites Closely Related to Onnamide A from Two Marine Sponges of the Genus *Theonella*. Tetrahedron **48**, 8369 (1992).

670. HONG, C.Y., and Y. KISHI: Total Synthesis of Onnamide A. J. Amer. Chem. Soc. **113**, 9693 (1991).

671. HONG, C.Y., and Y. KISHI: Total Synthesis of Mycalamides A and B. J. Organ. Chem. (USA) **55**, 4242 (1990).

672. DEBITUS, C., M. CESARIO, J. GILHEM, C. PASCARD, and M. PAIS: Corallistine, a New Polynitrogen Compound from the Sponge *Corallistes fulvodesmus* L. & L. Tetrahedron Letters **30**, 1535 (1989).

673. KASHMAN, Y., S. HIRSH, O.J. MCCONNELL, I. OHTANI, T. KUSUMI, and H. KAKISAWA: Ptilomycalin A: A Novel Polycyclic Guanidine Alkaloid of Marine Origin. J. Amer. Chem. Soc. **111**, 8925 (1989).

674. OHTANI, I., T. KUSUMI, H. KAKISAWA, Y. KASHMAN, S. HIRSH: Structure and Chemical Properties of Ptilomycalin A. J. Am. Chem. Soc. **114**, 8472 (1992).

675. OHTANI, I., T. KUSUMI, and H. KAKISAWA: An Insight into the Conformation of Ptilomycalin A. The NMR Properties of Trifluoroacetylated Spermidine Analogues. Tetrahedron Letters **33**, 2525 (1992).

676. BERLINCK, R.G.S., J.C. BRAEKMAN, D. DALOZE, K. HALLENGA, R. OTTINGER, I. BRUNO, and R. RICCIO: Two New Guanidine Alkaloids from the Mediterranean Sponge *Crambe crambe*. Tetrahedron Letters **31**, 6531 (1990).

677. BERLINCK, R.G.S., J.C. BRAEKMAN, D. DALOZE, I. BRUNO, R. RICCIO, D. ROGEAU, and P. AMADE: Crambines C1 and C2: Two Further Ichthyotoxic Guanidine Alkaloids from the Sponge *Crambe crambe*. J. Nat. Prod. (Lloydia) **55**, 528 (1992).

678. BERLINCK, R.G.S.: Chromatographic Approach to Polar Compounds: Isolation of Hydrophyllic Constituents of the Marine Sponge *Crambe crambe*. Química Nova **17**, 167 (1994).

679. BERLINCK, R.G.S., J.C. BRAEKMAN, D. DALOZE, I. BRUNO, R. RICCIO, D. ROGEAU, and P. AMADE: Biologically Active Guanidine Alkaloids from the Mediterranean Sponge *Crambe crambe*. In: Guanidino Compounds in Biology and Medicine (P. P. DE DEYN, B. MARESCAU, V. STALON, and I.A. QURESHI, eds.), p. 53. London: Libbey. 1992.

680. JARES-ERIJMAN, E.A., A.A. INGRUM, F. SUN, and K.L. RINEHART: On the Structures of Crambescins B and C1. J. Nat. Prod. (Lloydia) **56**, 2186 (1993).

681. JARES-ERIJMAN, E.A., R. SAKAI, and K.L. RINEHART, Jr.: Crambescidins: New Antiviral and Cytotoxic Compounds from the Sponge *Crambe crambe*. J. Organ. Chem. (USA) **56**, 5712 (1991).

682. BERLINCK, R.G.S, J.C. BRAEKMAN, D. DALOZE, I. BRUNO, R. RICCIO, S. FERRI, S. SPAMPINATO, and E. SPERONI: Polycyclic Guanidine Alkaloids from the Marine Sponge *Crambe crambe*, and Ca^{++} Channels Blocker Activity of Crambescidin 816. J. Nat. Prod. (Lloydia) **56**, 1007 (1993).

683. JARES-ERIJMAN, E.A., A.L. INGRUM, J.R. CARNEY, K.L. RINEHART, and R. SAKAI: Polycyclic Guanidine-Containing Compounds from the Mediterranean Sponge *Crambe crambe*: The Structure of 13,14,15-Isocrambescidin 800 and the Absolute Stereochemistry of the Pentacyclic Guanidine Moieties of the Crambescidins. J. Organ. Chem. (USA) **58**, 4805 (1993).

684. SNIDER, B.B., and Z. SHI: Biomimetic Synthesis of the Bicyclic Guanidine Moieties of Crambines A and B. J. Organ. Chem. (USA) **57**, 2526 (1992).

685. SNIDER, B.B., and Z. SHI: Biomimetic Synthesis of the Central Tricyclic Portion of Ptilomycalin A. Tetrahedron Letters **34**, 2099 (1993).

686. SNIDER, B.B., and Z. SHI: Biomimetic Synthesis of the Pentacyclic Nucleus of Ptilomycalin A. J. Amer. Chem. Soc. **116**, 549 (1994).

687. SNIDER, B.B., and Z. SHI: Biomimetic Synthesis of (±)-Crambines A, B, C1 and C2. Revision of the Structures of Crambines B and C1. J. Organ. Chem. (USA) **58**, 3828 (1993).

688. FUSETANI, N., S. MATSUNAGA, H. MATSUMOTO, and Y. TAKEBAYASHI: Cyclotheonamides, Potent Thrombin Inhibitors, from a Marine Sponge *Theonella* sp. J. Amer. Chem. Soc. **112**, 7053 (1990).

689. WIPF, P., and H. KIM: Total Synthesis of Cyclotheonamide A. J. Organ. Chem. (USA) **58**, 5592 (1993).

690. LEE, A.Y., M. HAGIHARA, R. KARMACHARYA, M.W. ALBERS, S.L. SCHREIBER, and J. CLARDY: Atomic Structure of the Trypsin-Cyclotheonamide A Complex: Lessons for the Design of Serine Protease Inhibitors. J. Amer. Chem. Soc. **115**, 12619 (1993).

691. SUN, H.H., and S. SAKEMI: A Brominated (Aminoimidazolinyl)indole from the Sponge *Discodermia polydiscus*. J. Organ. Chem. (USA) **56**, 4307 (1991).

692. KOURANY-LEFOLL, E., M. PAIS, T. SÉVENET, E. GUITTET, A. MONTAGNAC, C. FONTAINE, G. GUÉNARD, M.T. ADELINE, and C. DEBITUS: Phloeodictines A and B: New Antibacterial and Cytotoxic Bicyclic Amidinium Salts from the New Caledonian Sponge, *Phloeodictyon* sp. J. Organ. Chem. (USA) **57**, 3832 (1992).

693. FUSETANI, N., Y. KAKAO, and S. MATSUNAGA: Nazumamide A, a Thrombin-Inhibitory Tetrapeptide from a Marine Sponge, *Theonella* sp. Tetrahedron Letters **32**, 7073 (1991).

694. HAYASHI, K., Y. HAMADA, and T. SHIOIRI: Synthesis of Nazumamide A, a Thrombin-Inhibitory Linear Tetrapeptide, from a Marine Sponge, *Theonella* sp. Tetrahedron Letters **33**, 5075 (1992).

695. VANWAGENEN, B.C., R. LARSEN, J.H. CARDELLINA II, D. RANDAZZO, Z.C. LIDERT, and C. SWITHENBANK: Ulosantoin, a Potent Insecticide from the Sponge *Ulosa ruetzleri*. J. Organ. Chem. (USA) **58**, 335 (1993).

696. CASAPULLO, A., E. FINAMORE, L. MINALE, and F. ZOLLO: A Dimeric Peptide Alkaloid of a Completely New Type, Anchinopeptolide A, from the Marine Sponge *Anchinoe tenacior*. Tetrahedron Letters **34**, 6297 (1993).

697. CHAN, G.W., S. MONG, M.E. HEMLING, A.J. FREYER, P.H. OFFEN, C.W. DEBROSSE, H.M. SARAU, and J.W. WESTLEY: New Leukotriene B$_4$ Receptor Antagonist: Leucettamine A and Related Imidazole Alkaloids from the Marine Sponge *Leucetta microraphis*. J. Nat. Prod. (Lloydia) **56**, 116 (1993).

698. FATTORUSSO, E., V. LANZOTTI, S. MAGNO, and E. NOVELLINO: Tryptophan Derivatives from a Mediterranean Anthozoan, *Astroides calycularis*. J. Nat. Prod. (Lloydia) **48**, 924 (1985).

699. FUSETANI, N., M. ASANO, S. MATSUNAGA, and K. HASHIMOTO: Bioactive Marine Metabolites, XV: Isolation of Aplysinopsin from the Scleratinian Coral *Tubastrea aurea* as an Inhibitor of Development of Fertilized Sea Urchin Eggs. Comp. Biochem. Physiol. **85B**, 845 (1986).

700. OKUDA, R.K., D. KLEIN, R.B. KINNEL, M. LI, and P.J. SCHEUER: Marine Natural Products: The Past Twenty Years and Beyond. Pure Appl. Chem. **54**, 1907 (1982).

701. GUELLA, G., I. MANCINI, H. ZIBROWIUS, and F. PIETRA: 160. Aplysinopsin-Type Alkaloids from *Dendrophyllia* sp., a Scleratinian Coral of the Family Dendrophyllidae of the Philippines. Facile Photochemical (*Z/E*) Photoisomerization and Thermal Reversal. Helv. Chim. Acta **72**, 1444 (1989).

702. SAKAI, R., and T. HIGA: Tubastrine, a New Guanidino Styrene from the Coral *Tubastrea aurea*. Chem. Lett. **1987**, 127.

703. DOMISSE, R.A., E.E. ESMANS, B. MARESCAU, F.C. ALDERWEIRELDT, and Y. ROBIN: Structure of Actiniamine a New Guanidino Compound from Sea Anemones of the Genus *Actinia*. Abstracts of the 16th IUPAC Congress on the Chemistry of Natural Products, Kyoto, 29 May–4 June 1988.

704. CHENG, M.T., and K.L. RINEHART, Jr.: Polyandrocarpidines: Antimicrobial and Cytotoxic Agents from a Marine Tunicate (*Polyandrocarpa* sp.) from the Gulf of California. J. Amer. Chem. Soc. **100**, 7409 (1978).

705. CARTÉ, B., and D.J. FAULKNER: Revised Structures for the Polyandrocarpidines. Tetrahedron Letters **23**, 3863 (1982).

706. RINEHART, K.L., Jr., G.R. HARBOUR, M.D. GRAVES, and M.T. CHENG: Synthesis of Hexahydropolyandrocarpidine (A Revised Structure). Tetrahedron Letters **24**, 1593 (1983).

707. GUSTAFSON, K., and R.J. ANDERSEN: Triophamine, a Unique Diacylguanidine from the Dorid Nudibranch *Triopha catalinae*. J. Organ. Chem. (USA) **47**, 2167 (1982).

708. GUSTAFSON, K., and R.J. ANDERSEN: Chemical Studies of British Columbia Nudibranchs. Tetrahedron **41**, 1101 (1985).

709. PIERS, E., J.M. CHONG, K. GUSTAFSON, and R.J. ANDERSEN: A Total Synthesis of (±)-Triophamine. Canad. J. Chem. **62**, 1 (1984).

710. FONT QUER, P.: Plantas Medicinales, el Dioscórides Renovado, p. 374. Barcelona: Editorial Labor, S.A. 1962.

711. TANRET, G.: Sur la Galégine, Alcaloïde Retiré du *Galega officinalis*. Bull. Soc. Chim. **15**, 613 (1914).

712. TANRET, G.: Sur un Alcaloïde Retiré du *Galega officinalis*. C. R. séances hebd. Acad. Sci. **158**, 1182 (1914).

713. TANRET, G.: Sur la Constitution de la Galégine. C. R. séances hebd. Acad. Sci. **158**, 1426 (1914).

714. BARGER, G., and F.D. WHITE: The Constitution of Galegine. Biochem. J. **17**, 827 (1923).

715. TANRET, G.: Sur la Constitution de la Galégine. Bull. Soc. Chim. **35**, 404 (1924).

716. SPÄTH, E., and S. PROKOPP: Über das Galegin. Ber. **57**, 474 (1924).

717. SPÄTH, E., and W. SPITZY: Die Synthese des Galegins. Ber. **58**, 2273 (1925).

718. SIMMONET, H., and G. TANRET: Sur les Propriétés Hypoglycémiantes du Sulfate de Galégine. C. R. séances hebd. Acad. Sci. **184**, 1600 (1927).

719. BRAUN, C.E.: The Chemistry of Some Suggested Insulin Substitutes. J. Chem. Education **8**, 2175 (1931).

720. PUFAHL, K., and K. SCHREIBER: Isolierung eines neuen Guanidin-Derivates aus der Geißraute, *Galega officinalis* L. Experientia **17**, 302 (1961).

721. THOAI, N.V., and G. DESVAGES: Sur la Nouvelle Guanidine Biologique Végétale, la 4-Hydroxy-Galégine. Bull. Soc. Chim. Biol. (Paris) **45**, 413 (1963).

722. SCHREIBER, K., K. PUFAHL, and H. BRÄUNIGER: Über ein neues Guanidinderivat aus der Geißraute, *Galega officinalis* L. Liebigs Ann. Chem. **671**, 147 (1969).
723. SCHREIBER, K., K. PUFAHL, and H. BRÄUNIGER: Synthese von 4-Hydroxy-dihydrogalegin und verwandten Guanidinderivaten. Liebigs Ann. Chem. **671**, 154 (1964).
724a. RUBINSHTEIN, M.M., and G.P. MENSHIKOV: The Alkaloids of *Sphaerophysa salsula*, I. J. Gen. Chem. (USSR) **14**, 161 (1944).
724b. RUBINSHTEIN, M.M., and G.P. MENSHIKOV: The Alkaloids of *Sphaerophysa salsula*, II: The Structure of Spherosine and the Partial Synthesis of Dihydrospherosine and Isodihydrospherosin. J. Gen. Chem. (USSR) **14**, 172 (1994).
724c. BIRCH, A.J., D.G. PETTIT, and R. SCHOFIELD: Studies in Relation to Biosynthesis, Part IX: The Structure of Spherophysine. J. Chem. Soc. (London) **1957**, 410.
725. CORRAL, R.A., O.O. ORAZI, and M.F. DE PETRUCELLI: A New Guanidine Alkaloid. Experientia **25**, 1020 (1969).
726. CORRAL, R.A., O.O. ORAZI, and M.F. DE PETRUCELLI: Synthesis of Pterogynine and Isolation of Its Isomer Pterogynidine, a New Guanidine Alkaloid. Chem. Commun. **1970**, 556.
727. HART, N.K., S.R. JOHNS, and J.A. LAMBERTON: Hexahydroimidazo-Pyrimidines, a New Class of Alkaloids from *Alchornea javanensis*. Chem. Commun. **1969**, 1484.
728. KHUONG-HUU, F., J.-P. LE FORESTIER, G. MAILLARD, and R. GOUTAREL: L'Alchornéine, Alcaloïde Dérivé de la Tétrahydroimidazo-[1,2-a] Pyrimidine, Isolé de Deux Euphorbiacées Africaines, l'*Alchornea floribunda* Muell. Arg. et l'*Alchornea hirtella* Benth. C. R. séances hebd. Acad. Sci. **270**, 2070 (1970).
729. CÉSARIO, M., and J. GUILHEM: Structure Cristalline de Bromométhylate d'Alchornéine, Dérivé de la Tetrahydroimidazo-[1,2-a]pyrimidine. C.R. séances hebd. Acad. Sci. **271**, 1552 (1970).
730. CÉSARIO, M., and J. GUILHEM: Structure Cristalline and Configuration Absolue d'un Nouvel Alcaloïde, l'Alchornéine. Acta Crystallogr. **B28**, 151 (1972).
731. KHUONG-HUU, F., J.-P. LE FORESTIER, and R. GOUTAREL: Alchornéine, Isoalchornéine et Alchornéinone, Produits Isolés de l'*Alchornea floribunda* Muell. Arg. Tetrahedron **28**, 5207 (1972).
732. MAAT, L., and H.C. BEYERMAN: The Imidazole Alkaloids. In: The Alkaloids, Vol. 22 (A. BROSSI, ed.), p. 302. New York: Academic Press. 1983.
733. SIDDIQUI, S., and Z. AHMED: Alkaloids from the Seeds of *Cassia absus*, Linn. Proc. Indian. Acad. Sci. **2A**, 421 (1935).
734. GUHA, S.K., and J.N. RAY: The Constitution of Chaksine, Part I. J. Indian Chem. Soc. **33**, 225 (1956).
735. SINGH, G., G.V. NAIR, K.P. AGGARWAL, and S.S. SAKSENA: Alkaline Degradation of Chaksine. Chem. and Ind. **1956**, 739.
736. SIDDIQUI, S., G. HAHN, V.N. SHARMA, and A. KAMAL: Constitution of Chaksine. Nature **178**, 373 (1956).
737. SIDDIQUI, S., G. HAHN, V.N. SHARMA, and A. KAMAL: Pyrolysis of Chaksine. Constitution of the Acid of M.P. 116°C. Chem. and Ind. **1956**, 1525.
738. WIESNER, K., Z. VALENTA, B.S. HURLBERT, F. BICKELHAUPT, and L.R. FOWLER: The Structure of Chaksine, a Monoterpene Alkaloid. J. Amer. Chem. Soc. **80**, 1521 (1958).
739. FOWLER, L.R., Z. VALENTA, and K. WIESNER: Structure of Chaksine. Chem. and Ind. **1962**, 95.
740. VOELTER, W., and W. WINTER: The Revised Structure and Absolute Configuration of Chaksine. Angew. Chem. Int. Ed. Eng. **24**, 959 (1985).
741. MAAT, L., and H.C. BEYERMAN: The Imidazole Alkaloids. In: The Alkaloids, Vol. 22 (A. BROSSI, ed.), p. 306. New York: Academic Press. 1983.

742. HAYASHI, Y.: Studies on the Ingredients of *Leonurus sibiricus* L. I. J. Pharm. Soc. Japan **82**, 1020 (1962).

743. HAYASHI, Y.: Studies on the Ingredients of *Leonurus sibiricus*, II: Structure of Leonurine. J. Pharm. Soc. Japan **82**, 1025 (1962).

744. KISHI, Y., S. SUGIURA, S. INOUE, Y. HAYASHI, and T. GOTO: Synthesis of Leonurine. Tetrahedron Letters **1968**, 637.

745. S. SUGIURA, S. INOUE, Y. HAYASHI, Y. KISHI, and T. GOTO: Structure and Synthesis of Leonurine. Tetrahedron **25**, 5155 (1969).

746. CHENG, K.F., C.S. YIP, H.W. YEUNG, and Y.C. KONG: Leonurine, an Improved Synthesis. Experientia **35**, 571 (1979).

747. REUTER, G., and H.J. DIEHL: Guanidinederivate in *Leonurus sibiricus* L. Pharmazie **26**, 777 (1971).

748. KONG, Y.C., H.W. YEUNG, Y.M. CHEUNG, J.C. HWANG, Y.W. CHAN, Y.P. LAW, K.H. NG, and C.H. YEUNG: Isolation of the Uterotonic Principle from *Leonurus artemisia*, the Chinese Motherwort. Am. J. Chin. Med. **4**, 373 (1976).

749. LUDWIG, R.A., E.Y. SPENCER, and C.H. UNWIN: An Antifungal Factor from Barley of Possible Significance in Disease Resistance. Canad. J. Botany **38**, 21 (1960).

750. KOSHIMIZU, K., E.Y. SPENCER, and A. STOESSL: The Antifungal Factor Barley. Canad. J. Botany **41**, 744 (1963).

751. STOESSL, A.: The Antifungal Factors in Barley–The Constitutions of Hordatines A and B. Tetrahedron Letters **1966**, 2287.

752. STOESSL, A.: The Antifungal Factors in Barley–Isolation and Synthesis of Hordatine A. Tetrahedron Letters **1966**, 2849.

753. STOESSL, A.: The Antifungal Factors in Barley, III: Isolation of *p*-Coumaroylagmatine. Phytochem. **4**, 973 (1965).

754. STOESSL, A.: The Antifungal Factors in Barley, IV: Isolation, Structure and Synthesis of the Hordatines. Canad. J. Chem. **45**, 1745 (1967).

755. SMITH, T.A., and G.R. BEST: Distribution of the Hordatines in Barley. Phytochem. **17**, 1093 (1978).

756. BIRD, C.R., and T.A. SMITH: The Biosynthesis of Coumaroylagmatine in Barley Seedlings. Phytochem. **20**, 2345 (1981).

757. BIRD, C.R., and T.A. SMITH: Agmatine Coumaroyltransferase (Barley Seedlings). In: Methods in Enzymology, Vol. 94 (H. TABOR and C.W. TABOR, eds.), p. 344. New York: Academic Press. 1983.

758. RABARON, A., M. KOCH, M. PLAT, J. PEYROUX, E. WENKERT, and D.W. COCHRAN: Structure Analysis by Carbon-13 Nuclear Magnetic Resonance Spectroscopy. Arenaïne. J. Amer. Chem. Soc. **93**, 6270 (1971).

759. YOSHIDA, T.: A New Amine, Stizolamine, from *Stizolobium hassjoo*. Phytochem. **15**, 1723 (1976).

760. YOSHIDA, T.: Biosynthesis of Stizolamine in *Stizolobium hassjoo*. Phytochem. **16**, 1824 (1977).

761. ROWAN, D.D, and D.L. GAYNOR: Isolation of Feeding Deterrents Against Argentine Stem Weevil from Ryegrass Infected with the Endophyte *Acremonium loliae*. J. Chem. Ecol. **12**, 647 (1986).

762. DUMAS, D.J.: Total Synthesis of Peramine. J. Organ. Chem. (USA) **53**, 4650 (1988).

763. ROWAN, D.D., and B.A. TAPPER: An Efficient Method for the Isolation of Peramine, an Insect Feeding Deterrent Produced by the Fungus *Acremonium lolii*. J. Nat. Prod. (Lloydia) **52**, 193 (1989).

764. ROWAN, D.D., J.J. DYMOCK, and M.A. BRIMBLE: Effect of Fungal Metabolite Peramine and Analogs on Feeding and Development of Argentine Stem Weevil (*Listronotus bonariensis*). J. Chem. Ecol. **16**, 1683 (1990).

765. NGAMGA, D., S.N.Y.F. FREE, Z.T. FOMUM, A. CHIARONI, C. RICHE, M.T. MARTIN, and
     B. BODO: Millaurine and Acetylmillaurine: Alkaloids from *Millettia laurentii*. J. Nat.
     Prod. (Lloydia) **56**, 2126 (1993).
766. FISCHER, F.G., and H. BOHN: Die Giftsekrete der Vogelspinnen. Liebigs Ann. Chem.
     603, 232 (1957).
767. WHITE, J.: Review of Clinical and Pathological Aspects of Spider-Bite in Australia. In:
     Progress in Venom and Toxin Research, Proceedings of the 1st Asia-Pacific Congress
     on Animal, Plant and Microbial Toxins (P. GOPALAKRISHNAKONE and C.K. TAN, eds.),
     p. 531. Singapore: National University of Singapore. 1987.
768. KAWAI, N., W. NIWA, and T. ABE: Spider Venom Contains Specific Receptor Blocker
     of Glutaminergic Synapses. Brain. Res. **247**, 169 (1982).
769. KAWAI, N., A. MIWA, and T. ABE: Effect of a Spider Toxin on Glutaminergic Synapses
     in the Mammalian Brain. Biomed. Res. **3**, 353 (1982).
770. ABE, T., N. KAWAI, and A. MIWA: Effects of a Spider Toxin on the Glutaminergic
     Synapse of Lobster Muscle. J. Physiol. (London) **339**, 243 (1983).
771. KAWAI, N., S. YAMAGISHI, M. SAITO, and K. FURUYA: Blockade of Synaptic Transmis-
     sion in the Squid Giant Synapse by a Spider Toxin, (JSTX). Brain. Res. **278**, 346 (1983).
772. KAWAI, N., A. MIWA, and T. ABE: Specific Antagonism of the Glutamate Receptor by
     an Extract from the Venom of the Spider *Araneus ventricosus*. Toxicon **21**, 438 (1983).
773. KAWAI, N., A. MIWA, and T. ABE: Block of Glutamate Receptor by a Spider Toxin. In:
     CNS Receptors from Molecular Pharmacology to Behaviour (P. MANDEL and F.V.
     DEFEUDIS, eds.), p. 221. New York: Raven Press. 1983.
774. KAWAI, N., A. MIWA, M. SAITO, H.S. PAN-HOU, and M. YOSHIOKA: Spider Toxin
     (JSTX) on the Glutamate Synapse. J. Physiol. (Paris) **79**, 228 (1984).
775. USHERWOOD, P.N.R., I.R. DUCE, and P. BODEN: Slowly-Reversible Block of Gluta-
     mate Receptor Channels by Venoms of the Spiders, *Argiope trifasciata* and *Araneus
     gemma*. J. Physiol. (Paris) **79**, 241 (1984).
776. TASHMUKHAMEDOV, B.A., P.B. USMANOV, I. KASAKOV, D. KALIKULOV, L.Y. YUKEL-
     SON, and B.U. ATAKUZIEV: Effects of Different Spider Venoms on Artificial and
     Biological Membrans. In: Toxins as Tools in Neurochemistry, p. 312. Berlin: de
     Gruyter. 1983.
777. USMANOV, P.B., D. KALIKULOV, N.G. SHADYEVA, and B.A. TASHMUKHAMEDOV: Effect
     of the Spider *Argiope lobata* Venom on the Glutaminergic and Cholinergic Synapses.
     Doklady Akad. Nauk (USSR) **273**, 1017 (1983).
778. BODEN, P., I.R. DUCE, and P.N.R. USHERWOOD: Activation-Induced Postsynaptic
     Block of Insect Nerve-Muscle Transmission by a Low Molecular Weight Fraction of
     Spider Venom. Brit. J. Pharmacol. **82**, 221 (1984).
779. MICHAELIS, E.K., N. GALTON, and S.L. EARLY: Spider Venoms Inhibit L-Glutamate
     Binding to Brain Synaptic Membrane Receptors. Proc. Natl. Acad. Sci. (USA) **81**,
     5571 (1984).
780. BATEMAN, A., P. BODEN, A. DELL, I.R. DUCE, D.L.J. QUICKE, and P.N.R. USHERWOOD:
     Postsynaptic Block of a Glutaminergic Synapse by Low Molecular Weight Fractions
     of Spider Venom. Brain Res. **339**, 237 (1985).
781. USHERWOOD, P.N.R., and I.R. DUCE: Antagonism of Glutamate Receptor Channel
     Complexes by Spider Venom Polypeptides. NeuroToxicology **6**, 239 (1985).
782. SAITO, M., N. KAWAI, A. MIWA, H. PAN-HOU, and M. YOSHIOKA: Spider Toxin (JSTX)
     Blocks Glutamate Synapse in Hippocampal Pyramidal Neurons. Brain Res. **346**, 397
     (1985).
783. USMANOV, P.B., D. KALIKULOV, N.G. SHADYEVA, A.B. NEMILIN, and B.A. TASHMUK-
     HAMEDOV: Postsynaptic Blocking of Glutaminergic and Cholinergic Synapses as a
     Common Property on Araneidae Spider Venoms. Toxicon **23**, 528 (1985).

784. MAGAZANIK, L.G., S.M. ANTONOV, I.M. FEDOROVA, T.M. VOLKOVA, and E.V. GRISHIN: Effects of the Spider *Argiope lobata* Crude Venom and Its Low Molecular Weight Component, Argiopin, on the Postsynaptic Membrane. Biologich. Membr. 3. 1204 (1986).

785. MIWA, A., N. KAWAI, M. SAITO, H. PAN-HOU, and M. YOSHIOKA: Effect of a Spider Toxin (JSXT) on Excitatory Postsynaptic Current at Neuromuscular Synapse of Spiny Lobster. J. Neurophysiol. **58**, 319 (1987).

786. GRISHIN, E.V., T.M. VOLKOVA, A.S. ARSENIEV, O.S. RESHETOVA, V.V. ONOPRIENKO, L.G. MAGAZANIK, S.M. ANTONOV, and I.M. FEDOROVA: Structure-Functional Characterization of Argiopine, an Ion-Channel Blocker from the Venom of the Spider *Argiope lobata*. Bioorg. Khim. **12**, 1121 (1986); Chem. Abstr. **105**, 186106 (1987).

787. ARAMAKI, Y., T. YASUHARA, T. HIGASHIJIMA, M. YOSHIOKA, A. MIWA, N. KAWAI, and T. NAKAJIMA: Chemical Characterization of Spider Toxin, JSTX and NSTX. Proc. Japan Acad. **62B**, 359 (1986).

788. ARAMAKI, Y., T. YASUHARA, T. HIGASHIJIMA, A. MIWA, N. KAWAI, and T. NAKAJIMA: Chemical Characterization of Spider Toxin, NSTX. Biomed. Res. **8**, 167 (1987).

789. ARAMAKI, Y., T. YASUHARA, K. SHIMAZAKI, N. KAWAI, and T. NAKAJIMA: Chemical Structure of Joro Spider Toxin (JSTX). Biomed. Res. **8**, 241 (1987).

790. HASHIMOTO, Y., Y. ENDO, K. SHUDO, Y. ARAMAKI, N. KAWAI, and T. NAKAJIMA: Synthesis of Spider Toxin (JSTX-3) and Its Analogs. Tetrahedron Letters **28**, 3511 (1987).

791. ADAMS, M.E., R.L. CARNEY, F.E. ENDERLIN, E.T. FU, M.A. JAREMA, J.P. LI, C.A. MILLER, D.A. SHOOLEY, M.J. SHAPIRO, and V.J. VENEMA: Structures and Biological Activities of Three Synaptic Antagonists from Orb Web Spider Venom. Biochem. Biophys. Res. Commun. **148**, 678 (1987).

792. TOKI, T., T. YASUHARA, Y. ARAMAKI, N. KAWAI, and T. NAKAJIMA: A New Type of Spider Toxin, Nephilatoxin, in the Venom of the Joro Spider, *Nephila clavata*. Biomed. Res. **9**, 75 (1988).

793. TOKI, T., T. YASUHARA, Y. ARAMAKI, K. OSAWA, A. MIWA, N. KAWAI, and T. NAKAJIMA: Isolation and Chemical Characterization of a Series of New Spider Toxins (Nephilatoxins) in the Venom of Joro Spider, *Nephila clavata*. Biomed. Res. **9**, 421 (1988).

794. GRISHIN, E.V., T.M. VOLKOVA, and A.S. ARSENIEV: Antagonists of Glutamate Receptors from the Venom of *Argiope lobata* Spider. Bioorg. Khim. **14**, 883 (1988); Chem. Abstr. **109**, 208561 (1988).

795. GRISHIN, E.V., T.M. VOLKOVA, and A.S. ARSENIEV: Isolation and Structure Analysis of Components from Venom of the Spider *Argiope lobata*. Toxicon **27**, 541 (1989).

796. TOKI, T., T. YASUHARA, Y. ARAMAKI, Y. HASHIMOTO, K. SHUDO, N. KAWAI, and T. NAKAJIMA: Molecular Structures of Spider Toxins (JSTX-1, 2, 3 and 4) in the Venom of *Nephila clavata* L. Koch. Jpn. J. Sanit. Zool. **41**, 9 (1990).

797. NAKAJIMA, T., and N. KAWAI: Spider Toxins, NSTX and JSTX, in the Venoms of *Nephila maculata* and *Nephila clavata*. In: Progress in Venom and Toxin Research, Proceedings of the 1st Asia-Pacific Congress on Animal, Plant and Microbial Toxins (P. GOPALAKRISHNAKONE and C.K. TAN, eds.), p. 573. Singapore: National University of Singapore. 1987.

798. BUDD, T., P. CLINTON, A. DELL, I.R. DUCE, S.J. JOHNSON, D.L.J. QUICKE, G.W. TAYLOR, P.N.R. USHERWOOD, and G. USOH: Isolation and Characterization of Glutamate Receptor Antagonists from Venoms of Orb-Web Spiders. Brain Res. **448**, 30 (1988).

799. SKINNER, W.S., M.E. ADAMS, G.B. QUISTAD, H. KATAOKA, B.J. CESARIN, F.E. ENDERLIN, and D.A. SHOOLEY: Purification and Characterization of Two Classes of

Neurotoxins from the Funnel Web Spider, *Agelenopsis aperta*. J. Biol. Chem. **264**, 2150 (1989).

800. JASYS, V.J., P.R. KELBAUGH, D.M. NASON, D. PHILLIPS, K.J. ROSNACK, N.A. SACCO-MANO, J.G. STROH, and R.A. VOLKMANN: Isolation, Structure Elucidation, and Synthesis of Novel Hydroxylamine-Containing Polyamines from the Venom of the *Agelenopsis aperta* Spider. J. Amer. Chem. Soc. **112**, 6696 (1990).

801. QUISTAD, G.B., S. SUWANRUMPHA, M.A. JAREMA, M.J. SHAPIRO, W.S. SKINNER, G.C. JAMIESON, A. LUI, and E.W. FU: Structures of Paralytic Acylpolyamines from the Spider *Agelenopsis aperta*. Biochem. Biophys. Res. Commun. **169**, 51 (1990).

801a. JASYS, V.J., P.R. KELBAUGH, D.M. NASON, D. PHILLIPS, K.L. ROSNACK, J.T. FORMAN, N.A. SACCOMANO, J.G. STROH, and R.A. VOLKMANN: Novel Quaternary Ammonium Salt-Containing Polyamines from the *Agelenopsis aperta* Funnel-Web Spider. J. Organ. Chem. (USA) **57**, 1814 (1992).

802. JACKSON, H., and P.N.R. USHERWOOD: Spider Toxins and Tools for Dissecting Elements of Excitatory Amino Acid Transmission. Trends in Neurosci. **11**, 278 (1988).

802a. USHERWOOD, P.N.R., and I.S. BLAGBROUGH: Spider Toxins Affecting Glutamate Receptors: Polyamines in Therapeutic Neurochemistry. Pharmac. Ther. **52**, 245 (1991).

802b. SCOTT, R.H., K.G. SUTTON, and A.C. DOLPHIN: Interactions of Polyamines with Neuronal Ion Channels. Trends in Neurosci. **16**, 153 (1993).

803. SACCOMANO, N.A., R.A. VOLKMANN, H. JACKSON, and T.N. PARKS: Polyamine Spider Toxins: Unique Pharmacological Tools. Ann. Rep. Med. Chem. **24**, 287 (1989).

804. TESHIMA, T., T. WAKAMIYA, Y. ARAMAKI, T. NAKAJIMA, N. KAWAI, and T. SHIBA: Synthesis of a New Neurotoxin NSTX-3 of Papua New Guinean Spider. Tetrahedron Letters **28**, 3509 (1987).

805. SHIH, T.L., J. RUIZ-SANCHEZ, and H. MROZIK: The Total Synthesis of Argiopine (Argiotoxin-636). Tetrahedron Letters **28**, 6015 (1987).

806. YELIN, E.A., B.F. DE MACEDO, V.V. ONOPRIENKO, N.E. OSOKINA, and O.B. TIKHO-MIROVA: Synthesis of Argiopine. Bioorg. Khim. **14**, 704 (1988).

807. JASYS, V.J., P.R. KELBAUGH, D.M. NASON, D. PHILLIPS, N.A. SACCOMANO, and R.A. VOLKMANN: The Total Synthesis of Argiotoxins 636, 659 and 673. Tetrahedron Letters **29**, 6223 (1988).

808. NASON, D.M., V.J. JASYS, P.R. KELBAUGH, D. PHILLIPS, N.A. SACCOMANO, and R.A. VOLKMANN: Synthesis of Neurotoxic Nephila Spider Venoms: NSTX-3 and JSTX-3. Tetrahedron Letters **30**, 2337 (1989).

809. TESHIMA, T., T. MATSUMOTO, M. MIYAGAWA, T. WAKAMIYA, T. SHIBA, N. NARAI, and M. YOSHIOKA: Total Synthesis of Clavamine, Insecticidally Active Compound Isolated from Venom of Joro Spider (*Nephila clavata*). Tetrahedron Letters **46**, 3819 (1990).

810. TESHIMA, T., T. MATSUMOTO, T. WAKAMIYA, T. SHIBA, Y. ARAMAKI, T. NAKAJIMA, and N. KAWAI: Total Synthesis of NSTX-3, Spider Toxin of *Nephila maculata*. Tetrahedron **47**, 3305 (1991).

811. MIYASHITA, M., H. SATO, A. YOSHIOKI, T. TOKI, M. MATSUSHITA, H. IRIE, T. YANAMI, Y. KIKUCHI, C. TAKASAKI, and T. NAKAJIMA: Synthetic Studies on Spider Neurotoxins (I): Total Synthesis of Nephilatoxins (NPTX-9 and NPTX-11), New Neurotoxins of Joro Spider (*Nephila clavata*). Tetrahedron Letters **33**, 2833 (1992).

812. MIYASHITA, M., H. SATO, M. MATSUSHITA, Y. KUSUMEGI, T. TOKI, A. YOSHIKOSHI, T. YANAMI, Y. KIKUCHI, C. TAKASAKI, T. NAKAJIMA, and H. IRIE: Synthetic Studies on Spider Neurotoxins (II): Total Synthesis of Nephilatoxins (NPTX-10 and NPTX-12), New Neurotoxins of Joro Spider (*Nephila clavata*). Tetrahedron Letters **33**, 2837 (1992).

813. MIYASHITA, M., M. MATSUSHITA, H. SATO, T. TOKI, T. NAKAJIMA, and H. IRIE: Total Synthesis of Nephilatoxin (NPTX-8) a New Neurotoxin of Joro Spider (*Nephila clavata*). Chem Lett. **1993**, 929.

814. SHUDO, K., Y. ENDO, Y. HASHIMOTO, Y. ARAMAKI, T. NAKAJIMA, and N. KAWAI: Newly Synthesized Analogues of the Spider Toxin Block the Crusteacean Glutamate Receptor. Neurosci. Res. **5**, 82 (1987).

815. TESHIMA, T., T. MATSUMOTO, T. WAKAMIYA, T. SHIBA, T. NAKAJIMA, and N. KAWAI: Structure-Activity Relationship of NSTX-3, Spider Toxin of *Nephila clavata*. Tetrahedron **46**, 3813 (1990).

816. KOVACS, L., and M. HESSE: Synthetic Analogues of Naturally Occurring Spider Toxins. Helv. Chim. Acta **75**, 1909 (1992).

817. FIEDLER, W.J., A. GUGGISBERG, and M. HESSE: Synthetische Analoga von niedermolekularen Spinnentoxinen mit Acyl-Polyamine-Struktur. Helv. Chim. Acta **76**, 1167 (1993).

818. PAN-HOU, H., and Y. SUDA: Molecular Action Mechanism of Spider Toxin on Glutamate Receptor: Role of 2,4-Dihydroxyphenylacetic Acid in Toxin Molecule. Brain Res. **418**, 198 (1987).

819. PIEK, T.: Site of Action of the Venom of the Digger Wasp *Philanthus triangulum* F. on the Fast Neuromuscular System of the Locust. Toxicon **3**, 191 (1966).

820. PIEK, T.: The Effect of the Venom of the Digger Wasp *Philanthus* on the Fast and Slow Excitatory and Inhibitory System in the Locust Muscle. Experientia **12**, 462 (1966).

821. PIEK, T., P. MANTEL, and E. ENGELS: Neuromuscular Block in Insects Caused by the Venom of the Digger Wasp *Philanthus triangulum* F. Comp. Gen. Pharmacol. **2**, 317 (1971).

822. MAY, T.E., and T. PIEK: Neuromuscular Block in Locust Skeletal Muscle Caused by the Venom Preparation Made from the Digger Wasp *Philanthus triangulum* F. from Egypt. J. Insect Physiol. **25**, 685 (1979).

823. PIEK, T., P. MANTEL, and H. JAS: Ion-Channel Block in Insects Caused by the Venom of the Digger Wasp *Philanthus triangulum* F. J. Insect Physiol. **26**, 315 (1980).

824. CLARK, R.B., P.L. DONALDSON, K.A.F. GRATION, J.J. LAMBERT, T. PIEK, R. RAMSEY, W. SPANJER, and P.N.R. USHERWOOD: Block of Locust Muscle Glutamate Receptors by δ-Philanthotoxin Occurs After Receptor Activation. Brain Res. **241**, 105 (1982).

825. PIEK, T.: δ-Philanthotoxin, a Semi-Irreversible Blocker of Ion-Channels. Comp. Biochem. Physiol. **72C**, 311 (1982).

826. PIEK, T., and W. SPANJER: Effects and Chemical Characterization of Some Paralysing Venoms of Solitary Wasps. In: Pesticide and Venom Neurotoxicity (D.L. SHANKLAND, R.M. HOLLINGWORTH, and T. SMYTH, eds.), p. 211. New York: Plenum Press. 1978.

827. PIEK, T., R.H. FOKKENS, H. HARST, C. KRUK, A. LIND, J. VAN MARLE, T. NAKAJIMA, N.M.M. NIBBERING, H. SHINOZAKI, and W. SPANJAR: Polyamine-Like Toxins—A New Class of Pesticides? In: Neurotoxicology '88, Molecular Basis of Drugs and Pesticides Action (G.G. LUNT, ed.), p. 61. Amsterdam-New York-Oxford: Excerpta Medica. 1988.

828. BRUNDELL, P., R. GOODNOW, Jr., C.J. KERRY, K. NAKANISHI, H.L. SUDAN, and P.N.R. USHERWOOD: Quisqualate-Sensitive Glutamate Receptors of the Locust *Schistocerca gregaria* Are Antagonised by Intracellularly Applied Philanthotoxin and Spermine. Neurosci. Lett. **131**, 196 (1991).

829. MATTHIES, H., Jr., P.T.H. BRACKLEY, P.N.R. USHERWOOD, and K.G. REYMANN: Philanthotoxin-343 Blocks Long-Term Potentiation in Rat Hippocampus. NeuroReport **3**, 649 (1992).

830. EDELFRAWI, A.T., M.E. EDELFRAWI, K. KONNO, N.A. MANSOUR, K. NAKANISHI, E.

OLTZ, and P.N.R. USHERWOOD: Structure and Synthesis of a Potent Glutamate Receptor Antagonist in Wasp Venom. Proc. Natl. Acad. Sci. (USA) **85**, 4910 (1988).

831. BRUCE, M., R. BUKOWNIK, A.T. EDELFRAWI, M.E. EDELFRAWI, R. GOODNOW, Jr., T. KALLINOPOULOS, K. KONNO, K. NAKANISHI, M. NIWA, and P.N.R. USHERWOOD: Structure-Activity Relationships of Analogues of the Wasp Toxin Philanthotoxin: Non-Competitive Antagonists of Quisqualate Receptors. Toxicon **28**, 1333 (1990).

832. NAKANISHI, K., R. GOODNOW, K. KONNO, M. NIWA, R. BUKOWNIK, T.A. KALLINO-POULOS, P.N.R. USHERWOOD, A.T. EDELFRAWI, and M.E. EDELFRAWI: Philantho-toxin-433 (PhTX-433), a Non-Competitive Glutamate Receptor Inhibitor. Pure Appl. Chem. **62**, 1223 (1990).

833. CHOI, S.-K., K. NAKANISHI, and P.N.R. USHERWOOD: Synthesis and Assay of Hybrid Analogs of Argiotoxin-636 and Philanthotoxin-433: Glutamate Receptor Antagonists. Tetrahedron **49**, 5777 (1993).

834. CHOI, S.-K., R.A. GOODNOW, A. KALIVRETENOS, G.W. CHILES, S. FUSHIYA, and K. NAKANISHI: Synthesis of Novel and Photolabile Philanthotoxin Analogs: Glutamate Receptor Antagonists. Tetrahedron **48**, 4793 (1992).

835. KALIVRETENOS, A.G., and K. NAKANISHI: Synthesis of Philanthotoxin Analogs with a Branched Polyamine Moiety. J. Organ. Chem. (EUA) **58**, 6596 (1993).

835a. BLAGBROUGH, I.S., P.T.H. BRACKLEY, M. BRUCE, B.W. BYCROFT, A.J. MATHER, S. MILLINGTON, H.L. SUDAN, and P.N.R. USHERWOOD: Arthropod Toxins as Leads for Novel Insecticides: An Assessment of Polyamine Amides as Glutamate Antagonists. Toxicon **30**, 303 (1992).

836. WIELAND, H., and F.J. WEIL: Über das Krötengift. Ber. **46**, 3315 (1913).

837. KODAMA, H.: Pharmacology of Sen-So. Chem. Abstr. **15**, 2500 (1921).

838. KONDO, H., and S. IKAWA: Bufotalin and Bufotoxin in "Senso" and Its Raw Material. J. Pharm. Soc. Jpn. **53**, 23 (1993); Chem. Abstr. **27**, 1887 (1933).

839. KONDO, H., and S. ONO: Constitution of Senso, IV. J. Pharm. Soc. Jpn. **58**, 37 (1938); Chem. Abstr. **32**, 3765 (1938).

840. CHEN, K.K., H. JENSEN, and A.L. CHEN: The Physiological Action of the Principles Isolated from the Secretion of *Bufo arenarum*. J. Pharmacol. **49**, 1 (1933); Chem. Abstr. **28**, 214 (1933).

841. CHEN, K.K., H. JENSEN, and A.L. CHEN: The Physiological Action of the Principles Isolated from the Secretion of the European Green Toad (*Bufo viridis viridis*). J. Pharmacol. **49**, 14 (1933); Chem. Abstr. **28**, 214 (1933).

842. CHEN, K.K., H. JENSEN, and A.L. CHEN: The Physiological Action of the Principles Isolated from the Secretion of the Japanese Toad (*Bufo formosus*). J. Pharmacol. **49**, 26 (1933); Chem. Abstr. **28**, 214 (1933).

843. CHEN, K.K., H. JENSEN, and A.L. CHEN: Action of Bufotoxins. Proc. Soc. Exp. Biol. Med. **29**, 907 (1931–1932).

844. CHEN, K.K., H. JENSEN, and A.L. CHEN: Action of Bufotenines. Proc. Soc. Exp. Biol. Med. **29**, 908 (1931–1932).

845. WIELAND, H.: Toad Venom. Sitzber. Bayr. Akad. Wiss. **1920**, 329; Chem. Abstr. **16**, 3073 (1921).

846. WIELAND, H., and R. ALLES: Über den Giftstoff der Kröte. Ber. **55**, 1789 (1922).

847. WIELAND, H., and F. VOCKE: Über die Giftstoffe der japanischen Kröte. Ann. **481**, 215 (1930).

848. JENSEN, H., and K.K. CHEN: Chemical Studies on Toads Poisons, III: The Secretion of the Tropical Toad, *Bufo marinus*. J. Biol. Chem. **87**, 755 (1930).

849. WIELAND, H., G. HESSE, and H. MITTASCH: Über basische Inhaltstoffe des Hautsekrets der Kröte. Ber. **64**, 2099 (1931).

850. JENSEN, H.: Chemical Studies on Toad Poisons, IV: Bufagin and Cinobufagin. Science 75, 53 (1932).
851. WIELAND, H., W. KONZ, and H. MITTASCH: Über Kröten-Giftstoffe, VII: Die Konstitution von Bufotenin und Bufotenidin. Ann. 513, 1 (1934).
852. WIELAND, H., G. HESSE, and R. HÜTTEL: Zur Kenntnis der Krötengiftstoffe, IX: Weiteres zur Konstitutionsfrage. Ann. 524, 203 (1936).
853. WIELAND, H., and T. WIELAND: Zur Kenntnis der Krötengiftstoffe, X: Die Konstitution des Bufothionins. Ann. 528, 234 (1937).
854. WIELAND, H., H. BEHRINGER, G. HESSE, and K. GABELEIN: Über Krötengiftstoffe, XI: Zur Konstitution des Bufotalins. Ann. 549, 209 (1941).
855. MEYER, K.: Über herzaktive Krötengifte (Bufogenine), 2. Mitteilung: Konstitution des Bufalins. Helv. Chim. Acta 32, 1238 (1949).
856. MEYER, K.: Über herzaktive Krötengifte (Bufogenine), 5. Mitteilung: Konstitution des Bufotalins. Helv. Chim. Acta 32, 1993 (1949).
857. MEYER, K.: Über herzaktive Krötengifte (Bufogenine), 7. Mitteilung: Resibufogenin und Artebufogenin aus Ch'an Su. Helv. Chim. Acta 35, 2444 (1952).
858. URSCHELER, H.R., C. TAMM, and T. REICHSTEIN: Die Giftstoffe der europäischen Erdkröte Bufo bufo bufo L. (Über Krötengifte, 8. Mitteilung). Helv. Chim. Acta 38, 883 (1955).
859. LINDE, H., and K. MEYER: Konstitution des Resibufogenins. Helv. Chim. Acta 42, 807 (1959).
860. LINDE, H., and K. MEYER: Zur Konstitution des Resibufogenins und Artebufogenins. Experientia 14, 238 (1958).
861. HOFER, P., H. LINDE, and K. MEYER: Konstitution des Cinobufagins. Experientia 15, 297 (1959).
862. HOFER, P., H. LINDE, and K. MEYER: Konstitution des Cinobufagins. Helv. Chim. Acta 43, 1955 (1960).
863. BERNOULLI, F., H. LINDE, and K. MEYER: Konstitution des Cinobufotalins. Helv. Chim. Acta 45, 240 (1962).
864. LINDE, H., P. HOFER, and K. MEYER: Cinobufaginol. Helv. Chim. Acta 49, 1243 (1966).
865. KAMANO, Y., H. YAMAMOTO, Y. TANAKA, and M. KOMATSU: The Isolation and Structure of New Bufadienolides, 3-(Hydrogen Suberates) of Resibufogenin, Cinobufagin and Bufalin, the Structure of the So-Called "Bufotoxins". Tetrahedron Letters 1968, 5673.
866. MEYER, K: Cardiotoxic Steroids from Toads. Mem. Inst. Butantan Simp. Internac. 33, 433 (1966).
867. LINDE-TEMPEL, H.O.: Konstitution der Bufotoxine. Helv. Chim. Acta 53, 2188 (1970).
868. PETTIT, G.R., and Y. KAMANO: The Structure of the Steriod Toad Venom Constituent Bufotoxin. Chem. Commun. 1972, 45.
869. PETTIT, G.R., Y. KAMANO, P. DRASAR, H. INOUE, and J.C. KNIGHT: Synthesis of Bufalitoxin and Bufotoxin. J. Organ. Chem. (USA) 52, 3573 (1987).
870. SHIMADA, K., Y. FUJII, and T. NAMBARA: Synthesis of Bufotoxin. Chem. and Ind. 1972, 258.
871. SHIMADA, K., and T. NAMBARA: Synthesis of 3-Suberoylalanine Ester of Digitoxigenin. Chem. Pharm. Bull. (Japan) 19, 1073 (1971).
872. CHEN, K.K., and A. KOVARIKOVA: Pharmacology and Toxicology of Toad Venom. J. Pharm. Sci. 56, 1535 (1967).
873. MARSHALL, P.G.: Steroids: Cardiotonic Glycosides and Aglycons; Toad Poisons. In: Rodd's Chemistry of Carbon Compounds, 2. Pt.: C–D–E (Suppl.) (M.F. ANSELL, ed.), p. 205. Amsterdam: Elsevier. 1974.

874. HÖRINGER, N., H.H.A. LINDE, and K. MEYER: Über neue Bufadienolide aus *Ch'an Su*. Helv. Chim. Acta **52**, 1097 (1969).

875. PETTIT, G.R., P. BROWN, F. BRUSCHWEILER, and L.H. HOUGHTON: Structure of the Bufadienolide Bufotalin. Chem. Commun. **1970**, 1566.

876. HÖRINGER, N., H.H.A. LINDE, and K. MEYER: Cardenolide aus dem Chinesischen Krötengift *Ch'an Su*. Helv. Chim. Acta **53**, 1503 (1970).

877. HÖRINGER, N., D. ZIVANOV, H.H.A. LINDE, and K. MEYER: Cardenolid-Korksäure-halbester und weitere Bufadienolid-Korksäurehalbester in *Ch'an Su*. Helv. Chim. Acta **53**, 1993 (1970).

878. HÖRINGER, N., D. ZIVANOV, H.H.A. LINDE, and K. MEYER: Weitere Cardenolide aus *Ch'an Su*. Helv. Chim. Acta **53**, 2051 (1970).

879. SHIMADA, K., Y. FUJII, and T. NAMBARA: Isolation of Bufotalin-3-sulfate from the Skin of *Bufo vulgaris formosus* Boulenger. Tetrahedron Letters **1974**, 2767.

880. SHIMADA, K., Y. FUJII, E. MITSUISHI, and T. NAMBARA: Occurrence of Novel Type Bufotoxin in Japanese Toad. Chem. Pharm. Bull. (Japan) **22**, 1673 (1974).

881. SHIMADA, K., Y. FUJII, E. YAMASHITA, Y. NIIZAKI, Y. SATO, and T. NAMBARA: Studies on Cardiotonic Steroids from the Skin of Japanese Toad. Chem. Pharm. Bull. (Japan) **25**, 714 (1977).

882. SHIMADA, K., Y. FUJII, E. MITSUISHI, and T. NAMBARA: Isolation of a New Type Bufotoxin from Skin of *Bufo vulgaris formosus* Boulenger. Tetrahedron Letters **1974**, 467.

883. SHIMADA, K., Y. FUJII, E. MITSUISHI, and T. NAMBARA: Occurrence of Gamabufotali-toxin in the Skin of Japanese Toad. Chem. and Ind. **1974**, 342.

884. SHIMADA, K., Y. FUJII, and T. NAMBARA: Gamabufotalitoxin Homologs from the Skin of Japanese Toad. Chem. and Ind. **1974**, 963.

885. SHIMADA, K., Y. FUJII, Y. NIIZAKI, and T. NAMBARA: Isolation of Gamabufotalin 3-Pimeloylarginine Ester from the Skin of Japanese Toad. Tetrahedron Letters **1975**, 653.

886. FUJII, Y., K. SHIMADA, Y. NIIZAKI, and T. NAMBARA: Cardenobufotoxin: Novel Conjugated Cardenolide from Japanese Toad. Tetrahedron Letters **1975**, 3017.

887. SHIMADA, K., Y. SATO, Y. FUJII, and T. NAMBARA: Occurrence of Bufalitoxin, Cinobufotoxin and Their Homologs in Japanese Toad. Chem. Pharm. Bull. (Japan) **24**, 1118 (1976).

888. SHIMADA, K., and T. NAMBARA: Isolation and Characterization of Cardiotonic Steroids Conjugates from the Skin of *Bufo marinus* (L.) Schneider. Chem. Pharm. Bull. (Japan) **27**, 1881 (1979).

889. SHIMADA, K., J.S. RO, K. OHISHI, and T. NAMBARA: Isolation and Characterization of Cinobufagin 3-Glutaroyl-L-arginine Ester from *Bufo bufo gargarizans* Cantor. Chem. Pharm. Bull. (Japan) **33**, 2767 (1985).

890. SHIMADA, K., N. ISHII, and T. NAMBARA: Occurrence of Bufadienolides in the Skin of *Bufo viridis* Laur. Chem. Pharm. Bull. (Japan) **34**, 3454 (1986).

891. SHIMADA, K., Y. SATO, and T. NAMBARA: Occurrence of Marinobufotoxin and Telocinobufotoxin Homologs in the Skin of *Bufo bakorensis* Borbour. Chem. Pharm. Bull. (Japan) **35**, 2300 (1987).

892. SHIMADA, K., K. OHISHI, and T. NAMBARA: Isolation and Characterization of New Bufotoxins from the Skin of *Bufo melanosticus* Schneider. Chem. Pharm. Bull. (Japan) **32**, 4396 (1984).

893. SHIMADA, K., and T. NAMBARA: Isolation and Characterization of a New Type of Bufotoxin from the Skin of *Bufo americanus*. Chem. Pharm. Bull. (Japan) **28**, 1559 (1980).

894. SHIMADA, K., and T. NAMBARA: Isolation of Marinobufagin 3-Suberoyl-L-glutamine Ester from the Skin of *Bufo americanus.* Tetrahedron Letters **1979**, 163.

895. SHIMADA, K., K. OHISHI, and T. NAMBARA: Isolation of Bufotalin 3-Suberoyl-histidine and 3-Methyl-histidine Esters from the Skin of *Bufo melanosticus.* Tetrahedron Letters **24**, 551 (1984).

896. NAMBARA, T., K. SHIMADA, and Y. FUJII: Synthesis of 3-Suberoylamino Acid Esters of Digitoxigenin. Chem. Pharm. Bull. (Japan) **20**, 1424 (1972).

897. SHIMADA, K., Y. FUJII, and T. NAMBARA: Syntheses of Bufotoxin Analogs. Chem. Pharm. Bull. (Japan) **21**, 2183 (1973).

898. SHIMADA, K., M. HASEGAWA, K. HASEBE, Y. FUJII, and T. NAMBARA: Studies on Steroids, CXIV: Separation of Cardiotonic Steroid Conjugates by High-Performance Liquid Chromatography. J. Chromatogr. **124**, 79 (1976).

899. NATORI, S., N. IKEKAWA, and M. SUZUKI, eds.: Isolation of Cardiotonic Constituents from Japanese Toads. In: Advances in Natural Products Chemistry: Extraction and Isolation of Biologically Active Compounds, p. 537. New York: Wiley. 1981.

900. KAMANO, Y., N. SATOH, H. NAKAYOSHI, G.R. PETTIT, and C.R. SMITH: Rhinovirus Inhibition by Bufadienolides. Chem. Pharm. Bull. (Japan) **36**, 326 (1988).

901. BERLINCK, R.G.S.: Alcaloïdes Guanidiniques et Autres Metabolites Sécondaires de *Crambe crambe* (Thiele). Thèse presentée pour l'obtention du grade de Docteur en Sciences, Faculté des Sciences, Université Libre de Bruxelles, 1992.

(*Received May 10, 1994*)

Author Index

Page numbers printed in *italics* refer to References

Subject Index

A-269A 124
A-269A' 124
A37812 247
Acanthella aurantiaca 194
Acanthella carteri 195
3,5-Acarnidine 198, 199
Acarnidines 197
Acarnus erithacus 197
Acer ginnala 5
Acer nikoense 28
Acer sp. 7, 28
Aceraceae 15
Acertannin 5, 6
[1-^{13}C]-Acetate 125
Acetic acid 131, 158, 159
Acetone 28, 29
N-Acetylagmatine 220
Acetylmillaurine 234, 235
Acremonium loliae 233
Actinia equina 220
Actinia fragacea 220
Actiniamine 220
Actinomadura sp. 176
Actinomycetes 123
Actissimin A 42
Adenine 5'-diphosphate 91
Agelas cf. *nemoechinata* 195
Agelas clathrodes 207
Agelas conifera 196
Agelas sp. 195, 206
Agelasidine A 206–208
Agelasidine B 206, 207
Agelasidine C 206, 207
(+)-Agelasidine C 207
(−)-Agelasidine C 207
(−)-Agelasidine D 207
Agelenopsis aperta 241
Ageline A 206
[U-^{14}C]-Agmatine 232

Agrimonia pilosa 13, 16, 31, 33
Agrimonia sp. 43, 46
Agrimonic acid A 13, 14, 16
Agrimonic acid B 13, 14, 16
Agrimoniin 31, 33, 43, 45, 46, 68, 94–96
Agrobacterium rhizogenes 88
Alanine 138, 139, 167, 232
L-Alanine 125
Albothricin 124, 247
Albumin 91
Alchornea floribunda 226, 227
Alchornea javanensis 225
Alchornea nintella 226
Alchornea sp. 228
Alchornéine 226, 227
Alchornéinone 227
Alchornidine 225, 226
Alchornine 225, 226
Aldisin 194
Alienanin A 41, 60, 61
Alienanin B 41, 60, 61
Alkyl hexahydroxydiphenate 90
Alnus fordii 16
Alnusiin 12, 13, 16
α-Amino-3,5-dichloro-4-hydroxyphenyl-
    acetic acid 158
1-[3-(2-Amino-4,6-
    dimethylpyrimidinyl)ureido]-4-(*N*-
    methylacetamido)butane
    derivative 189
Aminoiminomethanesulfonic acid 190
3-Amino-9-methoxy-2,6,8-trimethyl-10-
    phenyl-4,6-decadienoic acid 187
3-Amino-2-methylaminopropionic
    acid 163
Amycolatopsis azurea 175
AN-201 I 124
AN-201 II 124
AN-201 III 124

SpringerChemistry

Fortschritte der Chemie organischer Naturstoffe

Progress in the Chemistry

of Organic Natural Products

Founded by L. Zechmeister
Edited by W. Herz, G. W. Kirby, R. E. Moore, W. Steglich,
and Ch. Tamm

Volume 65

1995. 2 figures. IX, 618 pages. Cloth DM 440,–, öS 3080,–
Subscription price: Cloth DM 396,–, öS 2772,–. ISBN 3-211-82576-2

Contents:
Y. Asakawa: Chemical Constituents of the Bryophytes.

Volume 64

1995. 22 partly coloured figures. VII, 216 pages. Cloth DM 250,–, öS 1750,–
Subscription price: Cloth DM 225,–, öS 1575,–. ISBN 3-211-82533-9

Contents:
A. G. González and J. Bermejo Barrera: Chemistry and Sources
of Mono- and Bicyclic Sesquiterpenes from Ferula Species.
G. Prota: The Chemistry of Melanins and Melanogenesis.
H. J. M. Gijsen, J. B. P. A. Wijnberg, and Ae. de Groot: Structure,
Occurrence, Biosynthesis, Biological Activity, Synthesis, and
Chemistry of Aromadendrane Sesquiterpenoids.

 SpringerWienNewYork

P.O.Box 89, A-1201 Wien • New York, NY 10010, 175 Fifth Avenue
Heidelberger Platz 3, D-14197 Berlin • Tokyo 113, 3-13, Hongo 3-chome, Bunkyo-ku

SpringerChemistry

Fortschritte der Chemie organischer Naturstoffe

Progress in the Chemistry

of Organic Natural Products

Founded by L. Zechmeister
Edited by W. Herz, G. W. Kirby, R. E. Moore, W. Steglich,
and Ch. Tamm

Volume 63

1994. VII, 216 pages. Cloth DM 220,–, öS 1540,–
Subscription price: Cloth DM 198,–, öS 1386,–. ISBN 3-211-82443-X

Contents:
A. B. Ray and M. Gupta: Withasteroids, a Growing Group of
Naturally Occurring Steroidal Lactones.
L. Rodríguez-Hahn, B. Esquivel, and J. Cárdenas: Clerodane
Diterpenes in Labiatae.

Volume 62

1993. 52 figures. VIII, 330 pages. Cloth DM 280,–, öS 1960,–
Subscription price: Cloth DM 252,–, öS 1764,–. ISBN 3-211-82402-2

Contents:
Sujata V. Bhat: Forskolin and Congeners.
L. Minale, R. Riccio and F. Zollo: Steroidal Oligoglycosides and
Polyhydroxysteroids from Echinoderms.

 SpringerWienNewYork

P.O.Box 89, A-1201 Wien • New York, NY 10010, 175 Fifth Avenue
Heidelberger Platz 3, D-14197 Berlin • Tokyo 113, 3-13, Hongo 3-chome, Bunkyo-ku